Practice of Petri Nets in Manufacturing

Practice of Petri Nets in Manufacturing

F. DiCesare

Rensselaer Polytechnic Institute, Troy, USA

G. Harhalakis

Department of Mechanical Engineering and the Institute for Systems Research, University of Maryland, USA

J.M. Proth

INRIA-Lorraine, Metz, France

M. Silva

University of Zaragoza, Spain

F.B. Vernadat

INRIA-Lorraine, Metz, France

CHAPMAN & HALL
London · Glasgow · New York · Tokyo · Melbourne · Madras

Published by Chapman & Hall, 2–6 Boundary Row, London SE1 8HN

Chapman & Hall, 2–6 Boundary Row, London SE1 8HN, UK

Blackie Academic & Professional, Wester Cleddens Road, Bishopbriggs, Glasgow G64 2NZ, UK

Chapman & Hall Inc., 29 West 35th Street, New York NY10001, USA

Chapman & Hall Japan, Thomson Publishing Japan, Hirakawacho Nemoto Building, 6F, 1-7-11 Hirakawa-cho, Chiyoda-ku, Tokyo 102, Japan

Chapman & Hall Australia, Thomas Nelson Australia, 102 Dodds Street, South Melbourne, Victoria 3205, Australia

Chapman & Hall India, R. Seshadri, 32 Second Main Road, CIT East, Madras 600 035, India

---

First edition 1993

Typeset in 10/12pt Times by Graphicraft Typesetters Ltd, Hong Kong
Printed in Great Britain by Clays Ltd, Bungay, Suffolk

ISBN 0 412 41230 6

A catalogue record for this book is available from the British Library

Library of Congress Cataloging-in-Publication Data
Practice of Petri nets in manufacturing / F. DiCesare ... [et al.]. – 1st ed.
p. cm.
Includes bibliographical references and index.
ISBN 0–412–41230–6
1. Flexible manufacturing systems—Mathematical models. 2. Petri nets. 3. production control—Mathematical models. I. Dicesare, Frank.
TS155.6.P712 1993 93–3345
670′.42′7—dc20 CIP

Contents

Introduction

M. Silva

Significant changes have been occurring in industrialized countries since the Second World War. Production is moving towards sophisticated high-quality products, economy of scale has been replaced by economy of scope, jerky demands are progressively replacing steady demands, and competitiveness is becoming a worldwide phenomenon. These trends require highly automated manufacturing systems with small set-up times and high flexibility. As a consequence, implementation and running costs of modern manufacturing systems are drastically increasing, whereas their fields of application remain limited, and every day become even narrower, which increases the risk of early obsolescence. This is the reason why designers are trying to improve the preliminary design phase, also known as the 'paper study phase'.

The preliminary design phase includes, but is not limited to, the functional specification, and the evaluation of the system.

Many tools exist to support the functional specification of manufacturing systems. IDEF0 is one of these tools. It leads, using a top-down approach, to a precise functional description of the required system. However, its use cannot be extended further. In general, the evaluation starts with a modeling step, which depends on the evaluation tool used, and ends by applying the model to find out its main dynamic characteristics. Two main approaches can be used to perform this task, namely simulation and mathematical approach. Using simulation, the modeling tool is either a classical computer language, or a simulation language. The latter is much more helpful for common applications. Various mathematical approaches are also available. Queuing theory, for instance, makes possible the representation of the behavior of the system by means of differential equations whose solutions are the behavioral characteristics of the manufacturing system. However, neither computer languages, nor common mathematical tools (e.g. Operations Research), can be used to specify the system.

Furthermore, specification on one hand, and modeling and evaluation on the other hand, usually appear as totally disconnected activities from the point of view of the tools utilized. It is certainly a drawback for the integration of the activities performed at the preliminary design level.

We claim that Petri net theory is the only tool which can support not only the specification activity, but also the evaluation of the behavior of the system starting from the model provided by the specification.

The goal of this book is to provide the foundations of Petri nets to the designer and to give several applications which demonstrate the capability of this tool for designing complex systems. The authors deliberately restrict themselves to the general nets which have the most powerful properties. The size of the models obtained by using these kinds of nets is not a drawback if, as suggested by the authors, a modular approach is used. Furthermore, such an approach makes it possible, in some cases, to derive easily the properties of the manufacturing system from those of the model.

The authors

1

Introducing Petri nets

M. Silva

1.1 INTRODUCTION

Modern manufacturing systems are highly parallel and distributed. They need to be analyzed from qualitative and quantitative points of view. Qualitative analysis looks for properties like the absence of deadlocks, the absence of (store) overflows, or the presence of certain mutual exclusions in the use of shared resources (e.g. a robot). Its ultimate goal is to prove the correctness of the modeled system. Quantitative analysis looks for performance properties (e.g. throughput), responsiveness properties (e.g. average completion times) or utilization properties (e.g. average queue lengths or utilization rates). In other words, the quantitative analysis concerns the evaluation of the efficiency of the modeled system.

As in many engineering fields, the design of manufacturing systems can be carried out using **models**. Petri nets allow the construction of models amenable both for correctness and efficiency analysis. Moreover they can be implemented using many different techniques (hardware, microprogrammed, software). Because of the graphical nature of net models, they are mostly self-documented specifications, making easier the communication among designers and users. Net models can be used during the entire life cycle of manufacturing systems.

A Petri net (PN), like a differential equation, is a mathematical formalism. Petri nets find their basis in a few simple objects, relations and rules, yet can represent very complex behaviors. More precisely, Petri nets can be considered as a graph theoretic tool specially suited to model and analyze **discrete event dynamic systems** (*DEDS*) which exhibit parallel evolutions and whose behaviors are characterized by synchronization and sharing phenomena. Their suitability for modeling this type of system has led to their application in a wide range of fields. Examples of such DEDS are communication networks, computer systems and, the purpose of this book, discrete part manufacturing systems.

To be able to use a Petri net for modeling a given type of application, we must enrich it with an adequate interpretation. That is, we must associate a semantics (i.e. a 'physical' meaning), to the net's entities (places, transitions, tokens), evolution conditions and, eventually, define the actions generated by the evolutions. Broadly speaking, the interpretation gives a meaning to the net system and defines its relationships with the external world (i.e. the interpretation considers the environment in which the net model will be exercised).

The interpretation of graph theoretic tools is nothing new. A graph (in its theoretical sense) is a set of objects (nodes) with relations (see, for example, Deo (1974); Gibbons (1985)). With a graph the connectivity between sites (towns, points in a circuit, . . .) can be represented using obvious interpretations. Another kind of interpretation on graphs allow us to model discrete and finite dynamic systems: the nodes represent the states of the system, the arcs represent transitions between states. Particularizing a little more the state-based interpretation, **state diagrams** (SD) (see, for example, Breeding 1989), and **state transition diagrams** (STD) (see, for example, Ajmone *et al.* (1987)) are widely used interpreted graphs: SDs allow the modeling of finite state sequential switching systems, while STDs allow the modeling of homogeneous finite Markov chains. For both formalisms, SDs and STDs, the evolution of the system can be done in **continuous time** (asynchronous state diagrams; state transition rate diagrams) or in **discrete time** (synchronous state graphs; discrete time state transition diagrams).

Provided with adequate interpretations, PNs are able to model 'distributed state diagrams', the control flow of concurrent programs or queuing networks with synchronizations, among other possibilities. The evolution of a fully uninterpreted net system is said to be **autonomous**. An interpreted net system is said to be **non-autonomous** because its evolution depends also on the **state of the environment** considered by the associated interpretation. For example, the **timing** of a net is a particular interpretation by which its evolution depends also on time.

There exists a very rich body of knowledge around Petri nets theory and applications. The purpose of this chapter is to briefly overview in a semi-formal and illustrative way the basic modeling concepts and the main techniques for qualitative analysis. It can be said that Petri nets are suited for parallel systems even more than are automata for sequential systems. Anyhow, the main practical argument for employing PNs should be the use of a graphical, easy to understand single family of formalisms through all the different stages from the design until the implementation and operation.

Although many recent results are integrated in this text, the main line of argument closely follows Silva (1985). The chapter is basically organized in two parts. The first one (up to section 1.5) is devoted to different modeling issues. The second part, sections 1.6 and 1.7, is mainly devoted to qualitative analysis. More precisely, the chapter is structured as follows. Net structure and the dynamics of net systems are introduced in section 1.2. Some interesting features for the modeling of concurrent discrete event dynamic systems are presented in section 1.3. Section 1.4 introduces in an illustrative way some possible interpretations of net systems models. The existence of pathological behaviors on concurrent systems leads to the introduction of some basic qualitative properties in section 1.5. Their analysis is done in section 1.6, overviewing reachability graph (section 1.6.1), net system reductions (section 1.6.2) and linear algebra techniques (section 1.6.3). Section 1.7 is devoted to some basic net subclasses and their analysis. Obviously, the more restrained the net subclass is, the more powerful the analysis techniques are. Concluding and bibliographical remarks end this introductory presentation.

1.2 NETS AND NET SYSTEMS

A Petri net model of a dynamic system consists of two parts:

1. a **net structure**, a weighted-bipartite directed graph, that represents the static part of the system; and
2. a **marking**, representing a **distributed overall state** on the structure.

The above separation allows one to reason on net-based models at two different levels: **structural** and **behavioral**. Reasoning at the structural level we can derive some 'fast' conclusions on the behavior of the modeled system, relating when possible structural and behavioral properties. Purely behavioral reasonings are computationally very complex.

1.2.1 Net structure (what is a Petri net?)

To model a discrete-event-dynamic system we need to take into account its states and the events leading to the state-evolutions. In net systems the state is described by means of a set of **state variables** representing local conditions. Moreover, net models make explicit the existence of **state-transitions**. Therefore net structures are built on two disjoint sets of objects: **places** (represented as circles), and **transitions** (represented as bars or boxes). Places are the support of the state variables.

Places and transitions are related through a **weighted flow relation**, described by an unweighted flow relation, F, and a weighting function on F, W. Let us now give the formal definitions and see some examples.

Definition 1.1. A **Petri net** is a four-tuple:

$$N = \langle P, T, F, W \rangle$$

where:

P is a finite non-empty set of $n = |P|$ **places**
T is a finite non-empty set of $m = |T|$ **transitions**
$P \cap T = \emptyset$; i.e. places and transitions are disjoint sets
$F \subset (P \times T) \cup (T \times P)$ is the **flow relation** (set of directed arcs):
  $dom(F) \cup range(F) = P \cup T$
$W : F \rightarrow \mathbf{N}^+$ assigns a **weight** to each arc.

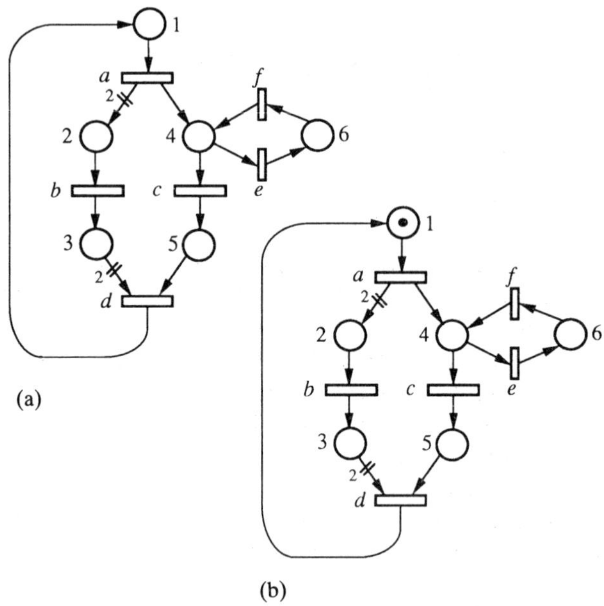

Figure 1.1 Net structure and net system: (a) N, net structure; (b) $\langle N, M_0 \rangle$, net system.

Figure 1.1(a) shows a net structure. Arcs are labeled with natural numbers, $W(p_i, t_j)$ or $W(t_i, p_j)$, the **arc weights**. As will be seen, non-unitary arc weights allow us to model **bulk arrivals** or **bulk services**. By convention, unlabeled arcs are weighted one. All the arc weights in the net of Fig. 1.1 are 1, except for arcs (a, p_2) and (p_3, d) whose weights are 2. In many practical cases there exists neither bulk arrival nor bulk service. Therefore all the arc weights are one. In this case the net is said to be **ordinary**.

A place p is an input (output) place of transition t if there exists an arc going from p to t (output respectively from t to p). In Fig. 1.1, $\{p_3, p_5\}$ are input places of d while $\{p_2, p_4\}$ are output places of a.

An alternative way to see Petri nets is to define the weighted flow relation through two *incidence functions*:

Definition 1.1(a). A **Petri net** is a four-tuple:

$$N = \langle P, T, Pre, Post \rangle$$

where:

P and T are disjoint, finite, non-empty sets of places and transitions, respectively

Pre: $P \times T \to N$ is the **pre-incidence** or **input** function

Post: $T \times P \to N$ is the **post-incidence** or **output** function

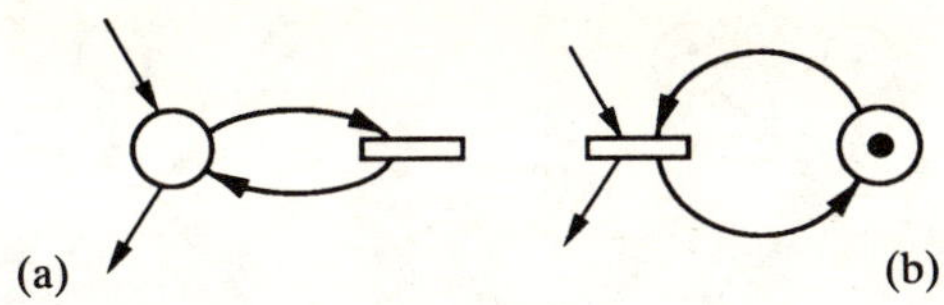

Figure 1.2 Two self-loops.

There is an arc going from the place p_i to the transition t_j iff $Pre(p_i, t_j) \neq 0$. Similarly, there is an arc going from transition t_k to place p_i iff $Post(t_k, p_i) \neq 0$. The arc weight, $Pre(p_i, t_j) = W(p_i, t_j)$ or $Post(t_k, p_i) = W(t_k, p_i)$, labels the corresponding arc. The **pre-** and **post-set** of transition $t \in T$ are defined respectively as $^{\bullet}t = \{p \mid Pre(p, t) > 0\}$ and $t^{\bullet} = \{p \mid Post(t, p) > 0\}$. The **pre-** and **post-set** of a place $p \in P$ are defined respectively as $^{\bullet}p = \{t \mid Post(t, p) > 0\}$ and $p^{\bullet} = \{t \mid Pre(p, t) > 0\}$.

A practical way of representing the net structure is to use **incidence matrices.** The incidence functions can be represented by means of **pre-** and **post-incidence** matrices, *Pre-* and *Post-*, both having $n = |P|$ rows and $m = |T|$ columns.

The pre- and post-incidence matrices of the net in Fig. 1.1(a) are as follows:

$$Pre = \begin{array}{c} \\ p_1 \\ p_2 \\ p_3 \\ p_4 \\ p_5 \\ p_6 \end{array} \begin{array}{c} \begin{array}{cccccc} a & b & c & d & e & f \end{array} \\ \begin{pmatrix} 1 & 0 & 0 & 0 & 0 & 0 \\ 0 & 1 & 0 & 0 & 0 & 0 \\ 0 & 0 & 0 & 2 & 0 & 0 \\ 0 & 0 & 1 & 0 & 1 & 0 \\ 0 & 0 & 0 & 1 & 0 & 0 \\ 0 & 0 & 0 & 0 & 0 & 1 \end{pmatrix} \end{array} \qquad Post = \begin{array}{c} \\ p_1 \\ p_2 \\ p_3 \\ p_4 \\ p_5 \\ p_6 \end{array} \begin{array}{c} \begin{array}{cccccc} a & b & c & d & e & f \end{array} \\ \begin{pmatrix} 0 & 0 & 0 & 1 & 0 & 0 \\ 2 & 0 & 0 & 0 & 0 & 0 \\ 0 & 1 & 0 & 0 & 0 & 0 \\ 1 & 0 & 0 & 0 & 0 & 1 \\ 0 & 0 & 1 & 0 & 0 & 0 \\ 0 & 0 & 0 & 0 & 1 & 0 \end{pmatrix} \end{array}$$

A pair of place p and transition t is called a **self-loop** if p is both an input and output place of t. A Petri net is said to be **pure** if it has no self-loops.

Figure 1.2 shows two self-loops. A self-loop can be easily eliminated (e.g. by expanding the transition into a sequence: initial transition – intermediate place – final transition). Pure nets are completely characterized by the (single) *incidence matrix*:

$$C = Post - Pre$$

Positive (negative) entries in C represent the post- (pre-) incidence function. If the net is not pure, the incidence matrix 'does not see' the self-loops.

1.2.2 Net systems: marking and token game

The structure of a net is something static. Assuming that the behavior of the system can be described in terms of system state and its changes, the

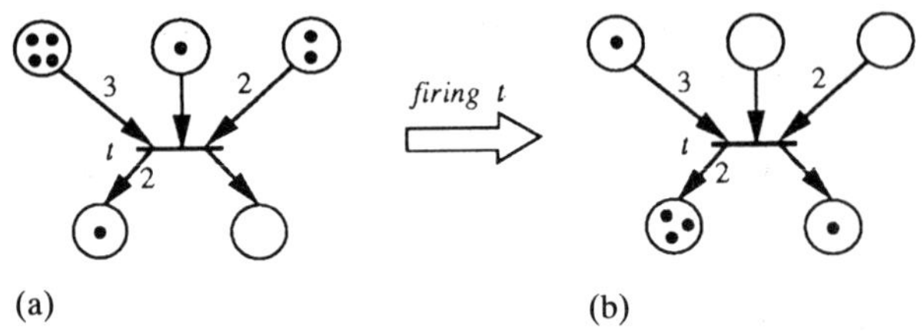

Figure 1.3 Firing transition t: marking evolution.

dynamics of a net structure are created by defining its **marking** and **marking evolution rule**.

Definition 1.2. The marking M of a net N is an application of P on $\mathbf{N}$, i.e. the assignment of a non-negative integer (number of tokens) to each place.

Definition 1.3. A **marked Petri net or net system** is the couple $\langle N, M_0 \rangle$, where N is a Petri net and M_0 is an initial marking.

The number of tokens at a place represents the **local state** of the place (i.e. the value of the state variable). The state of the overall net system is defined by the collection of local states of the places. A marking M is denoted as an $n = |P|$ vector whose pth component, $M(p)$, represents the number of tokens in place p. The vector M is the **state-vector** of the discrete event dynamic system described by the net system. Pictorially, we place $M(p)$ black dots (tokens) in the circle representing place p. Figure 1.1(b) represents a net system with an initial marking $M = (1, 0, 0, 0, 0, 0)^T$.

Once the distributed state is defined, the question is: how does a net system work? The evolution is defined through a **firing** or **occurrence rule**, informally named the '**token game**'. This is because net structures can be seen as 'special checkers', the tokens as 'markers' and the firing rule as the 'game rule'. Transitions represent potential moves in the 'token game'.

Definition 1.4 (token game). A marking in a net system evolves according to the following **firing** (or **occurrence**) **rule**:

1. A transition is said to be **enabled at a given marking** if each input place has at least as many tokens as the weight of the arc joining them.
2. The **firing** or **occurrence** of an enabled transition is an instantaneous operation that removes from (adds to) each input (output) place a number of tokens equal to the weight of the arc joining the place (transition) to the transition (place).

The pre-condition of a transition can be seen as the **resources required** for the transition to be fired. The post-condition represents the **resources produced** by the firing of the transition.

Transition t (Fig. 1.3(a)) is enabled. Its firing leads to the marking in Fig. 1.3(b), where t is no longer enabled. The only transition enabled in the net system of Fig. 1.1(b) is a. Its firing leads to the marking $M = (0, 2, 0, 1, 0, 0)^T$, where b, c and e are now enabled.

An important remark concerning the firing rule on our abstract model is that **enabled transitions are never forced to fire**. This is a form of **non-determinism**. In practical modeling the interpretation partially governs the firing of enabled transitions (e.g. depending on whether or not an external event associated to an enabled transition occurs). In section 1.3 we will come back to this important issue.

The enabling and firing of a transition can be represented in a very convenient way using incidence matrices and marking vectors. Let us denote the columns associated to t in the different incidence matrices as $Pre(t)$, $Post(t)$ and $C(t)$:

1. Transition t is enabled at M iff $M \geq Pre(t)$ (1.1)
2. Denoting as $M_1[t\rangle\, M_2$ the fact that M_2 is reached by firing t at M_1 (M_1 enables t):

$$M_2 = M_1 + Post(t) - Pre(t) = M_1 + C(t) \tag{1.2}$$

Assuming N to be pure (otherwise it can be easily transformed), it is not difficult to derive the following:

$$M_1[t\rangle\, M_2 \Leftrightarrow M_2 = M_1 + C \cdot e_t \geq 0 \tag{1.3}$$

where e_t is the **characteristic vector** of t: $e_t(x) :=$ **if** $x = t$ **then** 1 **else** 0

The right-hand side of the equivalence in eq. (1.3) is clearly a **state equation**: M_1 is the **present** state, M_2 the **next** state, e_t the input vector. Unfortunately classical control theory is not of great help to us when studying the dynamic behavior of net systems: the state (marking) and input vectors should take their values on non-negative integers.

Integrating the state equation from M_0 along a firing sequence $\sigma = t_i t_j \ldots$ leading to M_k (M_k is said to be reached from M_0 by means of σ) we can write:

$$M_0[\sigma\rangle\, M_k \Rightarrow M_k = M_0 + C \cdot \overline{\sigma} \geq 0,\ \overline{\sigma} \geq 0 \tag{1.4}$$

where $\overline{\sigma}$ is the **firing count vector** of σ: $\overline{\sigma}(t)$ is the number of times t has been fired in σ.

Equation (1.4) is called the **fundamental equation** or, more frequently, the **state equation** of the net system. (**Remark:** properly speaking the **state equation** is eq. (1.3), while eq. (1.4) is the **transition equation** in control theory terminology.)

The most important remark now is that only the right-hand implication exists in eq. (1.4). Otherwise stated, unfortunately a non-negative integer solution $\overline{\sigma} \geq 0$ of $M_k = M_0 + C \cdot \overline{\sigma} \geq 0$ does not imply there exists a σ such that M_k is reachable from M_0 (i.e. does not imply $M_0[\sigma\rangle\, M_k$). For example, assuming $M_0 = 0$ for the net in Fig. 1.1(a), $\overline{\sigma} = (1, 1, 1, 1, 0, 0)^T$ makes $M_k = M_0$. Nevertheless, there exists no σ firable from M_0! The integer couple $M_k = 0$, $\overline{\sigma} = (1, 1, 1, 1, 0, 0)^T$ is called a **spurious solution** of the state equation. The existence of spurious solutions is the main problem for the analysis of net systems using linear algebra techniques. However, many practical analysis results can be obtained using these techniques (see later sections 1.6.3 and 1.7.2).

1.3 ON MODELING FEATURES

Petri nets, as introduced thus far, are a mathematical formalism. This section presents a number of features which – in our opinion – make nets an interesting modeling formalism, specially suited for discrete-event dynamic systems with concurrent or parallel events and activities.

The considerations in this section are general, i.e. still on the abstract formalism, valid for any particular interpretation. Before concentrating on our main issue here, **on practical modeling**, it is important to highlight the fact that nets allow a natural **graphical** representation that makes them very much appreciated in engineering circles ('a picture is worth a thousand words!'). Nevertheless, big and not well-structured net models are difficult to understand and analyze. This means in practice that good modeling disciplines are very important.

As a preliminary remark on practical modeling, the reader can easily check the simplicity of representing with nets three basic modeling notions: **causal dependence** (e.g. sequence), **conflict** (decision, choice) and **concurrency**. Going back to our net system in Fig. 1.1(b) it is obvious that the firing of f must be done *after* (causal dependence) that of e. Also, it is clear that twice b and c must precede the firing of d. Moreover, c and e define a conflict. From any marking with $M(p_4) = 1$, transitions c and e are simultaneously enabled, but they cannot be simultaneously fired: **a decision must solve the conflict.**

As already mentioned, a major feature of nets is that they do not define in any way **how** and **when** (i.e. time independence) a given conflict should be solved, leading to **non-determinism** on its behavior.

Sequence and **conflict** are classical notions in sequential systems (e.g. in finite automata). **Concurrency** is a third concept that net systems represent in an extremely natural way. Informally speaking two transitions are concurrent at a given marking if they can be fired 'at the same time', i.e. simultaneously. Once transition a is fired in the net system at Fig. 1.1(b), the marking $M = (0, 2, 0, 1, 0, 0, 0)^T$ is reached. Then transitions b and c can be fired simultaneously. Moreover, because $M(p_2) = 2$, transition b can be fired concurrently to itself (idea of **re-entrancy**): **self-concurrency**.

Synchronizations are very important in the modeling of distributed and concurrent systems. How are synchronizations modeled with nets? Basically with those transitions with more than one input place (e.g. transition d in Fig. 1.1 or RV in Fig. 1.4(a)). A less obvious way of synchronization

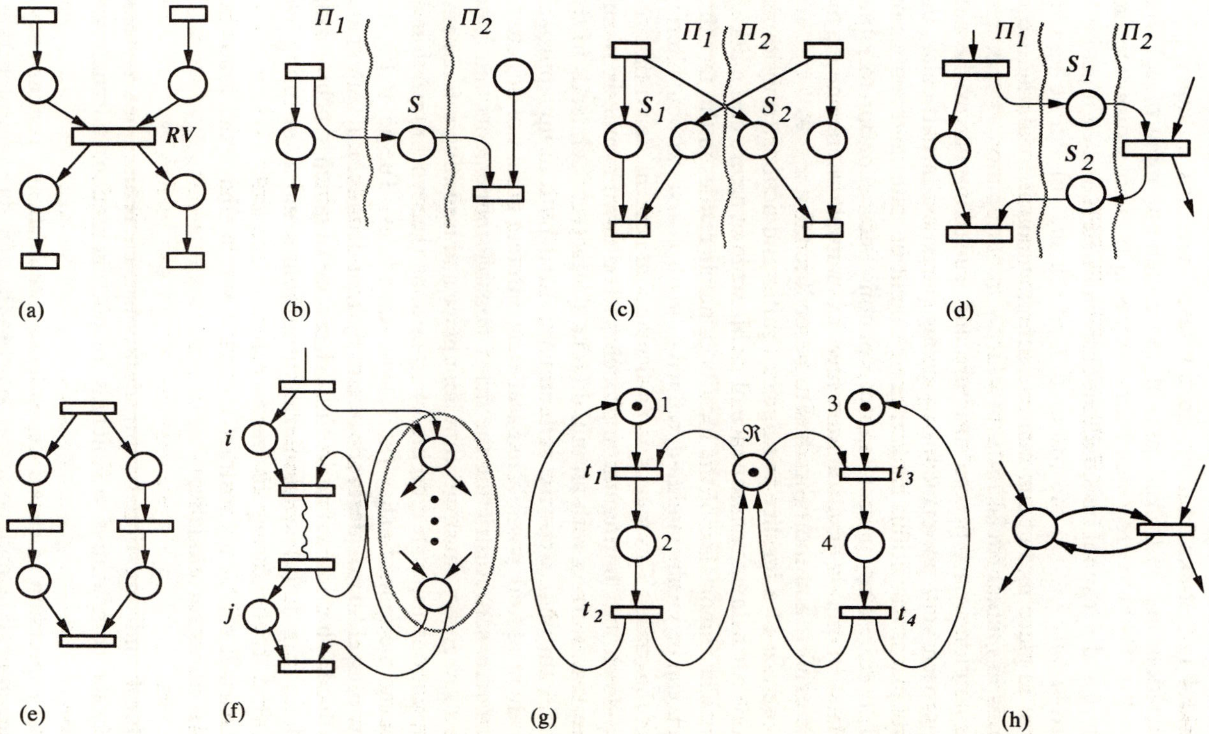

Figure 1.4 Typical synchronization schemes: (a) rendezvous, RV; (b) semaphore, S; (c) symmetric RV/semaphore; (d) asymmetric RV/semaphore (master/slave); (e) fork-join; (f) subprogram (p_i, p_j are in mutual exclusion, mutex); (g) shared-resource ($\mathfrak{R}$); (h) guard (condition reading).

may appear, even if a transition has only one input place, when the arc is weighted (e.g. assuming p_5 was not present in Fig. 1.1, transition d would also represent a synchronization because its firing would require the presence of two tokens on p_3).

Figure 1.4 is self-explanatory. In all cases nets are ordinary (thus synchronizations are on transitions with more than one input place). Just two remarks: (1) the correct behavior of schema (f) is based on the fact that both p_i and p_j cannot be simultaneously marked (i.e. they must be in *mutual exclusion*) and (2) the resource $\mathcal{R}$ of schema (g) can be used in place p_2 or in place p_4, but simultaneous use is impossible (i.e. the use in p_2 and p_4 is in mutual exclusion).

The separation in a bipartite structure and a marking makes the net-based approach very powerful for modeling purposes. In particular, the dichotomy places/transitions leads to a treatment of states and actions on an equal footing. This makes – in our opinion – nets superior to either purely state- or purely transition-oriented formalisms where one of the notions is explicit and the other has to be deduced.

The existence of a **locality principle** on states and actions (transitions) in net models is a direct consequence of its bipartite structure and marking definition. The importance of the locality principle resides in the fact that net models can be **locally modified**, **refined** or **made coarse**, without altering the rest of the model. This means, in particular, that nets can be synthesized using **top-down** and **bottom-up** approaches. Top-down synthesis is any procedure that, starting with an initial (very abstract) model, leads to the final model through **stepwise refinements**. In a bottom-up approach **modules** are produced, possibly in parallel by different groups of designers, and later **composed**. Restricting the many possible refinements and compositions strategies, we just mention here **place** and **transition refinements** and compositions through merging of transitions (i.e. **synchronization of modules**) and merging of places (i.e. **fusion of modules**). The net in Fig. 1.5 shows a two-level **hierarchical refinement**: p_5 (that defines local states) and θ_2 are refined. The net system in Fig. 1.1 can be obtained synchronizing two modules (Fig. 1.6(a), **synchronizing** transitions a and d) or fusing two modules (Fig. 1.6(b), **fusing** the places p_4).

Summarizing, still at an abstract level, net systems have the following practical features for modeling:

1. **Graphical** and **equational** representations. Therefore, net systems enjoy some comparative advantages for documentation and analytical studies.
2. Natural expression of **causal dependences**, **conflicts** and **concurrency**.
3. Simple, appealing and powerful **synchronization** mechanism making natural the construction of **mutual exclusion** constraints.
4. **Locality of states and actions** which allows the **hierarchical** and the **modular** construction of large net models.

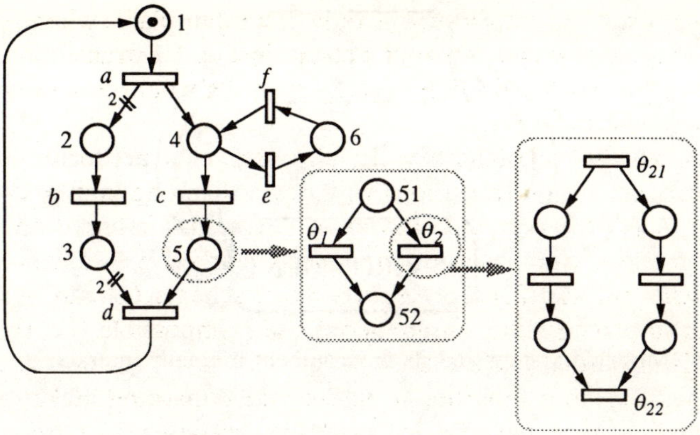

Figure 1.5 Hierarchical definition of a net system.

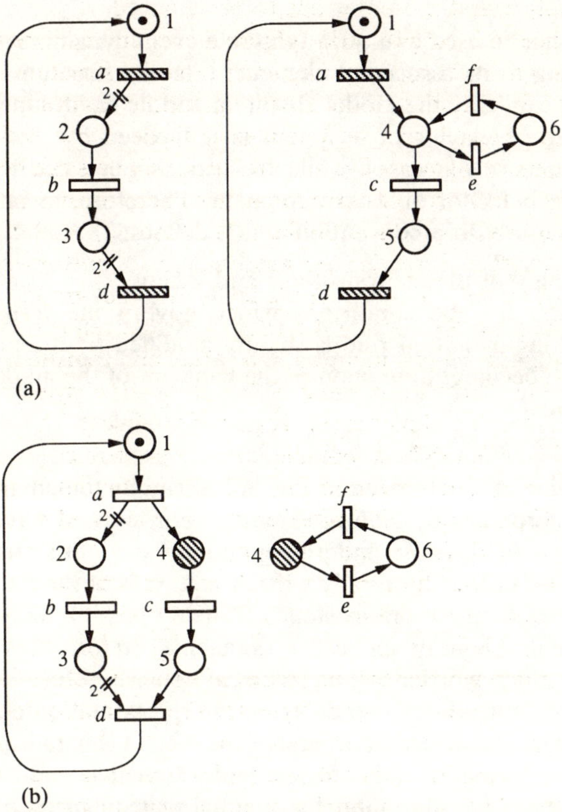

Figure 1.6 Two modular ways of constructing the net in Fig. 1.1: (a) synchronization; (b) fusion.

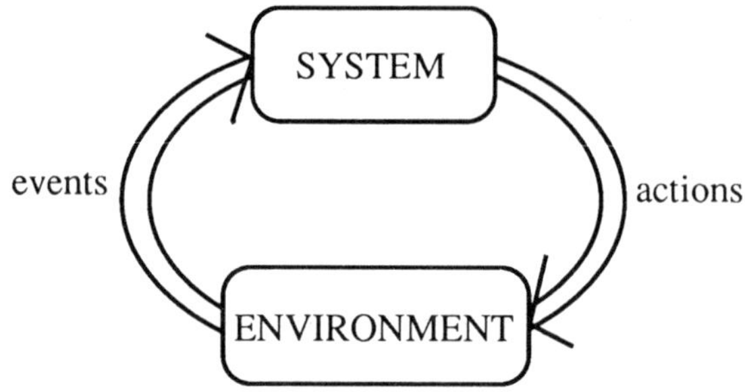

Figure 1.7 Modeled system and its environment mutually interact.

1.4 ON NET SYSTEMS INTERPRETATIONS

A Petri net can be used to model a discrete event dynamic systems assigning a **meaning** to its associated elements (places, transitions and tokens) and relating explicitly the modeled system and its **environment** (Fig. 1.7). In general the behavior of a system is influenced by the environment (through events in our case), while the actions generated by the system influence the behavior of its environment. Therefore, to interpret a net system is to establish a convention which defines:

1. The meaning of places, transitions and tokens.
2. A meaning for the conditions which govern the transition firing. The marking evolution rule is slightly modified by the interpretation, which also becomes a function of the behavior of the modeled system's **environment**.
3. The actions generated by the model.

If the behavior of a net system is not influenced by the environment, it is said to be **autonomous**. Non-autonomous net systems have more constrained behavior than the underlying autonomous net system.

The purpose of this informally written section is not to fix 'good' interpretations, but to show the existence of many possible ones, even for a given class of problems or application domain. Therefore, the reader should not be very much worried about technical details.

Section 1.4.1 introduces two different interpretations, one for modeling the control part of concurrent programs, the second generalizes the classical **state diagram** formalism (see, for example, Ercegovac and Lang (1985); Breeding (1989)) useful to model sequential switching systems. To clearly differentiate PNs as uninterpreted models from their different interpretations, **marking flow charts** and **marking diagrams** are the names given to the generalizations of flow charts and state diagrams.

1.4.1 Marking flow charts and marking diagrams

The environment of a net modeled system can affect the behavior of the net model and vice versa. Usually the interaction is done through:

(a) **Events** and/or **predicates** over some 'external states'. These 'guard' the firing of transitions (from the environment to the net modeled system).
(b) **Actions** that, generated by the net modeled system, cause the state of the environment to 'change'.

Depending more precisely on the application domain (software, hardware controllers, logical automatisms, etc.), many interpretations exist. In some cases, actions are associated with the **firing** of transitions (as in **Mealy-Automata**); in other cases actions are associated with the **marking** of places (as in **Moore-Automata**).

Even for a given application domain, there exist many possible interpretations. Thus no formal or rigid definition of interpretations will be given. Only two cases are considered for illustration purposes. First there is a collection of comments on what net interpretations look like to model the control part of concurrent software systems. Later a discrete-event-controller (a production cell with two machines, one robot and a store) is modeled using marking diagrams.

(a) Marking flow charts

In modeling **software** the more natural interpretation is based on the classical **control part (CP)–operative part (OP)** decomposition. Using the CP–OP decomposition principle, the state of a program (sequential or concurrent) can be considered as the concatenation of a **control state** (for the CP) and a **data state** (for the OP). Places will represent parts of the control state (defined by the marking of the net), while predicates over the data state are associated with the transitions. Predicates allow decisions to be taken (i.e. conflicts to be solved).

Therefore the firing of transitions is governed by the net marking and predicates on the data. Data transformations can be associated with the firing of a transition (usually implemented by means of sequential modules activated by the firing of the transition).

For a given program, the **execution of an instruction** (at a more abstract level a **block** of instructions or **module**) will be represented by the firing of a transition. The preincidence (input) function of each transition determines a condition which must be fulfilled for the instruction (or the block) to be executed. In general, with each transition a label, consisting of a 'predicate over data/data transformation' couple, is associated. If a transition is not conditioned by a predicate, that field is omitted in the label. If a transition does not represent any data transformation, the

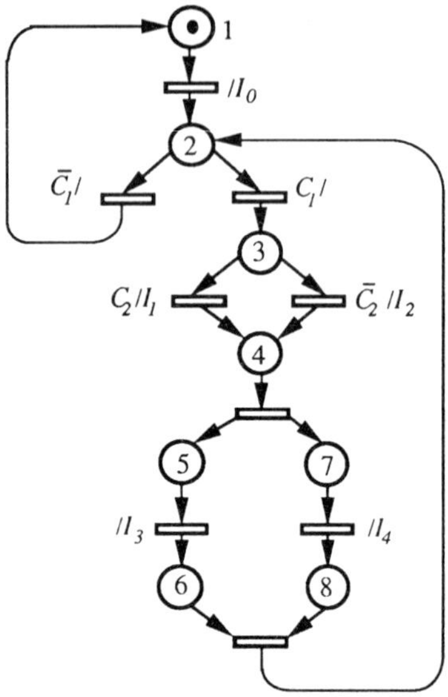


```
loop
  I0
  while C1 do (*p2*)
  | if C2 then I1 (*p3*)
  |         else I2
  | endif (*p4*)
  | par.begin  I3, I4 (*p5, p7*)
  | par.end  (*p6, p8*)
  endwhile (*p2*)
endloop
```

Figure 1.8 Flow control representation of a simple parallel-PASCAL-like program.

corresponding field is omitted. The complete condition for firing a transition is the intersection of the net system enabling condition and, possibly, the associated **predicate over data**.

According to the above comments, at a certain level of abstraction the interpreted net system models the control part, while the operative part is implemented through the modules associated with the firing of transitions. Figure 1.8 is practically self-explanatory.

(b) Marking diagrams

In modeling **discrete-event-controllers**, **state diagrams** (see, for example, Ercegovac and Lang (1985); Breeding (1989)) are classical interpretations of graphs allowing the modeling of **finite sequential switching automata**.

In state diagrams, the nodes of the graph represent the states, while the arcs are labeled with external events and external state conditions. Actions are associated with the transitions (arcs) or the states (nodes). Eventually actions may be conditioned by the external state. State diagram interpretations can be easily applied to Petri net systems leading to some marking diagrams (MD): (a) events and Boolean functions of external variables guard the firing of transitions, and (b) actions can be associated with transitions (level-actions) or to the marking of places.

Let us concentrate now on our manufacturing domain through an example. It consists of a **producer** (MACHINE 1) – **buffer** (a store of seven positions)–

consumer (MACHINE 2) schema with a **mutual exclusion semaphore** (*R*, a robot) (Fig. 1.9(a)). Many production systems can be constructed concatenating different stages of such schema.

The following behavior is assumed for the cell (Fig. 1.9(b)). Raw parts arrive through a conveyor. The arrival of a part is detected by a presence (e.g. photoelectric) sensor: $\Pi_1 = 1$ iff a part is present. When a raw part is present, MACHINE 1 is not loaded and the robot is free, it proceeds to load the machine (*load*; *el*: end_of_load). The machine performs operations op_1 and waits for deposit in the buffer (*wait-dep.*).

The *deposit* is done when there is an empty niche in the buffer and the robot is free again. End_of_deposit, *ed*, is represented by a transition. MACHINE 2 proceeds in an analogous way, but once op_2 has finished, eop_2 ($end\_of\_op_2$) waits for the robot to perform the *unloading*, assuming the second sensor detects that there is any part at the beginning of the finished parts conveyor ($\Pi_2 = 0$ iff the conveyor is free).

The net system model in Figure 1.9(b) (let us call these interpreted net models marking diagrams) specifies the above behavior. It is labeled with external conditions at transitions (Π_1 label t_1, $\overline{\Pi_2}$ label t_7), and actions at places (*load*, op_1, *deposit*, op_2, *unload*, *withdrawal*).

1.4.2 Timed net systems

Uninterpreted Petri nets do not include any notion of time and are aimed to model only the logical behavior of systems by describing the causal relations existing between events. The introduction of a timing specification is essential if we want to use this class of model to consider **performance**, **scheduling** or **real-time control** problems.

Timing and firing process. Since Petri nets are bipartite graphs, historically there have been two ways of introducing the concept of time in them, namely, associating a time interpretation with either places or transitions. Because transitions represent activities that change the state (marking) of the net, it seems 'natural' to associate a duration with these activities (transitions). In order to solve conflicts between transitions many authors tend to define a 'timed firing' of transitions in **three phases**: a first instantaneous phase in which an enabled transition removes tokens from its input places, then a **timed phase** in which the transitions are 'working', and a final instantaneous phase in which tokens are deposited into the output places. If we want to model **pre-emption** of activities after their starting, however, we are forced to associate an enabling time with transitions and define **atomic firing.** In this way conflicts should be solved at the end of the delay of enabled transitions. The solution is always in favor of the first transition that elapses its firing time among the conflicting transitions (Ajmone *et al.*, 1989).

In any case, from the above discussion it follows that the only effect of these different timing interpretations on the behavior of a model is due

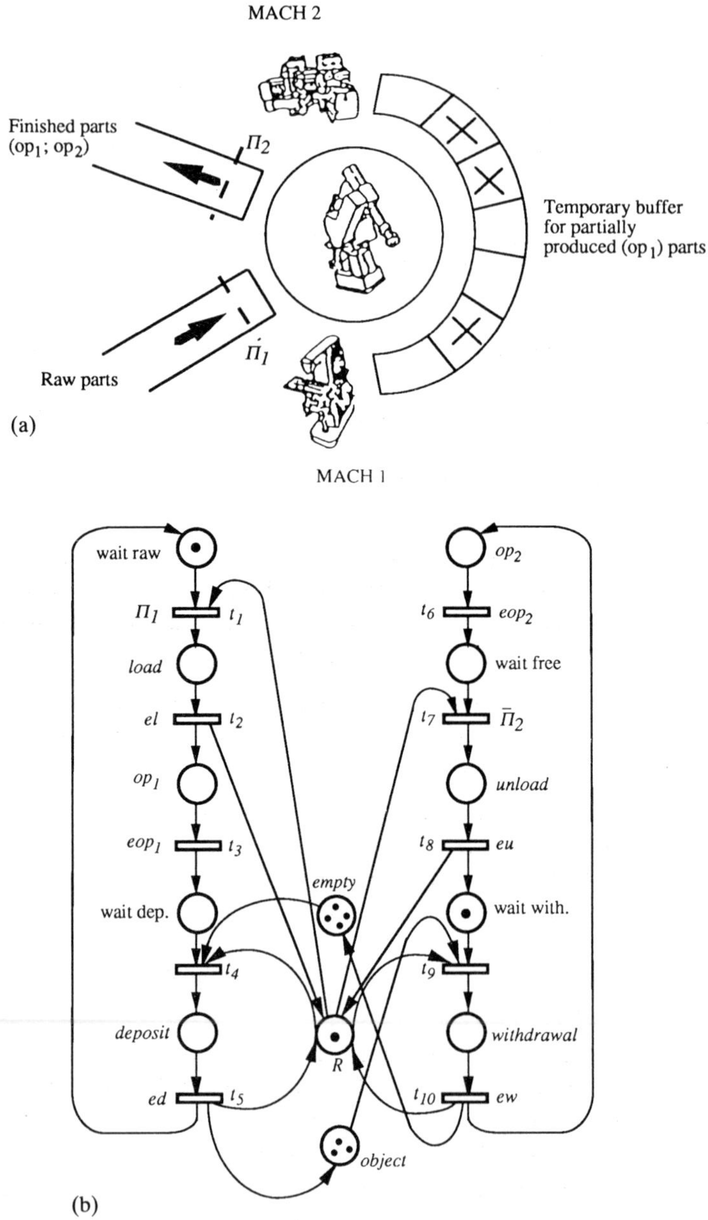

Figure 1.9 A production cell with two machines, one robot and a store: (a) Schema of a manufacturing cell; (b) Net system specifying the behavior.

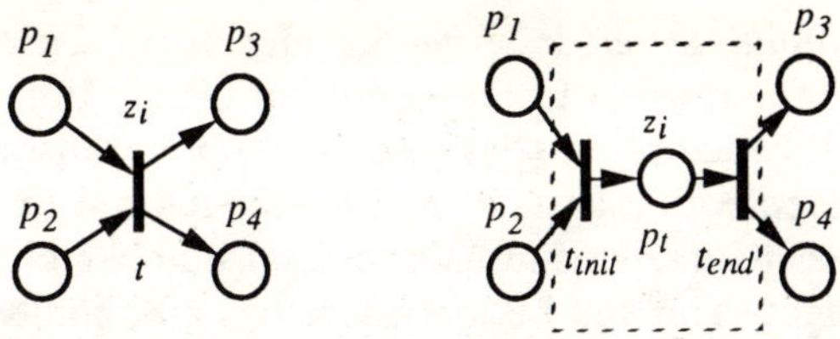

Figure 1.10 Visualization of the 3 phases firing of a timed transition: place p_t will hold each token for z_i time units.

to the different implications that the choices have on the **resolution of conflicts**.

Let us now give in a more precise way two alternative definitions for **deterministically** timed Petri nets. The reader is left to reflect on the relationship between the two. For illustration purposes only **three phases timed firing** is now considered.

Definition 1.5. A (deterministically) transition-timed Petri net (t-TPN) is a couple $\langle N, Z \rangle$ such that $N = \langle P, T, Pre, Post \rangle$ and Z is a function which assigns a non-negative real number, z_i, to each transition in the net: $Z{:}T \to \mathbf{R}^+$; $z_i = Z(t_i)$ is called the **firing time** of the transition t_i.

The marking evolution rule of t-TPNs is nearly identical to that of a PN. The only difference is that firing t_i takes z_i time units. Adopting the three phases approach (Fig. 1.10):

1. When the transition t is enabled, a firing is initiated. Conflicts are non-deterministically solved as in the untimed net system. In this phase $Pre(p, t)$ tokens disappear from each input place of t.
2. The firing process (which represents an operation) remains for z_i time units.
3. When the z_i time units have elapsed, the firing ends. In this phase $Post(t, p)$ tokens are added to each output place of t.

Definition 1.5(bis). A (deterministically) place-timed Petri Net (p-TPN) is a couple $\langle N, R \rangle$ such that $N = \langle P, T, Pre, Post \rangle$ and R is a function which assigns a non-negative real number r_i to each place in the net: $R{:}P \to \mathbf{R}^+$; $r_i = R(p_i)$ is the minimum **residence time** of a token in p_i.

A token in a p-TPN can be in either of two states: **ready** or **not ready**. If the tokens are ready the marking evolution rule is the same as for an autonomous net system. Not ready tokens do not enable transitions, as though they were not present yet. When a token reaches a place, it goes into the non-ready state, and becomes ready again after an interval of $r_i = R(p_i)$ time units.

In basic timed PN models, conflict resolution strategies are not specified: they are purely **non-deterministic**.

Single versus multiple server semantics: degree of self-concurrency. A possible source of confusion in the definition of any timed net model is related to the **self-concurrency** (or **re-entrance**) of a transition. In the case of timing associated with places, it seems quite natural to define **unavailability time** which is independent of the total number of tokens already present in the place. This can be interpreted as an '**infinite server**' policy from the **queuing theory** perspective. In the case of time associated with transitions, the adopted semantics is less obvious. Assume that transition t is k-enabled at a given marking. Then either one firing of t occurs at that time or k firings occur in parallel (i.e. the idea of one or more servers).

Thus **single server** and **infinite server** semantics can be considered in transition-timed net models. Of course an infinite server transition can always be constrained to a 'k-server' behavior by just adding a self-loop place around the transition (i.e. an input and output place) with k-tokens (e.g. the self-loop place in Fig. 1.2(b) constrains the transition to a single server). Therefore, the infinite server semantic appears to be the most general one. However, this generality of the infinite server assumption is usually paid in terms of complexity of the analysis algorithms (e.g. for computing performance figures).

Some comments on stochastic net systems. Definitions 1.5 and 1.5(bis) consider only deterministic timing of transitions. In many cases the timing is not deterministic, being characterized by the probability distribution function (PDF) of a random variable. In this case **stochastic net systems** are defined. The most usual approach is to consider one-phase firing **transition-timed models**. Stochastic Petri nets are defined in Chapter 4. The following informal comments (that can be skipped without affecting the comprehension of the rest of the material) try to point out that stochastic Petri nets can be viewed as queuing networks provided with a semantically simple and formal way of introducing synchronizations among queues.

The consideration of stochastic Petri nets as mentioned above is limited in practice by the fact that routing probabilities' are not naturally expressible when one-phase firing is assumed for transitions. The advantage of one-phase over three-phase firing is that tokens do not 'disappear' as in the second phase of the three-phase semantics; thus token (e.g. customers) conservation laws of the uninterpreted net system are preserved.

Generalized stochastic Petri nets (GSPN) have (one-phase) stochastically timed transitions and immediate transitions. Immediate transitions fire in zero time (i.e. instantaneous firing). Conflicts among immediate transitions are solved, among other mechanisms, using routing probability schemas. Because a single-phase firing semantics is used, immediate transitions are **prioritized** with respect to timed transitions (see Ajmone *et al.* (1984) for the seminal definition of GSPNs, and Ajmone *et al.* (1989)).

From a conceptual point of view the modeling power of GSPNs is greater than **extended queuing networks** (EQN), that were introduced to partially overcome the lack of expressibility of synchronizations in queuing networks.

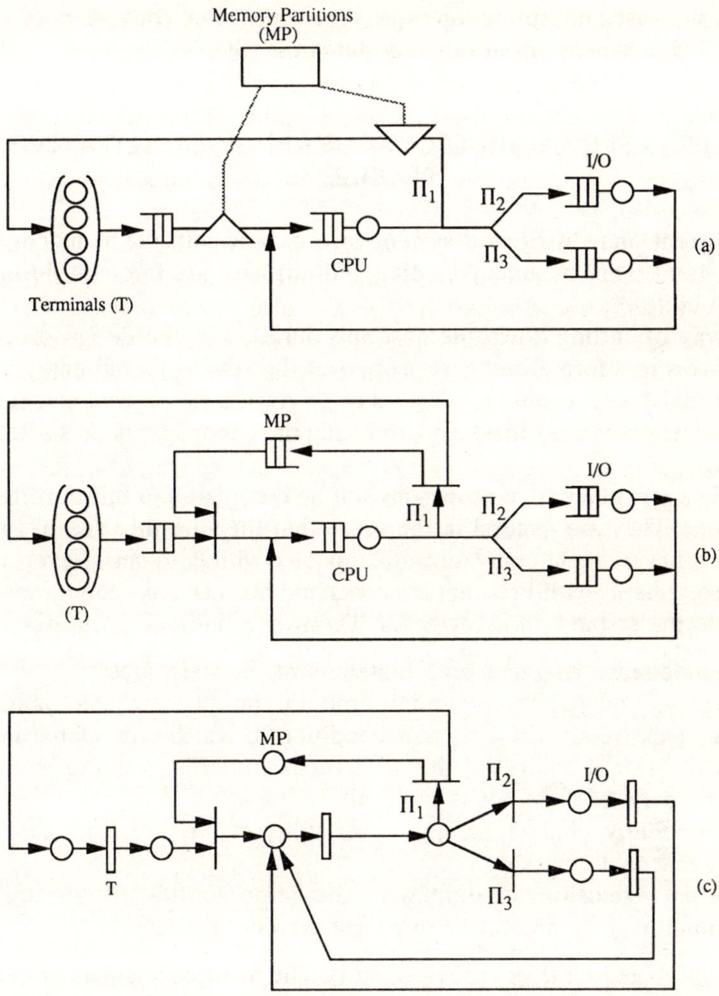

Figure 1.11 Central server representations (Mailles, 1987): (a) extended queuing network; (b) Descriptive queuing network; (c) Generalized stochastic Petri net (a free choice net with three conflicting immediate transitions: Π_1, Π_2, Π_3).

More precisely EQNs allow us to express particular synchronization schemas as:

- fork and joins
- passive resources
- replicated customers.

Figure 1.11 (Mailles, 1987) is almost self-explanatory. As a summary, stochastic Petri net systems can be viewed as Petri net systems provided with a stochastic interpretation or, as sometimes more convenient (Campos *et al.*, 1991), **synchronized queuing networks** (SQNs).

1.5 APPROACHING CONCURRENCY QUALITATIVE PROBLEMS

Concurrent and distributed systems are usually difficult to manage and understand. Thus misunderstanding and mistakes are frequent during the design cycle.

A way of cutting down the cost and duration of the design process is to express in a formalized way properties the system should enjoy and to use formal proof techniques. Errors could eventually be detected close to the moment they are introduced, reducing their propagation to subsequent stages.

Only a few qualitative properties will be considered in this introductory chapter.* They are general in the sense that they are meaningful for any concurrent system, not only for those modeled with Petri nets. Nevertheless, their statements with Petri net concepts and objects make them especially 'easy to understand' in many cases. The properties to be considered are:

1. **Boundedness**, characterizing finiteness of the state space.
2. **Liveness**, related to potential firability in all reachable markings. **Deadlock-freeness** is a weaker condition in which only global infinite activity (i.e. firability) of the net system model is guaranteed, even if some parts of it do not work at all.
3. **Reversibility**, characterizing recoverability of the initial marking from any reachable marking.
4. **Mutual exclusion**, dealing with the impossibility of simultaneous **submarkings** (p-mutex) or **firing concurrency** (t-mutex).

Let us consider the net in Fig. 1.12(a). The firing of t_2 allows us to reach the marking $M = (0, 0, 1, 1)^T$ (i.e. p_3 and p_4 have one token). Firing now t_4, $M^1 = (1, 0, 1, 0)^T$ is reached. Repeating ω times the sequence t_2t_4 the marking $M^\omega = (1, 0, \omega, 0)^T$ is reached. So the marking of p_3 can be arbitrarily high. In practice the capacity of the physical element represented by p_3 should be finite, so an **overflow** can appear. Place p_3 is said to be **unbounded**. Attention must be paid to the above situation because unboundedness can be a pathological situation. System boundedness (i.e. all places bounded) is a good behavioral property.

The maximum number of tokens a place may contain is its (marking) **bound**. A place is bounded if its bound is finite. A net system is bounded if each place is bounded.

* Quantitative properties (e.g. transitions throughput, utilization ratios, etc. . . .) are considered in Chapter 4, devoted to performance evaluation.

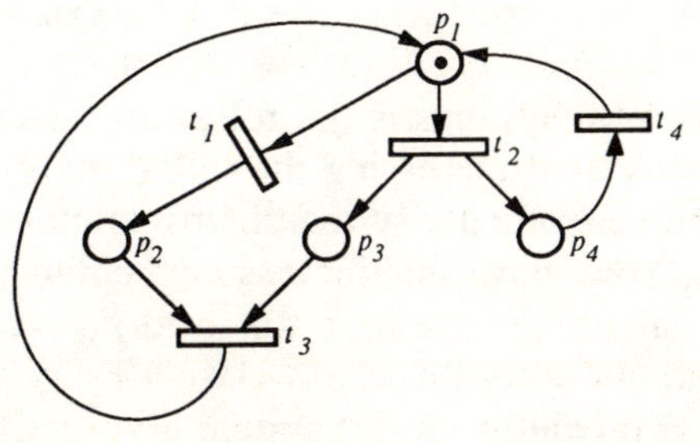

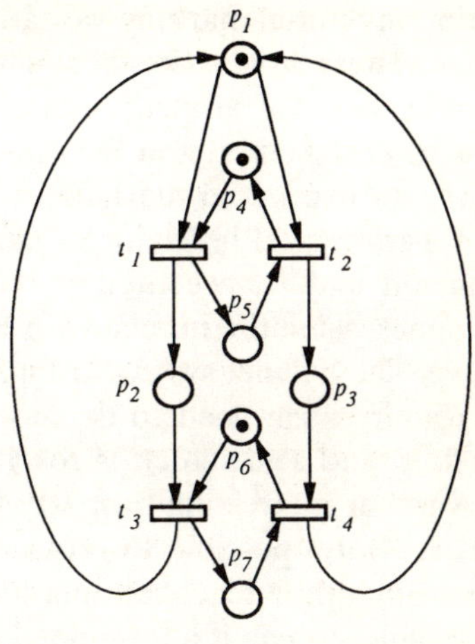

(b)

Figure 1.12 On qualitative pathological behaviors: (a) an unbounded, deadlockable (non-live), non-reversible net system; (b) Increasing the initial marking (e.g. $M_0(p_5) = 1$) the live net system is killed!

For any initial marking we can define on the net structure of Fig. 1.1(a) the following token conservation laws hold:

$$2M(p_1) + M(p_2) + M(p_3) = 2M_0(p_1) + M_0(p_2) + M_0(p_3) = K_1(M_0)$$
$$M(p_1) + M(p_4) + M(p_5) + M(p_6) = M_0(p_1) + M_0(p_4) + M_0(p_5) + M_0(p_6) = K_2(M_0)$$

where M_0 is the initial marking and M any reachable marking. Therefore:

$$M(p_1) \leq \min(K_1(M_0)/2, K_2(M_0))$$
$$M(p_i) \leq K_1(M_0) \quad i = 2, 3$$
$$M(p_j) \leq K_2(M_0) \quad j = 4, 5, 6$$

The above inequalities mean that *for any* M_0 the net system is bounded. This property, stronger than boundedness, is called **structural boundedness** because it holds independently of the initial marking (only finiteness of M_0 is assumed).

Let us now fire t_1 from the marking in Fig. 1.12(a). After that, no transition can be fired: a **total deadlock** has been reached. A net system is said to be **deadlock-free** (i.e. from any reachable marking) if at least one transition can always be fired. A stronger condition than deadlock-freeness is **liveness**. A transition t is potentially firable at a given marking M if there exists a transition firing sequence σ leading to a marking M' in which t is enabled (i.e. $M[\sigma\rangle M' \geq Pre(t)$). A transition is **live** if it is potentially firable in all reachable markings. In other words, a transition is live if it never loses the possibility of firing (i.e.of performing some activity). A net system is live if all the transitions are live.

For any initial marking we can define on the net structure in Fig. 1.12(a) non-liveness holds (in fact, a total deadlock can always be reached). Non-liveness for arbitrary initial markings reflects a pathology of the net structure: **structural non-liveness.** A net is structurally live if there exists at least one live initial marking.

A paradoxical behavior of concurrent systems is the following: at first glance it may be accepted as intuitive that increasing the initial marking (e.g. increasing the number of resources) of a net system 'helps' in making it live. The live net system in Fig. 1.12(b) shows that increasing the number of resources can lead to deadlock situations: adding a token to p_5, t_2 can be fired and a deadlock is reached!

Another interesting property is **reversibility**. A net system is reversible if it is always possible to return to the initial marking (i.e. it is reachable from any other reachable marking).The net system in Fig. 1.12(a) is not reversible. In fact if a total deadlock exists at some reachable marking, the net system cannot be reversible; the reverse is not true as is pointed out in Fig. 1.13(6), where the net system is not reversible but live, thus deadlock-free.

Liveness, boundedness and reversibility are just three different 'good' behavior properties that may be interesting to study in a net system. Figure 1.13 shows examples of the eight cases we may have. Therefore boundedness, liveness and reversibility are independent properties.

The last basic property we introduce in this section is **mutual exclusion.** This property captures constraints like the impossibility of a simultaneous access by two robots to a single store. Two places (transitions) are in mutual exclusion if they can never be simultaneously marked (fired). For the net system in Fig. 1.4(g) we can write: $M(p_2) + M(p_4) + M(\mathcal{R}) = 1$. Thus

$$M(p_2) = 1 \Rightarrow M(p_4) = M(\mathcal{R}) = 0$$
$$M'(p_4) = 1 \Rightarrow M'(p_2) = M'(\mathcal{R}) = 0$$

and $[M(p_2) = 0]$ or $[M(p_4) = 0]$ is true for every reachable marking (i.e. p_2 and p_4 are in mutual exclusion). Table 1.1 summarizes the definitions of the different properties we introduced in this section.

1.6 QUALITATIVE ANALYSIS OF NET SYSTEM MODELS

Techniques for analyzing net systems can be divided into the following groups:

1. analysis by enumeration
2. analysis by transformation
3. structural analysis
4. analysis by simulation.

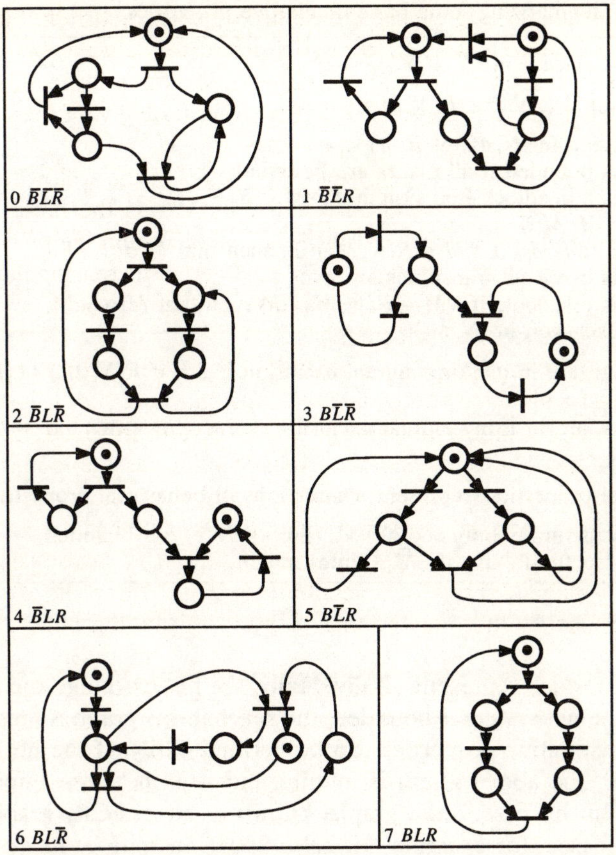

Figure 1.13 Boundedness (B), liveness (L) and reversibility (R) are independent properties.

The first three groups are called **static methods**, and their application to nets systems as abstract models leads to **exact results**. Simulation methods are called **dynamic** and proceed exercising the net system model under certain strategies. In this case some bugs can be detected (e.g. some deadlocks), allowing 'some confidence on the model', if problems are not manifested during the simulation process. However, in general, simulation methods do not allow properties to be proved, even if they might be of great help in understanding the modeled system. In particular, simulation methods are extremely useful when time is associated with the net evolution (**timed** systems), or when we wish to know the response of the system described with a net in an environment which is also defined by simulation. In this section we will only overview some static methods applied to autonomous nets.

Enumeration methods are based on the construction of a **reachability**

Table 1.1 Summarizing some basic qualitative properties

1. Bound of place p in $\langle N, M_0\rangle$

   $B(p) = sup\{M(p)/M \in R(N, M_0)\}$
2. p is bounded in $\langle N, M_0\rangle$ if $B(p) < \infty$
3. $\langle N, M_0\rangle$ is bounded if all places are bounded
4. $\langle N, M_0\rangle$ is a deadlock-free system if $\forall M \in R(N, M_0)\ \exists t \in T$ such that t is firable at M, $M[t\rangle$
5. t is live in $\langle N, M_0\rangle$ if $\forall M \in R(N, M_0)\ \exists\sigma$ such that $M[\sigma_t > M'$
6. $\langle N, M_0\rangle$ is live if all transitions are live
7. $\langle N, M_0\rangle$ is reversible if $\forall M \in \mathrm{R}(N, M_0)\ \exists\sigma$ such that $M[\sigma > M_0$
8. Mutual exclusion in $\langle N, M_0\rangle$:
   - p_i and p_j are in marking mutual exclusion if $\nexists M \in R(N, M_0)$ s.t.$[M(p_i) > 0] \wedge [M(p_j) > 0]$
   - t_i and t_j are in firing mutual exclusion if $\nexists M \in R(N, M_0)$ s.t. $M \geq Pre(t_i) + Pre(t_j)$
9. Structural properties (represent abstractions of behavioral properties):
   - N is structurally bounded if $\forall M_0$ (finite) $\langle N, M_0\rangle$ is bounded
   - N is structurally live if $\exists M_0$ (finite) making $\langle N, M_0\rangle$ a live system

graph (RG) which represents, **individually**, the net markings and transition firings. If the net system is bounded, the reachability graph is finite and the different qualitative properties can be verified easily. If the net system is unbounded, the above graph is infinite and it is therefore impossible to construct. In this case, finite graphs known as **coverability graphs** can be constructed (see, for example, Finkel (1990)). In spite of its power, enumeration is often difficult to apply, even in nets with few places, because of its computational complexity (it is strongly combinatorial).

The enumeration analysis technique for bounded stochastic Petri net systems consists of the generation of an underlying **Markov chain** (MC). All performance figures are later computed on the MC. Assuming exponential (i.e. memoryless) firing times, the MC is isomorphous to the RG.

Analysis by transformation is based on the following idea: given a net system $\langle N, M_0\rangle$ in which we wish to verify the set of properties Π, we transform it into the net system $\langle N', M'_0\rangle$ such that:

1. $\langle N', M'_0\rangle$ satisfies the properties Π iff $\langle N, M_0\rangle$ satisfies them (i.e. the transformation *preserves* the properties Π).
2. It is easier to verify the properties Π in $\langle N', M'_0\rangle$ than in $\langle N, M_0\rangle$.

Reduction methods are a special class of transformation methods in which a sequence of net systems preserving the properties to be studied is constructed. The construction is done in such a way that the net system $\langle N^{i+1}, M_0^{i+1}\rangle$ is 'smaller' (i.e. has less markings) than the previous in the sequence, $\langle N^i, M_0^i\rangle$.

The applicability of reduction methods is limited by the existence of **irreducible net systems**. Practically speaking, the reductions obtained are normally considerable, and can allow the desired properties to be verified directly. Because of the existence of irreducible systems, this method must be complemented by some other methods.

Finally, **structural analysis** techniques carefully consider the net structure (hence their name), while the initial marking acts basically as a parameter. Structural analysis techniques investigate the relationships between the behavior of a net system and the structure of the net. In this last class of analysis techniques, we can distinguish two subgroups:

1. **Linear algebra/Linear programming-based techniques**, which are based on the net system **state equation**. In certain analysis they permit a fast diagnosis without the necessity of enumeration.
2. **Graph-based techniques**, in which the net is seen as a bipartite directed graph and some *ad hoc* reasonings are applied. These methods are especially effective in analyzing restricted subclasses of ordinary nets.

The three groups of analysis techniques outlined above are by no means exclusive, but rather they are complementary. Normally the designer can use them according to the needs of the ongoing analysis process. Obviously, although we have distinguished between reduction and structural analysis methods, it must be pointed out that most popular reduction techniques act basically on the net structure level and thus can be considered also as structural techniques.

For what concerns the qualitative analysis of **interpreted systems**, it should be pointed out that the properties of the underlying autonomous model can be only sufficient (e.g. for boundedness), necessary (e.g. for reachability) or neither sufficient nor necessary (e.g. for liveness). For particular net subclasses (e.g. simple nets) liveness properties are preserved under 'reasonable assumptions' on the behavior of the environment (e.g. fair progress: i.e. not infinite delay in firing a continuously enabled transition; and local fairness on choices, i.e. all outcomes of a conflict are being chosen repeatedly).

1.6.1 Reachability graph

A marking is said to be reachable in a system $\langle N, M_0 \rangle$ if there exists a sequence σ applicable at M_0 such that $M_0[\sigma\rangle\, M$. If we were able to compute all reachable markings, $M \in R(N, M_0)$, and their reachability relationships, all qualitative behavioral properties should be analyzable. A major problem arises in systems in which the number of reachable markings is infinite (unbounded systems). Because infinite state (marking) systems cannot easily be represented by enumeration, finite representations have been proposed. But in this case it is possible that relevant information for a given property has been lost and some properties cannot be decided.

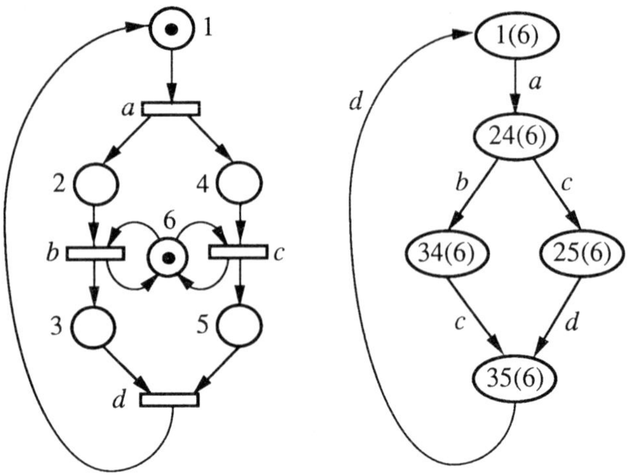

Figure 1.14 Bounded, live and reversible system and its reachability graph.

This presentation being more engineering oriented, let us restrict ourselves to the case of bounded systems.

Reachability analysis approach for bounded systems is based on the **exhaustive sequentialized simulation** of the possible marking evolutions. The main limitation of the approach is its computational complexity, so-called **state explosion problem**: the number of markings can be exponential with respect to the size of the net (measured, for example, by the number of places).

Definition 1.6. The **reachability graph** associated with a system $\langle N, M_0 \rangle$ is a graph $RG(N, M_0)$ in which each node represents a marking reachable form M_0 and each arc represents the firing of a transition. There exists an arc, labeled t_k, which goes from the node representing M_i to that representing M_j iff on firing t_k from M_i we reach M_j : $M_i[t_k\rangle\, M_j$.

If the marked net is bounded, the graph construction process is straightforward. It finishes when all the possible firings from the reachable markings have been explored.

Let us consider, for example, the system in Fig. 1.14, assuming p_6 is removed. Initially only a is enabled. Firing it, p_2 and p_4 are marked and b and c are enabled. If b is fired before, the marked places are p_3 and p_4, otherwise if c is fired before, p_2 and p_5 should be marked. From any of the last two markings p_3 and p_5 will be unavoidably marked and d being the only enabled transition, the initial marking should be recovered. Therefore our system has five markings (thus is **bounded**). From a direct inspection of the set of markings (i.e. the state space) it is easy to conclude that all places are **1-bounded**. A closer look allows us also to state that p_1,

p_2 and p_3 (p_1, p_4 and p_5) are in **mutual exclusion** (i.e. a pair of places is never simultaneously marked).

Moreover, considering the reachable markings and the net structure (the pre-function), **firing concurrency** between transitions b and c appears (in the reachability graph b and c seem to be conflicting!). Observe at this point that introducing p_6 in our system does **not** change the reachability graph, but transitions b and c become in **firing mutual exclusion**. This example shows that concurrency-mutual exclusion on firings cannot be studied on the reachability graph alone, because this gives a sequentialized vision of the concurrent evolutions.

Reversibility (i.e. recoverability of the initial marking) cannot be studied considering only the state space. The reachability relations among states (markings) must be taken into account. Because all markings are reachable from the initial one, a system should be reversible iff in the reachability graph there exists a path from any node to the original one. In other words:

Property 1.1. A net system is reversible iff its reachability graph is strongly connected.

If a system is reversible, any transition can be fired once and again iff that transition is firable at least once from the initial marking. This is true because the initial marking being always reachable, the transition can be fired again. More precisely, we are saying that in a reversible net system a transition is live iff it is firable in a sequence starting in the initial marking. If the system is not reversible but bounded, liveness is also characterizable but the condition is slightly more complex:

Property 1.2. Let $\langle N, M_0 \rangle$ be a bounded system. Transition t is live in $\langle N, M_0 \rangle$ iff t labels at least one arc of every strongly connected component of the reachability graph containing its transitive closure (i.e. such that the set of its successor nodes is included in the component itself).

In general, a system may be unbounded. Therefore the reachability graph construction process would never end. To avoid this, an abandon condition can be taken into account in the construction process:

Property 1.3. Place p is bounded in $\langle N, M_0 \rangle$ iff there exists no M_j, reached from M_i such that $M_j \geq M_i$ and $M_j(p) > M_i(p)$. Therefore, the system $\langle N, M_0 \rangle$ is unbounded (abandon condition) iff there exists M_j reachable from M_i such that $M_j \geq M_i$ and $M_j \neq M_i$.

If $M_i[\sigma\rangle M_j$, $M_j \geq M_i$ and $M_j \neq M_i$, the repetition of σ allows us to conclude on unboundedness. The proof that the condition is also necessary is based on a result from Karp–Miller (Karp and Miller, 1969).

A last and obvious consideration about the construction/analysis of the reachability graph gives an additional abandon condition:

Property 1.4. A system is deadlockable iff a marking not enabling any transition is found (i.e. a node without successor points out a deadlock).

Concluding, analysis techniques based on the reachability graph (theoretically possible for bounded systems) are very simple from a conceptual point of view. The problem that makes this approach impractical in many cases is its computational complexity: the **state explosion problem**. Figure 1.15 shows this on a very simple net system: parts are sent from store 1 to stores 2 and 3. The subnet generated by places $\{B, C, D, E\}$ imposes some restrictions on the way parts are distributed to the destination stores (i.e. partially schedule the distribution). The reachability graph is, even if it has been 'structured' for clearer presentation, difficult to understand and manage. The reader can try to check on the reachability graph (!) that the imposed distribution strategy is: parts are sent in a 1 : 1 relation to the destination stores, but allowing sometimes up to four consecutive dispatches to a given store (i.e. locally adjusting the possible demand, but maintaining the overall fair distribution).

Last, but not least, it is important to observe that reachability/coverability graphs are built for a given initial marking. If the number of resources (e.g. number of machines, size of stores, etc. . . .) changes, new (and completely different) graphs should be computed. Otherwise stated: reachability/coverability does not allow **parametric analysis** on the behavior of net systems.

1.6.2 Net system reductions

Even if reachability graph-based analysis techniques are complete for bounded systems, the computational complexity limits their applicability in practice. Net system reduction is a different analysis technique that allows the analysis of net models by producing transformations on its structure and, eventually, on its initial marking.

The approach is based on the definition of a **kit** or **catalog of reduction rules**, each one preserving the subset of properties (liveness, boundedness, reversibility, etc.) to be analyzed. The transformation procedure is **iterative** by nature: given the property (or properties) to be analyzed, the subset of rules that preserve it (them) is applied until the reduced system becomes **irreducible**. The irreducible system can be so simple that the property under study is trivially checked (see later, Fig. 1.17(d)). In other cases, the irreducible net is just 'more simple' to be analyzed, and other analysis techniques should be used: in other words, techniques to analyze net system models are **complementary**, not exclusive.

Reduction rules are transformation rules interesting for net analysis. When considered in the reverse sense they become **expansion** rules, interesting for net synthesis: **stepwise refinements** (or **top-down**) approach. Using this approach, with adequate expansion rules, the model will verify

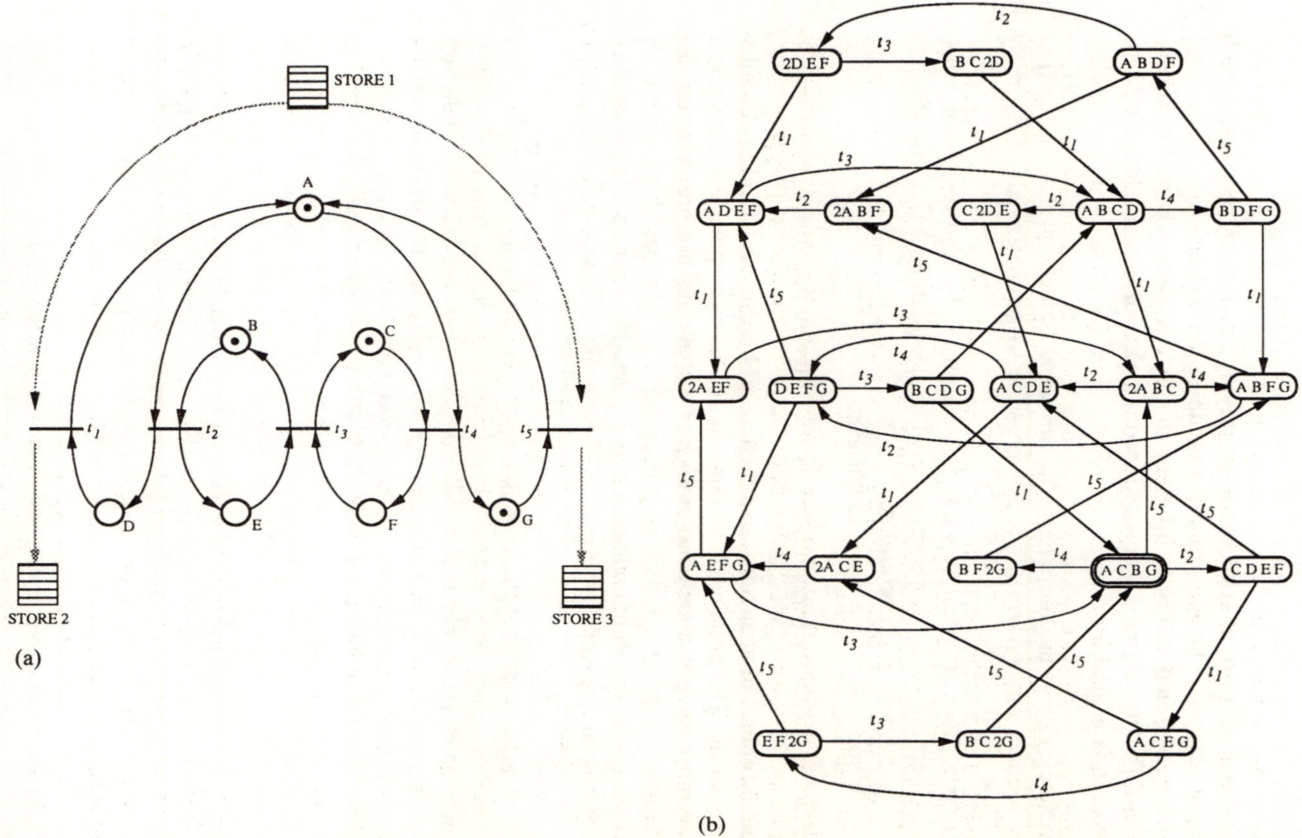

Figure 1.15 Parts of STORE 1 are sended to STORE 2 and STORE 3 according to the strategy defined by the subnet generated by {B, C, E, F}: (a) the net system; (b) the RG.

the specification by construction. This is interesting when comparing with the more classical approach based on the iteration of description and analysis. The iterative process has two basic disadvantages:

1. the **lack of general criteria** for modifying (correcting) a model which does not meet the requirements made in the validation.
2. the **operational difficulty** inherent to the validation phase. Obviously, this difficulty will be strongly reduced if a computer-aided design (CAD) system is available.

Nevertheless, because there exists no **universal** reduction rules kit (i.e. that fully reduces any system), it is not possible to synthesize all of them by stepwise refinements.

From a practical point of view, the design of the transformation rules catalog represents a **compromise** between **completeness** (i.e. transformation capabilities) and **usefulness**.

Reduction rules have a single pattern:

if an applicability precondition is true **then** reduce the net system.

Behavioral and/or **structural** statements can be done for the applicability precondition. The behavioral statements can be more powerful for a given initial marking, but their computation is usually much more complex. So the applicability preconditions presented here are based on **structural** considerations, the initial marking playing an auxiliary role as a **parameter**. According to this, reduction rules will have the following general pattern:

if *structural condition* and *initial marking condition* are true
then make *structural change* and *marking change*.

A very basic kit of reduction rules is presented. Additional details are given only for the rule of *implicit places*, which are redundancies in the net system model: if an implicit place is removed, then (illusory) synchronizations disappear and other reduction rules can be applied.

(a) A basic kit of reduction rules

Figure 1.16 presents graphically structural and marking conditions of a kit of six particular cases of reduction rules (Silva, 1985). It is not difficult to observe that they preserve such properties as liveness, the bounds of places (thus boundedness) and, if the second place in RA1 has only one input transition, reversibility:

- RA1 is a particular case of the **macroplace rule** (Silva, 1981).
- RA2 is a particular case of the *transition fusion rules* (Berthelot, 1987).
- RB1 and RC1 are particular cases of the **implicit place rule** (Silva, 1985; Silva and Colom, 1988) (to be considered later in more detail). Observe

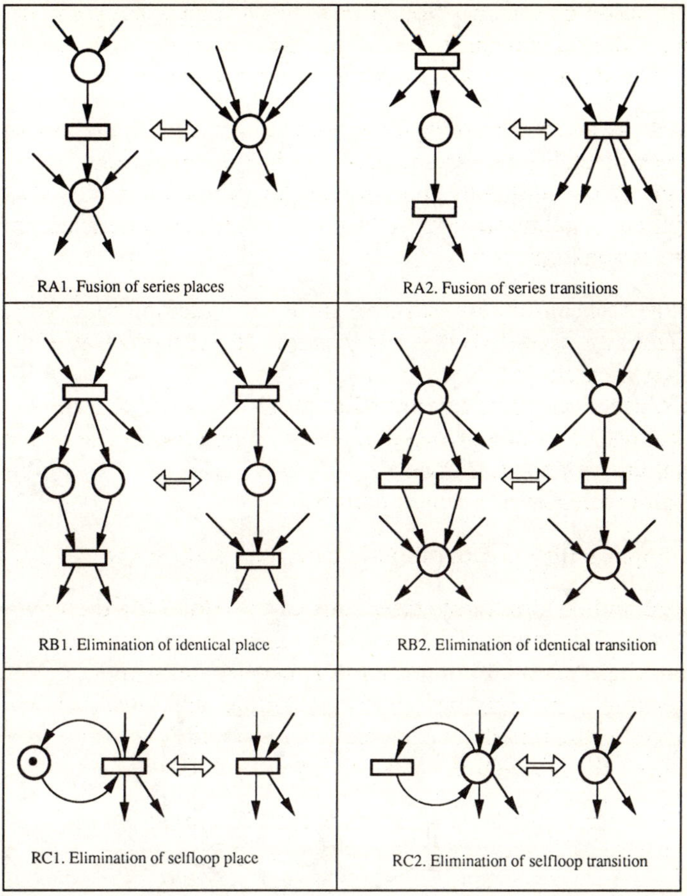

Figure 1.16 A basic reduction kit.

that RC1 can be trivially generalized creating **several self-loops** in which the place always appears. Liveness, the bound of places and reversibility are preserved. Moreover, if the place contains **several tokens**, liveness, boundedness (in general, not the bound of the net system) and reversibility are preserved.

- RB2 and RC2 are particular cases of **identical** and **identity transition rules** (Berthelot, 1987).

An interesting remark is the analogy between rules at the same level in Fig. 1.17: basically rules RX2 are obtained from rules RX1 by changing the role of places and transitions (**duality**) and **reversing** the arrows (something important only for rules RA). Duality (and reversing) are important concepts for a deep understanding and systematic presentation of many structural results on net theory.

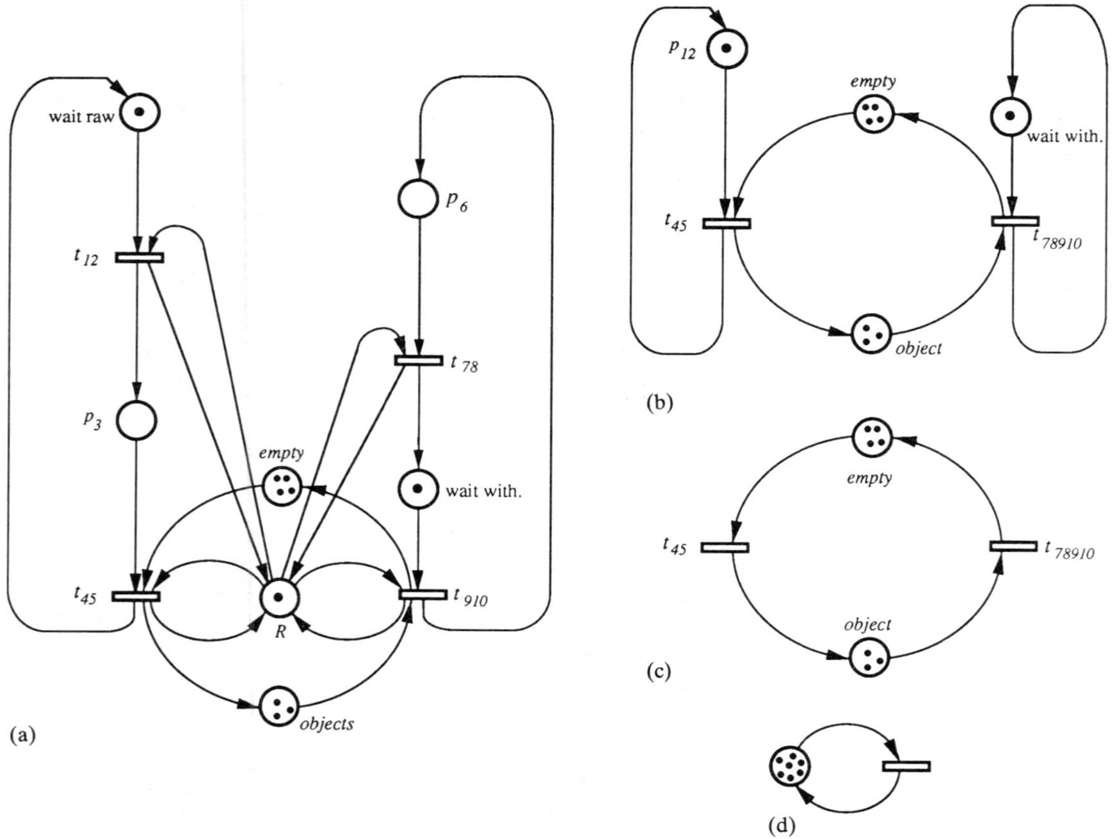

Figure 1.17 The reduction process shows (see (d)) that the net system in Fig. 1.11 is live, 7-bounded and reversible.

Let us now consider the net system in Fig. 1.9. The subnet defined by *op*$_1$–t_3–*wait dep.* verifies the precondition of rule RA1. Thus it can be reduced to a place, p_3 (Fig. 1.17(a)). The same holds for *op*$_2$–t_6–*wait free* that is reduced to p_6 (Fig. 1.17(a)). The subnets t_1–*load*–t_2, t_4–*deposit*–t_5, t_7–*unload*–t_8 and t_9–*withdraw*–t_{10} can be reduced according to RA2 (see t_{12}, t_{45}, t_{78} and t_{910} in Fig. 1.17(a)). Place R in Fig. 1.17(a) is implicit (one of the trivial generalizations mentioned for RC1). Thus it can be removed, and *wait raw*–t_{12}–p_3 and t_{910}–p_6–t_{78} can be reduced to p_{12} and t_{78910}, respectively (see Fig. 1.17(b)). Places p_{12} and *wait with.* are implicit (RC1) in Fig. 1.17(b), thus the net system in Fig. 1.17(c) is obtained. Playing the token game, a place (e.g. *object*) can became empty in Fig. 1.17(c) and t_{45}–*object*–t_{78910} can be reduced (RA2) to a single transition (Fig. 1.17(d)). Therefore, the original net system is live, 7-bounded and reversible.

(b) Implicit places

A place in a net system can only constrain the firable sequences. If a place, for an initial marking, never constrains the firable sequences, it can be removed without changing the **sequential observation** of the behavior of the net system (i.e. the set of firable sequences). These behaviorally defined places are called **(firing) implicit places**.

Let N be a net and N_p the net resulting from removing place p from N, Pre and Pre^p are the corresponding pre-incidence functions. If M_0 is an initial marking for N_p, $M_0 = (M_0^p, M_0(p))$ denotes the initial marking of N.

Definition 1.7. Given a system $\langle N, M_0 \rangle$, the place p is implicit (IP) if for any reachable marking in $\langle N_p, M_0^p \rangle$, (i.e. $\forall M^p \in R(N_p, M_0^p)$) and any output transition of p (i.e. $\forall t \in \{p^\bullet\}$) the following holds:

$$M^p \geq Pre^p(t) \Rightarrow M(p) \geq Pre(p, t)$$

The net system in Fig. 1.18 is unbounded (p_4 is the unique unbounded place) and non-reversible (also because of p_4). Place p_4 is implicit. Removing p_4 the system becomes bounded and reversible! Place p_6 in Fig. 1.14 imposes firing mutual exclusion between b and c. Since p_6 is an implicit place, the reduction rule does not preserve firing mutual exclusion. According to the definition, firable sequences are preserved. Thus the following is true:

Property 1.5. The elimination of implicit places:

1. **Preserves**: deadlock-freeness, liveness and marking mutual exclusions.
2. **Does not preserve**: boundedness, reversibility and firing mutual exclusion.

Sometimes it is practical to impose a second condition to the definition of implicit places: marking redundancy (i.e. computable from the other markings). The marking of p_4 in Fig. 1.18 cannot be computed from the other markings, thus it does not belong to the restricted case we are

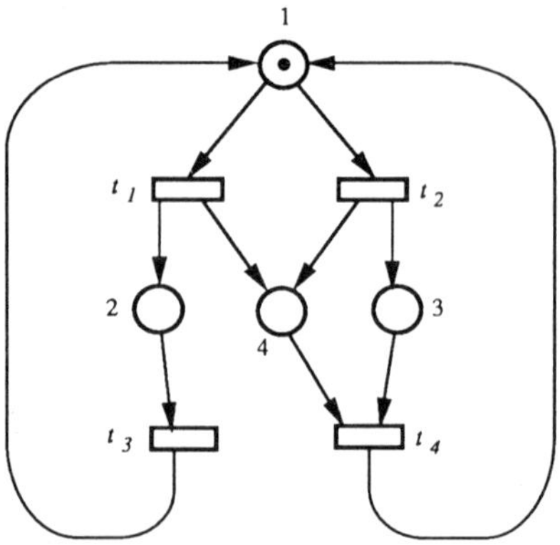

Figure 1.18 Place p_4 is firing implicit but not marking implicit. Removing p_4 the 'false' synchronization in t_4 disappears.

considering now: marking implicit places. Because of the additional redundancy, marking implicit places preserves the state space (i.e. the reachability graphs of the net system with and without p are isomorphous) and therefore preserves two already mentioned properties: **boundedness** and **reversibility**.

How can implicit places be detected? How complex is the process? The property is behavioral, so computationally complex behavior-based algorithms should be used. The next property gives a very simple algorithm, based on the solution of a linear programming problem (LPP1) to detect 'most of the practical cases'. Because, LPPs are of polynomial time complexity (Nemhauser *et al.*, 1989), the technique has this complexity. This approach derives from some relatively complex arguments (see Colom and Silva, 1991b).

Property 1.6. Let $\langle N, M_0\rangle$ be a net system, and z defined as follows:

$$\begin{aligned} z = \min \quad & Y^T \cdot M_0 + \mu \\ \text{subject to} \quad & Y^T \cdot C \leq C(p) \\ & Y^T \cdot Pre(t) + \mu \geq Pre(p,t) \quad \forall t \in p^\bullet \\ & Y \geq 0,\ Y(p) = 0 \end{aligned} \qquad \text{(LPP1)}$$

Assuming $M_0(p) \geq 0$, **if** $M_0(p) \geq z$ **then** p is implicit.

Remark if $Y^TC = C(p)$ constraint, then p is a marking implicit place.

The computation of LPP1 for p_9 (Fig. 1.19(a)) gives $z = 0$, for:

1. $[Y^T = (0\,0\,1\,1\,1\,0\,1\,0\,0)] \Leftrightarrow [C(p_9) = C(p_3) + C(p_4) + C(p_5) + C(p_7)]$
2. $\mu = -1$

Because $M_0(p) \geq z = 0$, p_9 is implicit and can be removed. Once p_9 is removed, a similar computation for p_2 can be done and p_2 is also shown to be implicit.

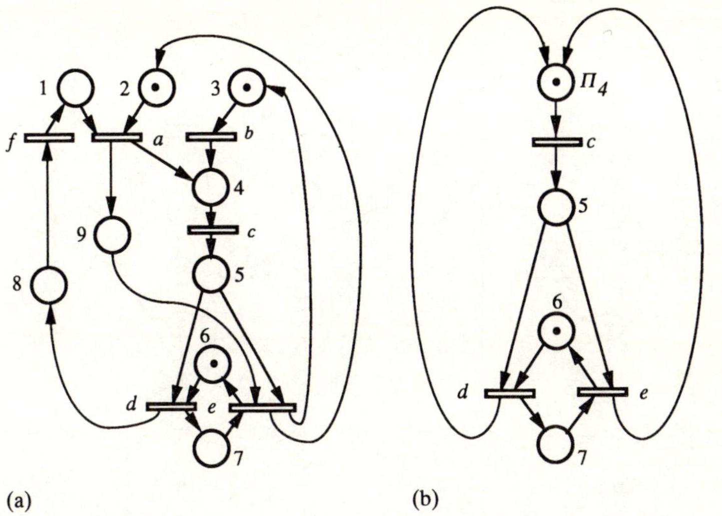

Figure 1.19 Places p_9 and p_2 (or p_2 and p_7) are implicits.

Figure 1.19(b) shows a reduced net system. It can be obtained reducing p_3–b–p_4 into a place (say p_{34}) (RA1) and finally p_8–f–p_1–a–p_{34} into Π_4. Now RA1 allows us to fuse Π_4 and p_5. The new place is implicit, so it can be removed. Then a cycle with p_6–d–p_7–e–p_6 remains. Finally it can be reduced to a basic net, p_6–t_{de}–p_6, with one token. Therefore the original net system is live and bounded (**Note**: it is also reversible, but we cannot guarantee this because of the fusion of p_3–b–p_4 in p_{34}).

1.6.3 Linear algebraic techniques

The behavior of a net system model is clearly non-linear, nevertheless the so-called state equation, $M = M_0 + C \cdot \overline{\sigma} \geq 0$, $\overline{\sigma} \geq 0$ (see eq. (1.4)), represents a nice **linear relaxation**. Unfortunately the existence of **spurious** solutions (section 1.1.2) leads usually to **semidecision algorithms** (i.e., only necessary or only sufficient conditions) to analyze such behavioral properties as reachability, boundedness, deadlock-freeness, mutual exclusion, liveness or reversibility. For example, $\tilde{M}_1 = (0,0,0,2,0)^T$ and $\tilde{M}_2 = (0,2,0,0,0)^T$ are two spurious solutions for the system in Fig. 1.20(a). The first allows us to say that p_4 is 2-bounded, while it is really 1-bounded (check it). $\tilde{M}_2$ is a deadlock. Then using the state equation we cannot conclude that the system in Fig. 1.20(a) is deadlock-free.

Spurious solutions can be removed using different approaches (Colom and Silva, 1991b). For example, it is clear that adding implicit places, a new system model with identical behavior is obtained. For certain net systems, if the implicit places are chosen carefully, the state equation of the new system may have no integer spurious solution preventing conclusion on the bound of a place or the deadlock freeness of the system. The net

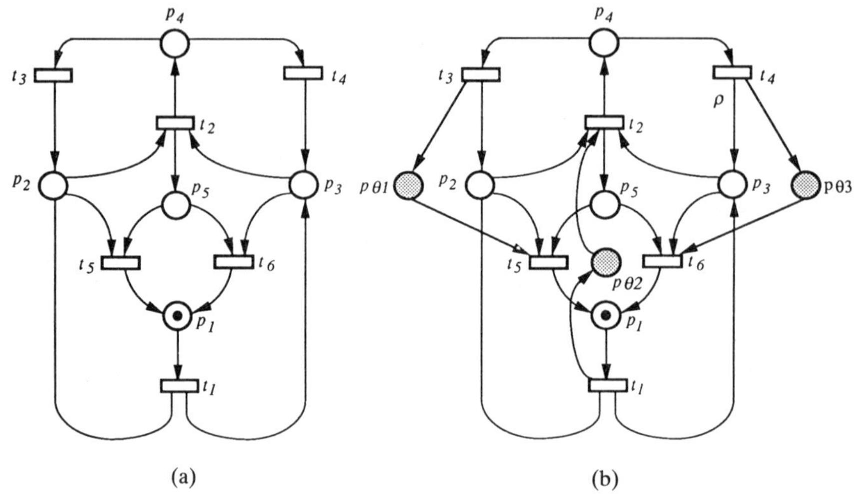

Figure 1.20 Two equal behavior 1-bounded and live net systems: dashed places, $\{p_{\theta 1}, p_{\theta 2}, p_{\theta 3}\}$ are implicit.

system in Fig. 1.20(b) has been obtained by adding implicit places $p_{\theta 1}$, $p_{\theta 2}$ and $p_{\theta 3}$ to that in Fig. 1.20(a). The above-mentioned spurious solutions, $\tilde{M}_1$ and $\tilde{M}_2$, are not projections on P of the solutions of the new state equation. Moreover, we can conclude now that the new (and original) net systems were 1-bounded for p_4 and deadlock-free!

Classical reasoning to prove logical properties uses **invariants** on the behavior of a system. The right and left non-negative annullers of the incidence matrix lead to two kinds of structural objects (p- and t-semiflows):

1. $Y \geq 0$, $Y^T \cdot C = 0 \Rightarrow Y^T \cdot M = Y^T \cdot M_0$ (token conservation law)
2. $X \geq 0$, $C \cdot X = 0 \Rightarrow \exists M_0$ such that $M_0[\sigma\rangle\, M_0$ and $\overline{\sigma} = X$ (cyclic marking behavior)

The token conservation laws are **marking invariants** induced by p-semiflows. Usually they are called ***p*-invariants**.

The computation of minimal p-semiflows (Y) and minimal t-semiflows (X) has been extensively studied. However, an **exponential number** of minimal semiflows may appear. Therefore the time complexity of this computation cannot be polynomial. In Colom and Silva (1991a) a study is carried out merging traditional techniques in convex geometry with those developed within Petri nets. From a conceptual point of view, the consideration of semiflows provides **decomposed views** of the structure of the net model. In Fig. 1.21 the decomposition induced by the minimal p-semiflows of the system in Fig. 1.9 is graphically presented. The induced minimal p-invariants are given in equations (1.5–8).

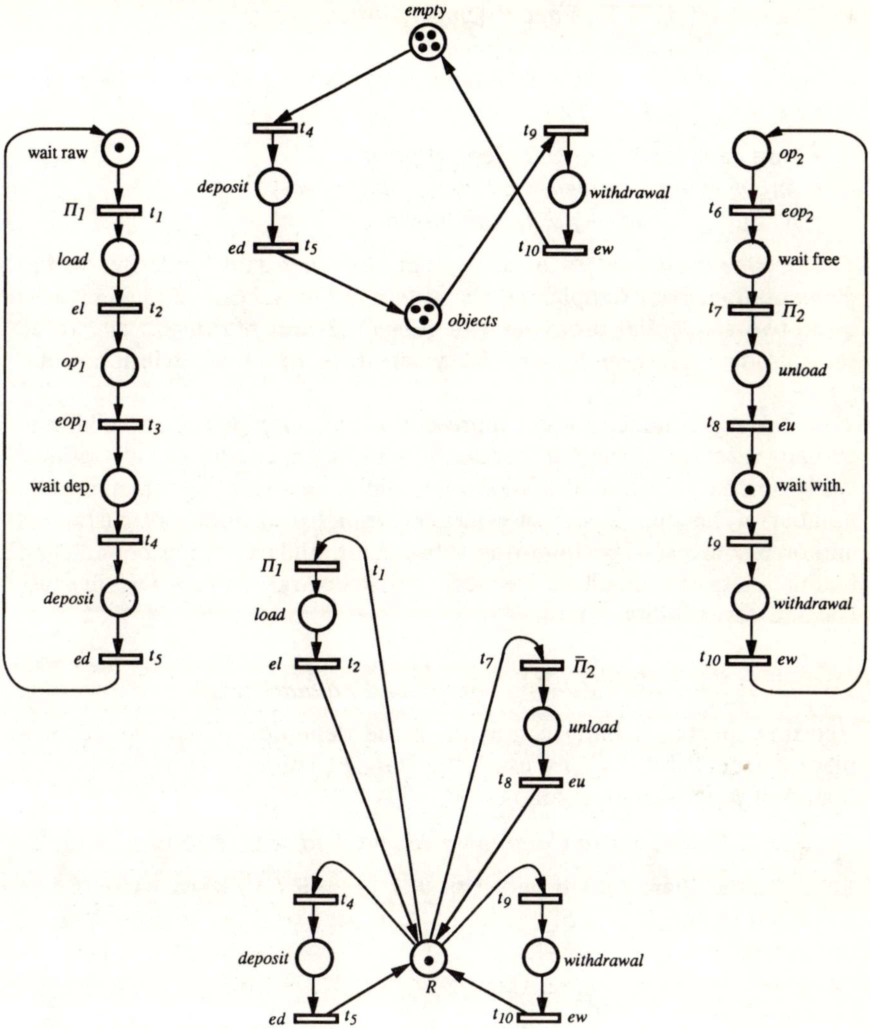

Figure 1.21 A decomposed view of the net system in Fig. 1.9.

$$M(\text{wait raw}) + M(\textit{load}) + M(\textit{op}_1) + M(\text{wait dep.}) + M(\textit{deposit}) = 1 \quad (1.5)$$

$$M(\textit{op}_2) + M(\text{wait free}) + M(\textit{unload}) + M(\text{wait with.}) + M(\textit{withdrawal}) = 1 \quad (1.6)$$

$$M(\textit{empty}) + M(\textit{deposit}) + M(\textit{object}) + M(\textit{withdrawal}) = 7 \quad (1.7)$$

$$M(R) + M(\textit{load}) + M(\textit{unload}) + M(\textit{deposit}) + M(\textit{withdrawal}) = 1 \quad (1.8)$$

Because markings are non-negative integers (i.e. $\forall p \in P, M(p) \geq 0$), the following can be easily stated:

1. Bounds: $M(p_i) \leq 1 \quad \forall p_i \in P \setminus \{empty, object\}$
   $M(empty) \leq 7, \quad M(object) \leq 7$
2. Marking mutual exclusions among the following subset of places (i.e. $M(p_i)M(p_j) = 0,\ i \neq j$):
   - wait raw, *load*, op_1, wait dep., *deposit*
   - op_2, wait free, *unload*, *wait with.*, *withdrawal*
   - *R*, *load*, *unload*, *deposit*, *withdrawal*

The decomposed view of a net system is even useful to derive an **implementation**. For example, the net system in Fig. 1.9 can be implemented using two sequential processes (for *Machine1* and *Machine2*) and three semaphores (*object*, *empty* and *R*), where *R* is a mutual exclusion semaphore.

Linear algebraic techniques represent *fast to compute* (polynomial time in many cases) *semidecision* (necessary or sufficient conditions) algorithms, easily amenable to initial marking *parametric* analysis (e.g. changing the number of customers, size of resources, initial distribution of customers and/or resources). The following subsections study marking bounds and boundedness (a), deadlock-freeness (b), structural liveness and liveness (c), and reversibility (d).

(a) Marking bounds and boundedness

Relaxing the reachability condition in the definition of the bound of a place p (see Table 1.1), by using the state equation (1.4), the structural bound of p is defined as follows:

$$SB(p) = sup\ \{M(p)/M = M_0 + C \cdot \bar{\sigma} \geq 0,\ \bar{\sigma} \geq 0\}$$

Let e_p be the **characteristic vector** of p: $e_p(f)$:= **if** $f = p$ **then** 1 **else** 0. The structural bound of p, $SB(p)$, can be expressed as a linear programming problem:

$$SB(p) = \max \quad e_p^T \cdot M \quad \text{subject to } M = M_0 + C \cdot \bar{\sigma} \geq 0,\ \bar{\sigma} \geq 0 \qquad \text{(LPP2)}$$

Therefore $SB(p)$ can be computed in polynomial time. In sparse-matrix problems (matrix C is usually sparse), good implementations of the classical **simplex method** lead to quasi-linear time complexities.

Because $SB(p)$ has been defined using a linear relaxation of the reachability in the system, then $SB(p) \geq B(p)$. Therefore, if we are investigating the k-boundedness of a place (i.e. $M(p) \leq k$), we have a sufficient condition in polynomial time:

if $SB(p) \leq k$ **then** $B(p) \leq k$ (i.e. p is k-bounded)

In the sequel we argue on classical results from linear programming and convex geometry theories. We must assume the reader is aware of these theories (see, for example, Murty (1983); Nemhauser *et al.* (1989); otherwise all the needed arguments are compiled and adapted in (Silva and Colom, 1988). The important point in this overview is to convey the idea that other theories are helpful to understand in a deep and general framework many sparse results on net systems' behaviors.

The **dual** linear programming problem of LPP2 is the following (see any text on linear programming to check it):

$$\begin{aligned} \tilde{SB}(p) = \min \quad & Y^T \cdot M_0 \\ \text{subject to } \quad & Y^T \cdot C \leq 0 \\ & Y \geq e_p \end{aligned} \qquad \text{(LPP3)}$$

LPP2 has always a feasible solution ($M = M_0$, $\overline{\sigma} = 0$). Using duality and boundedness theorems from linear programming theory, both LPP2 and LPP3 are bounded (thus p is structurally bounded) and $SB(p) = \tilde{SB}(p)$ iff there exists a **feasible solution** for LPP3:

$$\exists Y \geq e_p \quad \text{such that} \quad Y^T \cdot C \leq 0 \qquad (1.9)$$

The reader can easily check that LPP3 makes in polynomial time an 'implicit search' for the structural bound of p on a set of structural components including all the p-semiflows ($Y \geq 0$, $Y^T \cdot C = 0$).

From the above discussion and using the **alternatives theorem** (essentially the Minkowski–Farkas lemma in algebraic form) the following properties can be proved:

Property 1.7. The following three statements are equivalent:

1. p is structurally (i.e. for any M_0) bounded;
2. $\exists Y \geq e_p$ such that $Y^T \cdot C \leq 0$ {place-based characterization};
3. $\forall X \geq 0$ such that $C \cdot X \geq 0$, $e_p^T \cdot C \cdot X = C(p) \cdot X = 0$ is satisfied {transition-based characterization}.

Property 1.8. The following three statements are equivalent:

1. N is structurally (i.e. $\forall M_0$) bounded;
2. $\exists Y \geq 1$ such that $Y^T \cdot C \leq 0$ {place-based characterization};
3. $\nexists X \geq 0$ such that $C \cdot X \geq 0$ $C \cdot X = 0$ is satisfied (i.e. $\nexists X \geq 0$ such that $C \cdot X \geq 0$ and $C \cdot X \neq 0$) {transition-based characterization}.

(b) Deadlock-freeness (and liveness)

Token conservation laws are invariant properties of the behavior of net systems that may be very useful to prove **deadlock-freeness**. Using the invariants in eqs (1.5–1.8), we shall prove that our net system in Fig. 1.9 is deadlock-free.

If there exists a deadlock, no transition can be fired. Let us try to construct a marking in which no transition is firable. When a unique input place of a transition exists, that place must be unmarked. So $M(load) =$

$M(op_1) = M(deposit) = M(op_2) = M(unload) = M(withdrawal) = 0$, and the token conservation laws in eqs (1.5–1.8) reduce to:

$$M(\text{wait raw}) + M(\text{wait dep.}) = 1 \quad (1.5a)$$
$$M(\text{wait free}) + M(\text{wait with.}) = 1 \quad (1.6a)$$
$$M(empty) + M(object) = 7 \quad (1.7a)$$
$$M(R) = 1 \quad (1.8a)$$

Because R should always be marked at the present stage, to prevent the firing of t_1 and t_7, places 'wait raw' and 'wait free' should be unmarked. The token conservation laws are reduced once more, leading to:

$$M(\text{wait dep.}) = 1 \quad (1.5b)$$
$$M(\text{wait with.}) = 1 \quad (1.6b)$$
$$M(empty) + M(object) = 7 \quad (1.7b)$$
$$M(R) = 1 \quad (1.8b)$$

Since $M(\text{wait dep.}) = M(\text{wait with.}) = 1$, to avoid the firing of t_4 and t_9, $M(empty) + M(object) = 0$ is needed. This contradicts eq. (1.7tris), so the net system is deadlock-free.

A more compact, algorithmic presentation of the above deadlock-freeness proof is:

> **if** $M(load) + M(op_1) + M(deposit) + M(op_2) + M(unload) + M(withdrawal) \geq 1$
> **then** *one of* t_2, t_3, t_5, t_6, t_8, or t_{10} *is firable*
> **else if** $M(\text{wait raw}) + M(\text{wait free}) \geq 1$
>   **then** *one of* t_1 or t_7 *is firable*
>   **else** *one of* t_4 or t_9 *is firable*

Even if the above is an *ad hoc* proof, it can be fully automated (i.e. automatic proving). To prove deadlock-freeness by means of linear algebra we must express the condition 'transition t is not firable at marking M' using linear constraints. This can always be done. Nevertheless, it turns out to be very efficient in net systems where for each place the structural bound $SB(p)$, computed through LPP2 or LPP3 is equal to the weight of its output arcs: $SB(p) = W(p, t)\ \forall p \in P, \forall t \in p^{\bullet}$. The condition in linear form is:

$$\sum_{p_i \in {}^{\bullet}t} M(p_i) \leq \sum_{p \in P} Pre(p,t) - 1$$

In words: the amount of tokens in the input places of t is less than required. Therefore:

Property 1.9. Let $\langle N, M_0\rangle$ be a net system such that $SB(p) = W(p, t)\ \forall p \in P, \forall t \in p^{\bullet}$. A sufficient condition for $\langle N, M_0\rangle$ to be deadlock-free is that the following linear system has **no** solution ($dim\ (1_q^T) = q$):

$$M = M_0 + C \cdot \overline{\sigma} \geq 0, \overline{\sigma} \geq 0$$
$$M^T \cdot Pre \leq 1_n^T \cdot Pre - 1_m^T$$

For the system in Fig. 1.4(g), the system of inequalities $M^T \cdot Pre \leq 1^T \cdot Pre - 1$ looks as follows ($M(p) \geq 0\ \forall p \in p$, thus $M(p) \leq 0$ is equivalent to $M(p) = 0$):

$$\begin{array}{ll} M(p_1) + M(\mathcal{R}) \leq 1 & \{t_1\} \\ M(p_2) = 0 & \{t_2\} \\ M(p_3) + M(\mathcal{R}) \leq 1 & \{t_3\} \\ M(p_4) = 0 & \{t_4\} \end{array}$$

The above system of inequalities and equations together with the state equation form an **inconsistent** linear system. Thus the net system is deadlock-free.

The deadlock-free system in Fig. 1.9 does not verify the precondition of Property 1.9. Nevertheless the approach has been generalized (see Colom *et al.* (1990b)), and that conclusion can be obtained from the inconsistency of a single linear system.

As a final remark, we want to point out that liveness can be proved for the net system in Fig. 1.9. Liveness implies deadlock-freeness, but the reverse is not true in general. Nevertheless, if the incidence matrix, C, has a right-orthogonal space of dimension one (i.e. $C \cdot X = 0 \Rightarrow X = k \cdot A$) and $A \geq 1$ (i.e. the net is consistent), then any infinite behavior must contain all transitions with relative firings given by A (t_i). Thus deadlock-freeness implies, in this case, liveness. For the net in Fig. 1.9 the above property holds with $A = 1$. Therefore the net system being deadlock-free is also live.

(c) *Structural liveness and liveness*

A general approach to linearly analyze liveness needs the use of **invariants** and some **inductive** reasoning. A **necessary** condition for a transition t to be live in a system $\langle N, M_0 \rangle$ is its eventual infinite firability (i.e. the existence of a **firing repetitive sequence** σ_R containing t).

Using the state equation (1.4) as a linear relaxation of the reachability condition, an **upper bound** of the number of times t can be fired in $\langle N, M_0 \rangle$ is given by the following LPP (e_t (θ): > **if** $\theta = t$ **then** 1 **else** 0):

$$\begin{array}{rl} SR(t) = \max & e_t^T \cdot \overline{\sigma} \\ & \text{subject to } M = M_0 + C \cdot \overline{\sigma} \geq 0 \\ & \overline{\sigma} \geq 0 \end{array} \qquad \text{(LPP4)}$$

The dual of LPP4 is:

$$\begin{array}{rl} S\tilde{R}(t) = \min & Y^T \cdot M_0 \\ & \text{subject to } Y^T \cdot C \leq - e_t^T \\ & Y \geq 0 \end{array} \qquad \text{(LPP5)}$$

We are interested on characterizing when $SR(t)$ goes to infinity. The problem LPP4 has $M = M_0$ and $\overline{\sigma} = 0$ as a feasible solution. Using first duality and unboundedness theorems from linear programming and later the alternatives theorem, the following properties can be stated:

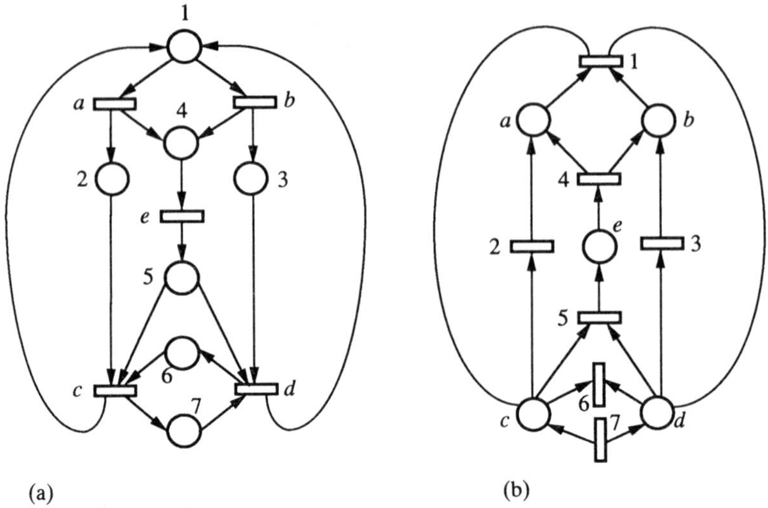

Figure 1.22 Two conservative and consistent, structurally non-live nets:
(a) $rank(C) = 4$, $/T/ - l - \delta = 5 - 1 - 1 = 3$, thus N is not struct-live;
(b) $rank(C) = 4$, $/T/ - l - \delta = 7 - 1 - 2 = 4$, thus no answer.

Property 1.10. The following three statements are equivalent:

1. t is structurally repetitive (i.e. there exists a 'large enough' M_0 such that t can be fired infinitely often);
2. $\nexists Y \geq 0$ such that $Y^T \cdot C \leq - e_t^T$ {place perspective};
3. $\exists X \geq e_t$ such that $C \cdot X \geq 0$ {transition perspective}.

Property 1.11. The following three statements are equivalent:

1. N is structurally repetitive (i.e. all transitions are structurally repetitive);
2. $\nexists Y \geq 0$ such that $Y^T \cdot C \leq 0$ and $Y^T \cdot C \neq 0$;
3. $\exists X \geq 1$ such that $C \cdot X \geq 0$.

Combining Properties 1.8 and 1.11 and considering that structural repetitiveness is a necessary condition for structural liveness, the following classical result can be obtained (Memmi and Roucairol, 1980; Brams, 1983; Silva, 1985):

if N is structurally live and structurally bounded
then $\exists X \geq 1$ such that $C \cdot X = 0$ (i.e. *N is consistent*)
$\exists Y \geq 1$ such that $Y^T \cdot C = 0$ (i.e. *N is conservative*)

Net structures in Fig. 1.22 are consistent and conservative, but there exists no live marking for them. A more careful analysis using the above result iteratively allows us to improve it with a **rank condition** on the incidence matrix of N, C. This and other results are summarized in the next property.

Let us say that t_i and t_j are in **equality conflict relation** if $Pre(t_i) = Pre(t_j) \neq 0$. Obviously, this is an *equivalence* relation, leading to a *partition* of transitions into *equivalence classes*. Let D_k be an equivalence class. We define the following quantities:

$$\delta_k = |D_k| - 1$$
$$\delta = \Sigma_k \delta_k$$

Property 1.12. Let C be the incidence matrix of N.

1. **if** N is structurally live **then** $\exists X \geq 1$ such that $C \cdot X \geq 0$
2. **if** N is structurally live and structurally bounded

   then $\exists X \geq 1$ such that $C \cdot X = 0$ (i.e. net consistency)
   $\exists Y \geq 1$ such that $Y^T \cdot C = 0$ (i.e. net conservativeness)
   $rank(C) \leq |T| - \delta - 1$

3. **if** N is connected, consistent and conservative **then** it is strongly connected.

The added rank condition allows us to state that the net in Fig. 1.22(a) is structurally non-live. Nevertheless, nothing can be said about structural liveness of the net in Fig. 1.22(b).

Property 1.12 is purely structural (i.e. the initial marking is not considered at all). Nevertheless it is clear that a small enough initial marking can make non-live a net system even if the net structure is well formed. A lower bound for the initial marking to make live a net system is based on p-invariants: if $t \in T$ is firable at least once, for any p-invariant Y, $Y^T \cdot M_0 \geq Y^T \cdot Pre(t)$. Therefore:

Property 1.13. If $\langle N, M_0 \rangle$ is a live system, then

$$\forall Y \geq 0 \text{ such that } Y^T \cdot C = 0, \quad Y^T \cdot M_0 \geq \max_{t \in T} (Y^T \cdot Pre(t)) \geq 1$$

Unfortunately no characterization of liveness exists in linear algebraic terms. The net in Fig. 1.12b is structurally live. Adding a token to p_5, all p-semiflows remain marked, but it is non-live.

(d) Reversibility (and liveness)

Let us now use a **Liapunov-stability-like** technique to prove that the net system in Fig. 1.9 is reversible. It serves to illustrate the use of invariants and some inductive reasonings.

As a preliminary consideration that makes the rest of the proof easier, the following simple property will be used: *let* $\langle N, M_1 \rangle$ *be a reversible system and* M_0 *reachable from* M_1 (i.e. $\exists \sigma$ such that $M_1[\sigma\rangle M_0$). Then $\langle N, M_0 \rangle$ is reversible.

Assume M_1 is like M_0 (Fig. 1.9), but with the following changes:

$$M_1(\text{wait raw}) = 0, \quad M_1(\text{wait dep.}) = 1$$
$$M_1(empty) = 0, \quad M_1(object) = 7$$

Let us prove first that $\langle N, M_1\rangle$ is reversible. Let L be a non-negative place weighting such that $L(p_i) = 0$ iff p_i is marked in M_1. Therefore, $L(\text{wait dep.}) = L(R) = L(object) = L(\text{wait with.}) = 0$ and $L(p_j) > 0$ for all the other places. The function $V(M) = L^T \cdot M$ has the following properties:

$$V(M) \geq 0 \quad \text{and} \quad V(M_1) = 0$$

For the system in Fig. 1.9 a stronger property holds: $V(M) = 0 \Leftrightarrow M = M_1$. This can be clearly seen because $L^T \cdot M = 0 \Leftrightarrow M(\text{wait raw}) = M(load) = M(op_1) = M(deposit) = M(empty) = M(op_2) = M(\text{wait free}) = M(unload) = M(withdrawal) = 0$. Even more, it is easy to check the following:

M_1 is the present marking $\Leftrightarrow$ t_9 is the unique firable transition

If there exists (warning: in Liapunov-stability criteria the universal quantifier is used!) a finite firing sequence (i.e. a finite trajectory) per reachable marking M_i such that $M_i[\sigma_k\rangle\, M_{i+1}$ and $V(M_i) > V(M_{i+1})$, in a finite number of transition firings $V(M) = 0$ is reached. Because $V(M) = 0 \Leftrightarrow M = M_1$, a proof that M_1 is reachable from any marking has been obtained (i.e. $\langle N, M_1\rangle$ is reversible).

Pre-multiplying the state equation (1.4) by L^T we obtain the following condition:

$$\textbf{if } \sigma_k = t_j \textbf{ then } [L^T \cdot M_{i+1} < L^T \cdot M_i] \Leftrightarrow L^T \cdot C\ (t_j) < 0$$

Now, removing in Fig. 1.9 the places marked at M_1 (i.e. wait dep., R, *object*, wait with.) and firable transitions (i.e. t_9) an *acyclic net* is obtained, so there exists an L such that $L^T \cdot C\ (t_j) < 0\ \forall_j \neq 9$.

For example, taking as weights the *levels* in the acyclic graph we have:

$$L(op_1) = L(unload) = 1$$
$$L(load) = L(\text{wait free}) = 2$$
$$L(\text{wait raw}) = L(op_2) = 3$$
$$L(deposit) = L(withdrawal) = 4$$
$$L(empty) = 5$$

and $L^T \cdot C = (-1, -1, -1, -1, -1, -1, -1, -1, +4, -1)$. In other words, the firing of any transition, except t_9, decreases $V(M) = L^T \cdot M$.

Using the algorithmic deadlock-freeness explanation in section (b), the reversibility of $\langle N, M_1\rangle$ is proven (observe that the p-invariants in eqs (1.5–1.8) remain for M_1):

if $M(load) + M(op_1) + M(deposit) + M(op_2) + M(unload) + M(withdrawal) \geq 1$
then $V(M)$ *can decrease firing* t_2, t_3, t_5, t_6, t_8 *or* t_{10}
else if $M(\text{wait raw}) + M(\text{wait free}) \geq 1$
then $V(M)$ *can decrease firing* t_1 *or* t_7

else *$V(M)$ can decrease firing t_4* **or**
t_9 is the unique firable transition {$\Leftrightarrow M_1$ is the present marking}

Because M_0 is reachable from M_1 (for example, by means of $\sigma = (t_9 \cdot t_{10} \cdot t_6 \cdot t_7 \cdot t_8)^5 \cdot t_4 \cdot t_5$), $\langle N, M_0 \rangle$ is a reversible system.

Once again liveness of the system in Fig. 1.9 can be proved, because the complete sequence (i.e. containing all transitions) $\sigma = t_1 \cdot t_2 \cdot t_3 \cdot t_4 \cdot t_5 \cdot t_9 \cdot t_{10} \cdot t_6 \cdot t_7 \cdot t_8$ can be fired. Since the system is reversible, no transition loses the possibility of firing (i.e. all transitions are live).

1.7 SOME NET SUBCLASSES AND THEIR ANALYSIS

Net subclasses will be defined exclusively by introducing constraints on the structure of ordinary nets. Therefore it is very easy to recognize if a net model belongs to a subclass (i.e. the **membership problem**). By restricting the generality of the model, it will be easier to study its behavior. In particular, powerful structural results allow us to fully characterize some (otherwise hard to study) properties such as liveness and reversibility.

1.7.1 Definition of four net subclasses

Let us introduce four important subclasses summarized in Fig. 1.23.

Definition 1.8. A **state machine** (SM) is an ordinary net such that:

$$\forall t \in T \; |{}^{\bullet}t| = 1 \quad \text{and} \quad |t^{\bullet}| = 1$$

that is, any transition has one input and one output place.

State machines allow the modeling of decisions (conflicts) and re-entrance. It is important to note that the concept of state machine, considered as a subclass of nets, is more general than the classical state diagram or state graph, since it can have more than one token. In any case, state machines do not 'create' tokens; thus can model only finite state systems.

Definition 1.9. A **marked graph** (MG) (also **event** or **synchronization graph**) is an ordinary net such that:

$$\forall p \in P \; |{}^{\bullet}p| = 1 \quad \text{and} \quad |p^{\bullet}| = 1$$

that is, any place has one input and one output transition.

MGs are a subclass of structurally decision-free nets. They can model systems for ordering activities as PERTs (program evaluation and review techniques) do. They are more general than PERTs in the sense that **recycling** is allowed and places can contain several tokens. Moreover, provided with an adequate stochastic interpretation, strongly connected MGs

State Machines

- Present neither synchronizations (JOINS) nor structural parallelism FORKS), but allow reentrancy and concurrency
- Only model finite state systems
- Analysis and synthesis theory is well understood

Marked Graphs

- Allow modeling synchronizations and structural parallelism but cannot model choices
- Can model non-finite state systems
- Allow modeling systems for ordering activities similar to PERTs, even integrating repetitive behaviours
- Analysis and synthesis theory is well understood

Free Choice Nets

- Allow modeling sequence, choice and concurrence relationships
- Choice and synchronizations cannot coincide on the same transition
- Do not allow modeling sequential subsystems synchronized through mutual exclusion semaphores
- Analysis and synthesis theory is well understood

Simple Nets

- Each transition has at most one input place shared with other transitions
- Allow modeling sequence, choice and concurrence relationships
- Choices are not free in general but can be solved locally
- Allow modeling sequential subsystems synchronized through mutual exclusion semaphores
- Analysis theory is partially understood

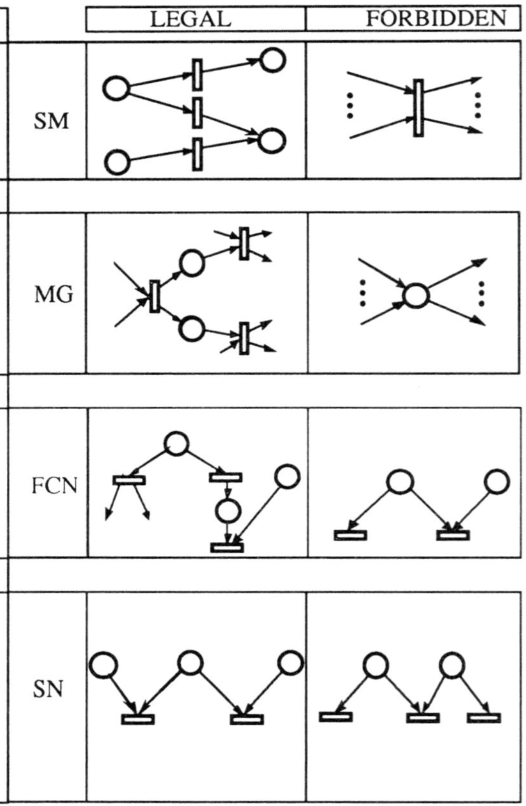

Figure 1.23 Four fundamental ordinary net subclasses.

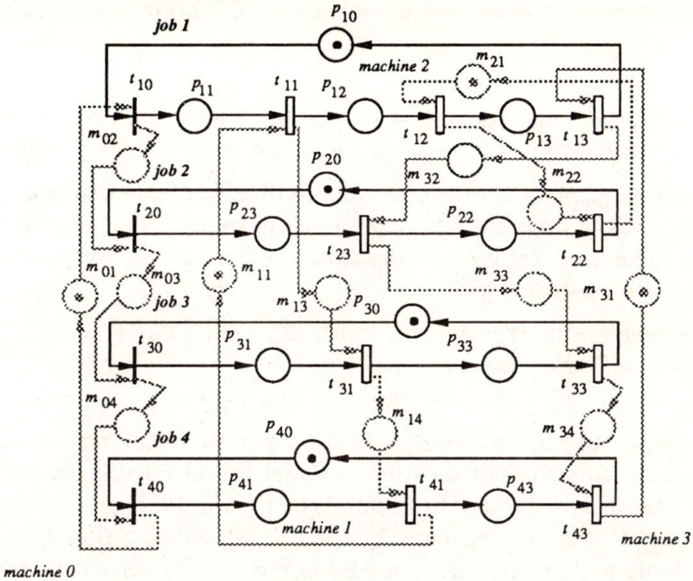

Figure 1.24 A job-shop system modeled with a marked graph (Hillion and Proth, 1989).

are equivalent (i.e. have the same descriptive power) to **fork/join queuing networks with blockings** (FJQN/B) (Dallery *et al.*, 1990).

Even if MGs are a very restrained class of nets, they may appear in the modeling of job-shop systems, once **production routing** of jobs and the **machine sequencing** are uniquely defined.

Let us assume that we want to model a 4-jobs–3-machines manufacturing system under the following fixed route control strategy:

(a) Product mix: 25% of *job i*, $\forall i = 1, 4$;
(b) Jobs should visit the machines as follows:

job 1: machine 1, machine 2, machine 3
job 2: machine 3, machine 2
job 3: machine 1, machine 3
job 4: machine 1, machine 3

(c) Machines should sequence the jobs as follows:

M_1*: job 1, job 2, job 3, job 4*
M_2*: job 1, job 2*
M_3*: job 1, job 2, job 3, job 4*

The net system in Fig. 1.24 models that specification.

The jobs to be performed are modeled with the horizontal circuits:

job 1: $p_{10} \cdot t_{10} \cdot p_{11} \cdot t_{11} \cdot p_{12} \cdot t_{12} \cdot p_{13} \cdot t_{13} \cdot p_{10}$
job 2: $p_{20} \cdot t_{20} \cdot p_{23} \cdot t_{23} \cdot p_{22} \cdot t_{22} \cdot p_{20}$
job 3: $p_{30} \cdot t_{30} \cdot p_{31} \cdot t_{31} \cdot p_{33} \cdot t_{33} \cdot p_{30}$
job 4: $p_{40} \cdot t_{40} \cdot p_{41} \cdot t_{41} \cdot p_{43} \cdot t_{43} \cdot p_{40}$

Because machines process only one job at a time, vertical circuits marked with a single token are added to determine the sequencing of the jobs on the corresponding machines (i.e. the token represents the availability of the machine to process a job):

machine 1: $m_{11} \cdot t_{11} \cdot m_{13} \cdot t_{31} \cdot m_{14} \cdot t_{41} \cdot m_{11}$
machine 2: $m_{21} \cdot t_{12} \cdot m_{22} \cdot t_{22} \cdot m_{21}$
machine 3: $m_{31} \cdot t_{13} \cdot m_{32} \cdot t_{23} \cdot t_{23} \cdot m_{33} \cdot t_{33} \cdot m_{34} \cdot t_{44} \cdot m_{31}$

Another class of examples where MGs are useful to model manufacturing problems is the following. A **flow line** is a tandem production system (i.e. a series of machines separated by buffers) (Gershwin, 1987). Material flows from the first to the last machine. Figure 1.25(a) shows a flow line with three machines and two buffers. The MG in Fig. 1.25(b) makes explicit the behavior of the model with **reliable** machines, assuming the so-called **blocking-after service mechanism** (i.e. machine processes even if there is no free-space in the output buffer).

Here it is worth noting that the modeling capacities of MGs and SMs are **dual** in the sense that SMs can model choices, but not synchronizations. On the other hand, MGs can model synchronizations, but cannot model choices.

The subclasses we introduce below contain SMs and MGs; therefore, they can model some restricted interleaving between choices and synchronizations, although not in all the cases that ordinary nets allow. Free-choice nets (FCNs) can be considered as an extension of:

(a) SMs by allowing MGs-type synchronization (i.e. if two places share a common output transition then this is their unique output transition); or
(b) MGs by allowing SMs-type conflicts (i.e. if two transitions share a common input place, then this is their unique input place).

It is not difficult to realize that both statements represent identical restrictions. FCNs are a common generalization in which choices and synchronizations do not directly interfere with each other.

Definition 1.10. A **free-choice net** (FCN) is an ordinary net such that:

$$\forall p \in P, |p^\bullet| > 1 \Rightarrow \forall t_k \in p^\bullet, |{}^\bullet t_k| = 1$$

That is, if two transitions, t_i and t_j, have a common input place p, it is the only input place of t_i and t_j.

From a behavioral point of view, FCNs are models in which either all or none of the output transitions of each place are enabled. In an FCN, if a

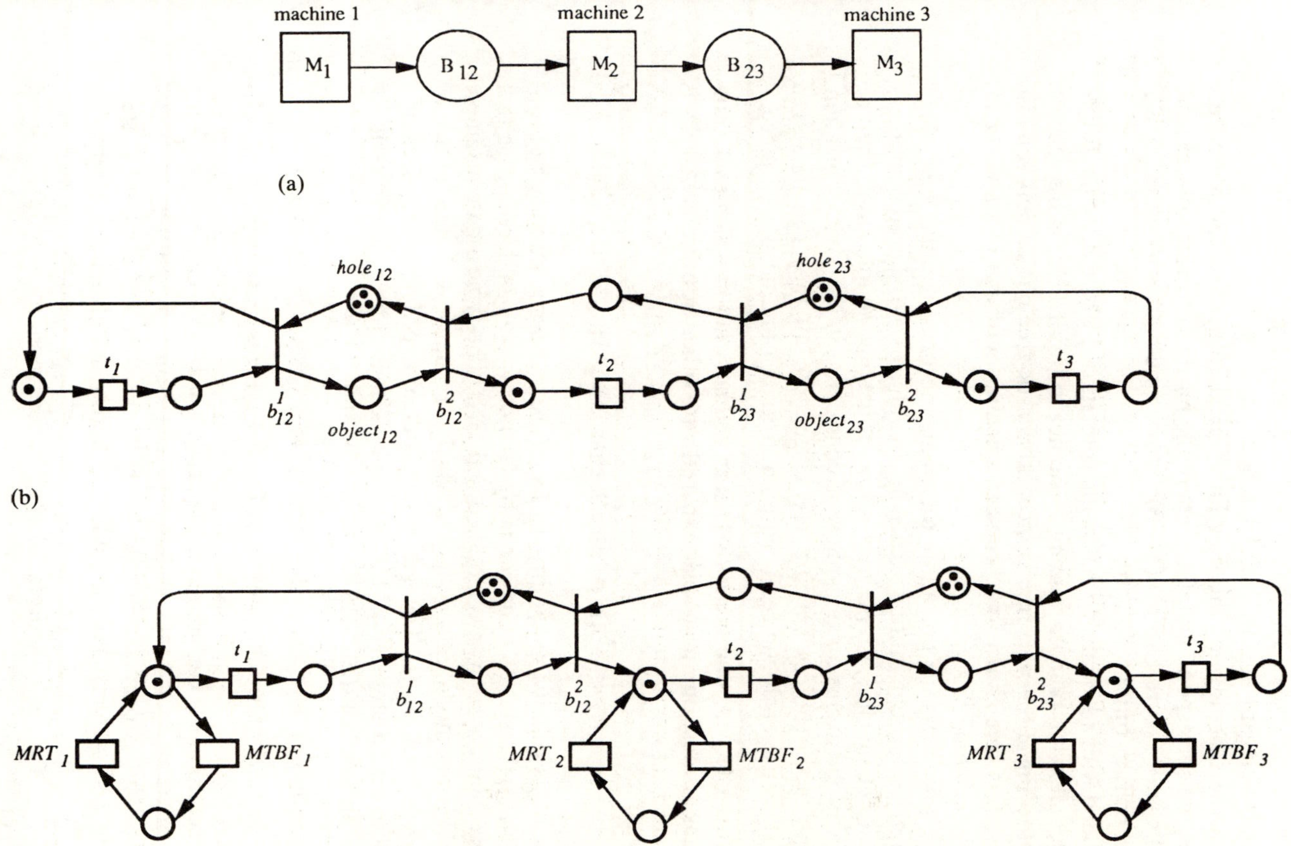

Figure 1.25 A flow line and two net systems representations: (a) a flow line with three machines and two buffers; (b) MG model assuming reliable machines; (c) Macroplace expansion for unreliable machines.

place is marked and has more than one output transition, the transition to be fired can be freely chosen (i.e. independently of the rest of the marking). Hence the name. The behavioral and structural analysis of FCNs is particularly elegant and well understood.

When the machines of the flow line (Fig. 1.25(a)) are considered **unreliable**, the corresponding refinement of the machine-working places leads to the FCN of Figure 1.25(c)) (MTBF: mean time between failures; MRT: mean repairing time).

Provided with adequate stochastic interpretation, FCNs extend the FJQN/B model allowing **random routings**. The net in Fig. 1.11(c) is free-choice. The net in Fig. 1.8 is also free-choice. Stochastic free-choice nets can also be viewed as **free-choice synchronized queuing networks** (Campos *et al.*, 1991).

Free-choice nets do not allow the modeling of sequential subsystems synchronized through mutual exclusion semaphores (i.e. shared resources). The next net subclass, simple nets, allows it in some simple case: when no more than one exclusion semaphore is considered in any synchronization. In simple nets, choices are not free in general but they can be **solved locally**, because each choice is centered around a unique shared place.

Definition 1.11. A **simple net** (SN) is an ordinary net in which each transition has at most one input place shared with other transitions. An SN is such that:

$$\forall t \in T \ |\{p \in {}^{\bullet}t| \text{ such that } |p^{\bullet}| > 1\}| \leq 1$$

In spite of its relative generality, this subclass has some interesting properties. Nevertheless, its behavior is by far not so well understood as that of FCNs. Many systems are modeled with SNs. The typical basic example of an SN is the model of a system in which a resource is shared by two or more users (Fig. 1.4(g)). The net of our production cell with two machines, one robot and a store (Fig. 1.9) is simple. Other simple non-free-choice nets are in Figs 1.12(b), 1.15, 1.17(a), 1.19 and 1.22(a).

1.7.2 On the analysis of the net systems subclasses

The structure of the considered net subclasses is rich enough to give plenty of information on the net systems we can define by putting an initial marking. This is particularly true for FCN systems and their subclasses, MG and SM systems. For SN systems there exist also some interesting results, but the stronger properties of free-choice systems cannot be extended (e.g. the rank theorem, or the liveness monotonicity with respect to the initial marking).

Our selection of results concentrates basically on liveness and reachability of live systems.

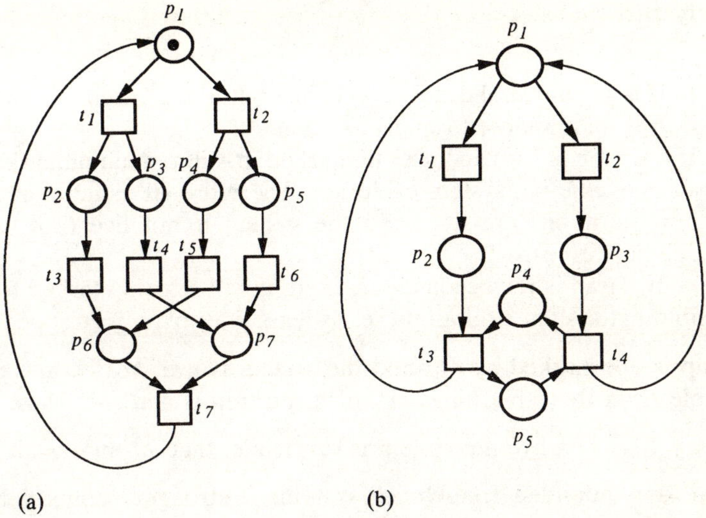

Figure 1.26 Two consistent and conservative free choice nets: (a) Structurally live ($r = \rho$), $r = rank(C) = 5$, $\rho = m + n - a - 1 = 7 + 7 - 8 - 1 = 5$; (b) Structurally non-live ($r \neq \rho$), $r = rank(C) = 3$, $\rho = m + n - a - 1 = 4 + 5 - 6 - 1 = 2$.

(a) Siphons, traps, liveness and reversibility

By means of graph-theory based reasoning it is possible to characterize many properties for net subclasses. **Siphons** (also called **structural deadlocks**, or more simply **deadlocks**) and **traps** are easily recognizable subsets of places that generate very particular subnets.

Definition 1.12. Let $N = \langle P, T, F \rangle$ be an ordinary net.

1. A **siphon** is a subset of places such that the set of its input transitions is contained in the set of its output transitions: $\Sigma \subseteq P$ is a siphon $\Leftrightarrow {}^\bullet\Sigma \subseteq \Sigma^\bullet$.
2. A **trap** is a subset of places such that the set of its output transitions is contained in the set of its input transitions: $\theta \subseteq P$ is a trap $\Leftrightarrow \theta^\bullet \subseteq {}^\bullet\theta$.

$\Sigma = \{p_1, p_2, p_4, p_5, p_6\}$ is a siphon for the net in Fig. 1.26(a): ${}^\bullet\Sigma = \{t_7, t_1, t_2, t_3, t_5\}$, while $\Sigma^\bullet = {}^\bullet\Sigma \cup \{t_6\}$. Σ contains a trap, $\theta = \Sigma \backslash p_5$. In fact θ is also a siphon (it is minimal: removing any number of places no siphon can be obtained).

Siphons and traps are reverse concepts: a subset of places of a net N is a siphon iff it is a trap on the reverse net, N^{-1} (i.e. that obtained reversing the arcs, its flow relation, F).

The following property 'explains' why 'siphons' (think of 'soda siphons') and 'traps' are the names of the above concepts.

Property 1.14

1. Siphons:
   (i) If M is a behavioral deadlock (i.e. dead-marking), then $D = \{p/M(p) = 0\}$ is an unmarked (empty) siphon.
   (ii) If a siphon is (or becomes) unmarked, it will remain unmarked for any possible net system evolution. Therefore all its input and output transitions are dead. So the system is not live (but can be deadlock-free).
2. Traps: If a trap is (or becomes) marked, it will remain marked for any possible net system evolution (i.e. at least one token is 'trapped').

If a trap is not marked by M_0, and the system is live, M_0 will not be recoverable from those markings in which the trap is marked. Thus:

Corollary 1.1. If a live net system is reversible, then M_0 marks all traps.

For live and bounded free-choice systems a stronger property holds: Marking all traps is a necessary and sufficient condition for reversibility (Best *et al.*, 1990). The net system in Fig. 1.26(a) is reversible. Nevertheless, if $M_0^T = (0\,1\,0\,0\,1\,0\,0)$, the new system is live and bounded but non-reversible: The trap $\theta = \{p_1, p_3, p_4, p_6, p_7\}$ is not marked under M_0.

A siphon which contains a marked trap will never become unmarked. So this more elaborate property can be helpful for some liveness characterizations.

Definition 1.13. Let N be an ordinary net. The system $\langle N, M_0\rangle$ has the **marked-siphon-trap property**, MST-property, if each siphon contains a marked trap under M_0.

A siphon (trap) is **minimal** if it is not contained in any other. Thus siphons in the above statement can be constrained to be minimal without any loss of generality.

The MST-property guarantees that all siphons will be marked. Thus no dead marking can be reached, according to property 1.14(1)(i). Therefore:

Property 1.15. If $\langle N, M_0\rangle$ has the MST-property, the system is deadlock-free.

Figure 1.27 presents some limitations of the MST-property for liveness characterization. Nevertheless, there exist the following interesting results on liveness.

Property 1.16. The MST-property is sufficient for liveness in simple net systems and necessary and sufficient for free-choice net systems.

As a corollary, the following **liveness monotonicity** result is true:

Corollary 1.2. Let $\langle N, M_0\rangle$ be a live free-choice system. Increasing M_0 preserves liveness $M_0' \geq M_0 \Rightarrow \langle N, M_0\rangle$ is live.

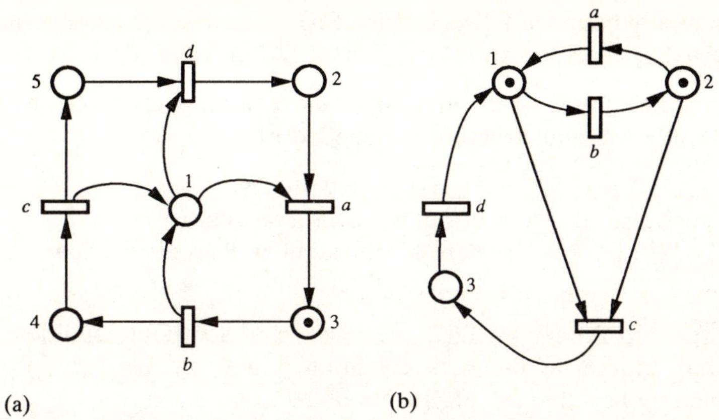

Figure 1.27 For liveness analysis the marked siphon-trap-property is not necessary for simple bounded nets nor sufficient for non-simple bounded nets: (a) the marked siphon-trap-property does not hold and the simple net is live and bounded; (b) the marked siphon-trap-property does not hold but the non-simple net is non-live (although deadlock-free) and bounded.

The above result does not apply to SN systems. The system in Fig. 1.12(b) is simple, $\Sigma = \{p_1, p_2, p_7\}$ is a siphon ($^\bullet\Sigma = \{t_3, t_4, t_1\}$, $\Sigma^\bullet = {}^\bullet\Sigma \cup \{t_2\}$) that does not contain any trap. If we assume $M_0(p_5) = 1$, t_2 can be fired and Σ becomes empty, leading to non-liveness.

SMs and MGs are FCNs. The set of places of an SM is a minimal siphon and a minimal trap iff it is **strongly connected**. Moreover, any **elementary circuit** of an MG is simultaneously a minimal siphon and a minimal trap. So the following can be stated.

Property 1.17

1. An SM is live for M_0 iff it is strongly connected and marked under M_0 (i.e. there is at least one token).
2. An MG is live for M_0 iff all circuits are marked under M_0.

Obviously, liveness is a **polynomial complexity** problem for SMs. By conventional logic equivalence the last property can be easily restated as follows: *an MG is live under M_0 iff there is no unmarked circuit.* From this statement it follows immediately that liveness can also be computed in polynomial time for MGs:

1. Remove from N all marked places under M_0 to obtain the net N_0;
2. The system $\langle N, M_0 \rangle$ is live iff N_0 is an acyclic graph.

The MG systems in Figs 1.24 and 1.25 are obviously live.

(b) Linear algebra, structural liveness and liveness

Structural boundedness is well characterized for general Petri nets (see Properties 1.7 and 1.8). In this section structural liveness is characterized for structurally bounded FCNs and for MGs. Liveness for those structurally live net systems is also fully characterized in linear algebraic terms.

Property 1.18. Let C be the incidence matrix of the free-choice net N. N is structurally live and structurally bounded iff:

$\exists X \geq 1$ such that $C \cdot X = 0$ (i.e. net consistency)
$\exists Y \geq 1$ such that $Y^T \cdot C = 0$ (i.e. net conservativeness)
$rank(C) = |T| + |P| - a - 1$, where a is the number of arcs in the flow relation *Pre*.

With respect to Property 1.12(2), necessary and sufficient conditions are now given. Moreover, the rank condition is now an *equality*, and δ is substituted by its value for FCNs: $\delta = a - |P|$.

The above property allows us to conclude on structural liveness (Fig. 1.26(a)) and structural non-liveness (Fig.1.26(b)) for consistent and conservative FCNs.

The particularization of Property 1.18 for SMs and MGs shows that in both cases the rank condition is redundant: it is satisfied when consistency and conservativeness hold. Even more, the following can be shown:

Property 1.19. A connected SM (MG) is consistent (conservative) if the net is strongly connected.

Connected SMs are conservative (then structurally bounded), thus structurally live iff strongly connected. Connected MGs are consistent, thus structurally bounded and structurally live iff strongly connected. This very last result can be generalized interpreting Property 1.17(2): with 'large enough' M_0, any MG is live. In other words, **MGs are structurally live nets.**

Property 1.18 has several important consequences (Esparza and Silva, 1991a): (1) Structural liveness can be decided in **polynomial time** (solving the consistency and conservativeness problems, and computing the rank of a matrix); (2) **duality theorem**; (3) a **kit of two reduction rules** (that is **complete**, i.e. is able to reduce all structurally live and structurally bounded FCNs).

Property 1.18 characterizes the lively and boundedly markable FCNs. Once we have one of these nets, we would like to know which are exactly the markings that make it live and bounded.

Property 1.20. Let N be a structurally live and structurally bounded FCN. $\langle N, M_0 \rangle$ is live iff all p-semiflows of N are marked at M_0 (i.e. $\nexists Y \geq 0$ such that $Y^T \cdot C = 0$ and $Y^T \cdot M_0 = 0$).

The necessary condition holds in general: if a p-semiflow (that is always a siphon and a trap) is unmarked at M_0, it remains unmarked forever. Liveness **monotonicity** for live and (structurally) bounded FCN systems is also a consequence of the above property. Moreover, liveness for structurally bounded FCNs is characterizable in **polynomial time**.

(c) Reversibility, reachability and marking bounds

Not all live and bounded FCNs are reversible (imagine $M_0^* = (0, 1, 0, 0, 1, 0, 0)^T$ for the net in Fig. 1.26(a). Nevertheless, it is always possible to reach from M_0 a marking such that a unique reversible system is obtained (i.e. there always exists a so-called **home state**). The following property characterizes the live, bounded and reversible FCN systems.

Property 1.21. Let $\langle N, M_0 \rangle$ be a live and bounded FCN system. It is reversible iff all traps are marked under M_0.

The reachability problem for the class of systems being considered is also solved (Desel and Esparza, 1990):

Property 1.22. Let $\langle N, M_0 \rangle$ be a live, reversible and bounded FCN system. The three following statements are equivalent:

(i) M is reachable from M_0 (i.e. $\exists \sigma$ such that $M_0[\sigma \rangle M$);
(ii) $M = M_0 + C \cdot \overline{\sigma}$, $\overline{\sigma} \geq 0$ where $M \in N^n$ marks all minimal traps;
(iii) $B^T \cdot M = B^T \cdot M_0$, where B is a basis of left annullers (i.e. $B^T \cdot C = 0$) and $M \in N^n$ marks all minimal traps.

Strongly connected SMs have one minimal trap ($\theta = P$) and $B = 1$. Therefore (Property 1.21) live SMs are reversible and the reachable markings are the non-negative integer solutions of $\Sigma_p M(p) = \Sigma_p M_0(p)$. For MGs minimal traps coincide with elementary circuits. Therefore liveness is equivalent to reversibility for strongly connected MGs. Even more, the above is true for MGs in general:

Property 1.23. Let $\langle N, M_0 \rangle$ be a live (possibly unbounded) MG system. The three following statements are equivalent:

(i) M is reachable from M_0 (i.e. $\exists \sigma$ such that $M_0[\sigma \rangle M$)
(ii) $M = M_0 + C \cdot \overline{\sigma}$, $\overline{\sigma} \geq 0$, where $M \in N^n$
(iii) $B_f^T \cdot M = B_f^T \cdot M_0$, where $M \in N^n$ and B_f is the **fundamental circuit** matrix of the graph.

Therefore it is easy to show that (possibly unbounded) MG systems are reversible iff live. Finally, using reachability characterizations it is not difficult to derive the following last property:

Property 1.24. Let $\langle N, M_0 \rangle$ be a live and bounded FCN system or a live (possibly unbounded) MG system. The behavioral and structural bounds of place p coincide:

$$\begin{aligned} B(p) &= \max\{M(p)/M \in R\ (N, M_0)\} = SB(p) \\ &= \max\{M(p)/M = M_0 + C \cdot \overline{\sigma} \geq 0,\ \overline{\sigma} \geq 0\} \end{aligned}$$

As an example, the above means (for the class of systems being considered) that the maximum required size of stores (buffers) can be computed exactly in **polynomial time** (solving LPP2 or LPP3).

The net system in Fig. 1.9 is simple but not free-choice (thus it is not a marked graph). As an example of **complementarity** among analysis techniques, let us prove its main global properties combining **reductions** and **structure theory** for net subclasses:

1. Reduction phase:
   - Use rule RA2 (Fig. 1.16) to fuse the following series of transitions of Fig. 1.9:

$$t_1\text{–}load\text{–}t_2 \Rightarrow t_{12}$$
$$t_4\text{–}deposit\text{–}t_5 \Rightarrow t_{45}$$
$$t_7\text{–}unload\text{–}t_8 \Rightarrow t_{78}$$
$$t_9\text{–}withdrawal\text{–}t_{10} \Rightarrow t_{910}$$

   - Use (marking) implicit place rule (only the basic extension of self-loop place, RC1 in Fig. 1.16) to remove place R.
2. The remaining net is a **marked graph**, therefore:
   - The MG is strongly connected, then conservative (Property 1.19). Conservativeness is a particular case of structural boundedness. Therefore the MG system is bounded.
   - Removing marked places (i.e. wait raw, *empty*, *object*, wait with.) an acyclic net is obtained. Then the MG system is live (Property 1.17(2)) and reversible (follows from Property 1.22 or Property 1.23).

Rules RA2 and RC1 preserve boundedness, liveness and reversibility. Therefore, the original manufacturing model is bounded, live and reversible.

1.8 CONCLUDING REMARKS

Petri nets theory and applications is a vast field. In this chapter a selection of **modeling issues** have been presented first. **Autonomous net systems** are abstract models. **Interpreted net systems** are 'natural' extensions of well-known formalisms like **state diagrams** (a formalism to model the functional behavior of sequential switching systems) or **queuing networks** (a formalism to model performance issues of systems where the sharing of resources plays an important role, but there are no synchronizations). **Marking diagrams** and **synchronized queuing networks** are examples of such (particular) net interpreted models.

The GRAFCET (see, for example, David and Alla (1989)), an international standard of the Commission Electrotechnique International (CEI #848, *Preparation of function charts for control systems*) is an interpreted bipartite graph, 1-bounded by definition, closely related in some aspects to marking diagrams. GRAFCETs are difficult to validate in general because too much information is put in the interpretation. They have been extensively used in industry at the local control level.

Qualitative (i.e. logical) analysis issues of net system models have been considered in the second part of the chapter. We tried to present a compilation of some practical results. The reader is referred to the literature where many more interesting analytical results are documented. Complementarity of qualitative analysis techniques has been emphasized. Moreover, quantitative (i.e. performance) modeling issues have just been introduced in the framework of net interpretations. A consideration in manufacturing, including quantitative analysis techniques, is delayed to Chapter 4. At the research level a main trend is the in-depth interleaving of qualitative and quantitative analysis techniques.

The behavior of non-autonomous net systems may be constrained by the environment. Therefore the interpreted net system behavior may be strictly included in the behavior of the underlying autonomous net system. Therefore, attention must be paid to the fact that the analysis of the underlying net system may be for the interpreted one **only necessary** (e.g. for marking reachability), **only sufficient** (e.g. for boundedness or mututal exclusion) or **neither necessary nor sufficient** (e.g. for liveness). In other words, the designer should interpret the properties of the underlying net system, given the constraints imposed by the environment: the results of the net system qualitative analysis can be considered as **elaborated warnings**. This generalizes from purely syntactical to a certain semantical level, those warnings in the compilation of programs.

Petri nets as introduced here represent a level in the hierarchy of net level models; namely **place/transition nets**. Subclasses have been considered to increase the practical decision power at the expense of practical modeling power. On the other side, extensions of place/transition net models allow us to increase the practical modeling power (e.g. **capacity** Petri nets: all places are bounded by a given capacity) or even the theoretical modeling power (e.g. **inhibitor arcs** or **transition priority** nets have the same descriptive power as **Turing machines**!). Therefore they are difficult to analyze in general. Nevertheless, in bounded systems inhibitor arcs or priority at transitions do not enlarge the modeling power. They are only practical modeling facilities.

Another kind of extension leads to the so-called **High-Level Nets** (HLN) (see Jensen and Rozenberg (1991) for a broad and recent perspective). In HLNs tokens have attributes (e.g. called colors) and arcs receive some inscriptions concerning the flow and transformation of the attributes of tokens. HLNs allow us to make much more compact models (see, for example, in manufacturing, Martinez *et al.* (1986) or Silva and Valette (1990)). Their analysis is presently not so mature as for place/transition nets.

Nets potentials for modeling discrete event dynamic systems (a consistent, graphical/analytical family of models covering the main modeling issues), and for its qualitative and quantitative analysis are not thc only basic arguments to use nets. **Real-time control systems** can be derived from nets in fairly systematic ways. This can be done at the **hardware**, **firmware** (i.e. microprogrammed) and **software** level (see Silva (1985)) for a broad overview of techniques). The generation of real-time software systems received particular attention in the past, putting nets close to **programmable logic controllers** (Colom *et al.*, 1986; Silva and Valette, 1990).

1.9 BIBLIOGRAPHICAL REMARKS

Petri nets were introduced in the Ph.D. of Carl Adam Petri (Petri, 1962). Today it is a very rich but relatively young field having an impact on many different industrial sectors. More than in seminal/historical papers we mainly (but not only) refer to books, tutorials or surveys, where the specialized contributions are explicitly pointed. A bibliography on Petri nets is periodically gathered by the *Gesellschaft für Mathematik und Datenverarbeitung* (GMD). The last issue was compiled in 1986 and published in Rozenberg, 1987, where some 2,634 entries are quoted.

Introductory texts to Petri nets and their applications have been written by Peterson (1981), Brams, (collective name of a group of French researchers) (Brams 1983), Silva (1985) and David and Alla (1989).

The material of two advanced courses on Petri nets is collected in Brauer (1980), Brauer *et al.* (1987a,b). There is a subseries of *Lecture Notes in Computer Science* (LNCS) entitled *Advances in Petri Nets* (Rozenberg, various dates). The *International Conference* (*European Workshop* until 1989) *on Application and Theory of Petri Nets* takes place every year, since 1980. The Proceedings are published by the organizers with the support of IBM-Germany. Published by the IEEE Computer Society Press, there are the *Proceedings of the International Workshops on Petri Nets and Performance Models* (PNPM) focusing on the time qualitative and quantitative aspects. Focused on *High-Level Petri Nets* (Jensen and Rozenberg, 1991) is a selection of papers.

There exist several surveys on Petri nets in general. Murata (1989) gives a broad perspective (containing 315 bibliographical entries) and received the IEEE Donald G. Fink Prize Award. Petri nets and manufacturing have also been the subject of several tutorials and surveys. Martinez *et al.* (1986) is a basic tutorial for manufacturing engineers introducing concepts on modeling with Petri nets. More survey-oriented are Valette (1987) and Silva and Valette (1990). The last one assumes knowledge on Petri nets basic concepts. Basically it focuses on some manufacturing features and how Petri nets may be helpful (contains more than 150 bibliographical entries). Invited or regular sessions on Petri nets and manufacturing appear at several conferences (e.g. IEEE International Conference on Robotics and Automation, World Congress of IMACS, and IEEE International Symposium on Circuits and Systems).

Finkel (1990) is a recent paper on the construction of **coverability graphs**. **Transformation techniques** for the analysis of net models are surveyed in

Berthelot (1987). An overview of seminal works on this topic is Berthelot *et al.* (1980). Silva (1981) studies the macroplace reduction rule, while Silva (1985) and Colom and Silva (1991b) generalize the implicit (or redundant) place concept from Berthelot *et al.* (1980). The reverse process of reduction is **stepwise refinement** for which Valette (1979) and Suzuki and Murata (1983) can be taken into account. A more general/abstract perspective of the topic is summarized in Brauer *et al.* (1991).

A seminal overview on the **state equation** based analysis of net models is Memmi and Roucairol (1980). The bridge between Petri nets and linear programming is covered in Silva and Colom (1988) and Colom and Silva (1991b). The identification of **minimal invariants** as **extreme directions of a cone** allows us to derive fast algorithms to compute the sets of minimal *p*- and *t*-semiflows of a net (Colom and Silva, 1991a). Additional results on linear algebra and Petri nets are in Colom *et al.* (1990b). **Structure theory** for net subclasses is surveyed in several works. Best (1987), Best and Thiagarajan (1987) and Esparza and Silva (1991a) are mainly concerned with the **free choice** subclass. Commoner *et al.* (1971) and Murata (1977) develop the basic theory of **marked graphs**. In Thulasiraman and Comeau (1987) linear programming is considered for MGs.

BIBLIOGRAPHY

Ajmone, M., Balbo, G. and Conte, G. A class of generalized stochastic Petri nets for the performance analysis of multiprocessor systems. *ACM Transactions on Computer Systems*, 2(2), 93–122, May 1984.

Ajmone, M., Balbo, G. and Conte, G. *Performance Models of Multiprocessor Systems*. MIT Press, Cambridge, MA, 1987.

Ajmone, M., Balbo, G., Bobbio, A., Chiola, G., Conte, G. and Cumani, A. The effect of execution policies on the semantic and analysis of stochastic Petri nets. *IEEE Transactions on Software Engineering*, 15(7), 832–846, July 1989.

Al-Jaar, R. Y. and Desrochers, A. A survey of Petri nets in automated manufacturing systems. In *IMACS World Congress*, vol. 2, pp. 503–510, Paris, June 1988.

Berthelot, G. Transformations and decompositions of nets. In W. Brauer *et al.* (1987a), pp. 359–376, 1987.

Berthelot, G., Roucairol, G. and Valk, R. Reductions of nets and parallel programs. In Brauer (1980), pp. 277–290, 1980.

Best, E. Structure theory of Petri nets: the free choice hiatus. In W. Brauer *et al.* (1987a), pp. 168–205, 1987.

Best, E. and Thiagarajan, P. S. Some classes of live and save Petri nets. In Voss *et al.* (1987), pp. 71–94, 1987.

Best, E., Cherskasova, L., Desel, J. and Esparza, J. Characterization of home states in free choice systems. *Hildesheimer Informatik-Berichte*, no. 7/90, July 1990.

Brams, G. W. *Réseaux de Petri: théorie et pratique* (2 vols). Paris, Masson, 1983.

Brauer, W. (ed.) *Net Theory and Applications*. LNCS '84, Springer-Verlag, Berlin, 1980.

Brauer, W., Reisig, W. and Rozenberg, G. *Petri Nets: Central Models and their Properties*. LNCS 254, Springer-Verlag, Berlin, 1987a.

Brauer, W., Reisig, W. and Rozenberg, G. *Petri Nets: Applications and Relationships to other Models of Concurrency*. LNCS 255, Springer-Verlag, Berlin, 1987b.

Brauer, W., Gold, R. and Vogler, W. A survey of behaviour and equivalence preserving refinements of Petri nets. *Advances in Petri Nets '90*. (G. Rozenberg, ed.), LNCS 483, pp. 1–46. Springer-Verlag, Berlin, 1991.

Breeding, K. I. *Digital Design Fundamentals*. Prentice-Hall, Englewood Cliffs, NJ, 1989.

Bruno, G. and Marchetto, G. Process translatable Petri-nets for the rapid prototyping of control systems. *IEEE Transactions on Software Engineering*, vol. SE-12, no. 2, pp. 346–357, February 1985.

Campos, J., Colom, J. M. and Silva, M. Performance evaluation of repetitive automated manufacturing systems. In *Proceedings of the 2nd International Conference on Computer Integrated Manufacturing* (IEEE Computer Society Press). Troy, New York, pp. 78–91, May 1990.

Campos, J., Chiola, G. and Silva, M. Properties and performance bounds for closed free choice synchronized monoclass queueing networks. *IEEE Transactions on Automatic Control*, 36(12), December 1991. [Special issue on *Multidimensional Queueing Networks*.]

Colom, J. M. and Silva, M. Convex geometry and semiflows in P/T nets. *Advances in Petri Nets '90* (G. Rozenberg, ed.). LNCS 483, Springer-Verlag, Berlin, pp. 79–112, 1991a.

Colom, J. M. and Silva, M. Improving the linearly based characterization of P/T nets. *Advances in Petri Nets '90* (G. Rozenberg, ed.). LNCS 483, Springer-Verlag, Berlin, pp. 113–145, 1991b.

Colom, J. M., Silva, M. and Villarroel, J. L. On software implementations of Petri nets and colored Petri nets using high-level concurrent languages. In *Proceedings of the 7th European Workshop on Application and Theory of Petri Nets*. Oxford, pp. 207–241, July 1986.

Colom, J. M., Campos, J. and Silva, M. On liveness analysis through linear algebraic techniques. *Dpto. Ing. Eléctrica e Informática, Research Report RR-90-11* (16 pp.). Also in *Deliverables of Esprit Basic Research Action Demon*, June 1990.

Commoner, F., Holt, A. W., Even, S. and Pnueli, A. Marked directed graphs. *Journal of Computer and System Sciences*, vol. 9, no. 2, pp. 72–79, 1971.

Dallery, Y., Liu, Z. and Towsley, D. Equivalence, reversibility and symmetry properties in fork/join queueing networks with blocking. *Université Pierre et Marie Curie, MASI Technical Report 90-32*, Paris, June 1990.

David, R. and Alla, H. *Du grafcet aux réseaux de Petri*. Hermes, Paris, 1989.

Deo, N. *Graph Theory with Applications to Engineering and Computer Science*. Prentice-Hall, Englewood Cliffs, NJ, 1974.

Desel, J. and Esparza, J. Reachability in reversible free-choice systems. *Technical University of Munich, SFB-Bericht*, No. 342/11/90A, June 1990.

Ercegovac, M. and Lang, T. *Digital Systems and Hardware, Firmware Algorithms*. John Wiley & Sons, New York, 1985.

Esparza, J. and Silva, M. On the analysis and synthesis of free choice systems. *Advances in Petri Nets '91* (G. Rozenberg, ed.). LNCS 483, Springer-Verlag, Berlin, pp. 243–286, 1991a.

Esparza, J. and Silva, M. Top-down synthesis of live and bounded free-choice nets. In *Proceedings of the 11th International Conference on Applications and Theory of Petri Nets*, pp. 63–83. Paris, June 1991b.

Finkel, A. A minimal coverability graph for Petri nets. In *Proceedings of the 11th International Conference on Applications and Theory of Petri Nets*, pp. 1–21. Paris, June 1990.

Genrich, H. J. and Lautenbach, K. Synchronisationsgraphen. *Acta Informatica* 2, pp. 143–161.

Gershwin, S. B. Representation and analysis of transfer lines with machines that have different processing rates. *Annals of Operations Research*, 9, 511–530, 1987.

Gibbons, A. *Algorithmic Graph Theory*. Cambridge University Press, London, 1985.

Girault, C. and Reisig, W. (eds). *Application and Theory of Petri Nets*. Informatik-Fachberichte 52, Springer-Verlag, Berlin, 1982.

Hack, M. T. Analysis of production schemata by Petri nets. *MIT, TR-94*. Boston, 1972 (corrected June 1974).

Hillion, H. P. and Proth, J. M. Performance evaluation of job-shop systems using timed event-graphs. *IEEE Transactions on Automatic Control*, 34(1), 3–9, January 1989.

Jensen, K. and Rozenberg, G. (eds). *High-Level Petri Nets. Theory and Application*. Springer-Verlag, Berlin, 1991.

Karp, R. and Miller, R. Parallel program schemata. *Journal of Computer and System Science*, 3(4), 167–195, May 1969.

Mailles, D. Files d'attente descriptives pour la modélisation de la synchronisation dans les systèmes informatiques. *Université P. et M. Curie. These d'Etat.* September 1987.

Martínez, J., Alla, H. and Silva, M. Petri nets for the specification of FMSs. In *Modelling and Design of Flexible Manufacturing Systems* (A. Kusiak, ed.). Elsevier, pp. 389–406, 1986.

Memmi, G. and Roucairol, G. Linear algebra in net theory. In Brauer (1980), pp. 213–223, 1980.

Molloy, M. K. Performance analysis using stochastic Petri nets. *IEEE Transactions on Computers*, 31(9), 913–917, September 1982.

Murata, T. Circuit theoretic analysis and synthesis of marked graphs. *IEEE Transactions on Circuits and Systems*, 24(7), 400–405, 1977.

Murata, T. Petri nets: properties, analysis and applications. *Proceedings of the IEEE*, 77(4), 541–580, April 1989.

Murty, K. G. *Linear Programming*. John Wiley & Sons. New York, 1983.

Nelson, R. Haibt, L. and Sheridan, P. Casting Petri-nets into programs. *IEEE Transactions on Software Engineering*, 9(5), 590–602, September 1983.

Nemhauser, G. L., Rinnoy Kan, A. H. G. and Todd, M. J. *Optimization*. Volume I of *Handbook in Operations Research and Management Science*. North-Holland, Amsterdam, 1989.

Peterson, J. L. *Petri Net Theory and the Modelling of Systems*. Prentice Hall, Englewood Cliffs, NJ, 1981.

Petri, C. A. *Kommunikation mit Automaten*. Bonn: Institut für Instrumentelle Mathematik, Schriften des IIM No. 2, 1962.

Proceedings of the European Workshop on Applications and Theory of Petri Nets (*EWPN '81*: Bad Honnef, Germany; *EWPN '82*: Varenna, Italy; *EWPN '83*: Toulouse, France; *EWPN '84*: Aarhus, Denmark; *EWPN '85*: Helsinki, Finland; *EWPN '86*: Oxford, UK; *EWPN '87*: Zaragoza, Spain; *EWPN '88*: Venezia, Italy; *EWPN '89*: Bonn, Germany).

Proceedings of the International Conference on Applications and Theory of Petri Nets (*ICPN '89*: Bonn, Germany; *ICPN '90*: Paris, France; *ICPN '91*: Gjern, Denmark).

Proceedings of the International Workshop on Petri Nets and Performance Models (*PNPM '87*: Madison, WI, USA, August 1987; *PNPM '89*: Kyoto, Japan, December 1989; *PNPM '91*: Melbourne, Australia, December 1991). IEEE Computer Society Press.

Proceedings of the International Workshop on Timed Petri Nets. Torino, Italy IEEE-Computer Society Press, July 1985.

Rozenberg, G. (ed.) *Advances in Petri Nets*. Lecture Notes in Computer Science: 188 (*APN '84*), 222 (*APN '85*), 266 (*APN '87*), 340 (*APN '88*), 424 (*APN '89*) and 483 (*APN '90*). Springer-Verlag, Berlin, various years.

Silva, M. Sur le concept de macroplace et son utilisation pour l'analyse des reseaux de Petri. *RAIRO-Systems Analysis and Control*, 15(4), 57–67, 1981.

Silva, M. *Las redes de Petri en la Automática y la Informática*. Editorial AC, Madrid, 1985.

Silva, M. and Velilla, S. Programmable logic controllers and Petri nets. In *Proceedings of the International Symposium of the IFAC – IFIC on Software for Computer Control*, SOCOCO '82, pp. 29–34 (G. Ferraté and E. A. Puente, eds). Pergamon Press, Oxford, 1982.

Silva, M. and Colom, J. M. On the computation of structural synchronic invariants in P/T nets. *Advances in Petri Nets '88* (G. Rozenberg, ed.). LNCS 340, pp. 386–417, Springer-Verlag, Berlin, 1988.

Silva, M. and Valette, R. Petri nets and flexible manufacturing. *Advances in Petri Nets '89* (G. Rozenberg, ed.). LNCS 424, pp. 375–417. Springer-Verlag, Berlin, 1990.

Suzuki, I. and Murata, T. A method for stepwise refinement and abstraction of Petri nets. *Journal of Computer and Systems Sciences*, 27(1), 51–76, August 1983.

Thulasiraman, K. and Comeau, M. Maximum-weight marking in marked graphs: algorithms and interpretations based on the simplex method. *IEEE Transactions on Circuits and Systems*, 34(12), 1535–1545, December M1987.

Valette, R. Analysis of Petri nets by stepwise refinements. *Journal of Computer and Systems Sciences*, vol. 18, pp. 35–46, 1979.

Valette, R. Nets in production systems. In Brauer (1987b), pp. 191–217, 1987.

Valette, R., Courvoisier, M., Bigou, J. M. and Alburkerque, J. A. A Petri net based programmable logic controller. *First International Conference on Computer Applications in Production and Engineering*, CAPE '83, Amsterdam, April 1983.

Voss, K., Genrich, H. J. and Rozenberg, G. (eds). *Concurrency and Nets*. Springer-Verlag, Berlin, 1987.

2

Principles of system modeling

J. M. Proth

2.1 MANUFACTURING SYSTEM MODELING: BASIC CONCEPTS

A manufacturing system is composed of two main parts: the **physical system** and the **management system**. Hereafter, the latter is also referred to as the **control** system or the **decision-making** system (DMS).

The physical system is the set of resources which operate on the raw material and/or on the work-in-process (WIP). For instance, machines, cells, transportation systems (conveyors, automated guided vehicles (AGVs), cranes, etc.), workers, storage devices, ovens, loading–unloading stations, quality control stations, belong to the physical system.

The DMS allows one to take advantage of the indetermination (i.e. the degree of freedom) of the physical system to make it work in a way which optimizes some criteria like productivity (which has to be maximized), WIP level (which has to be reduced as much as possible), total tardiness (which has to be as close to zero as possible).

The DMS can be divided into two subsystems: the part of the DMS which computes the decisions starting only from the state of the physical system (hereafter referred to as PSS-DMS for Physical System State-based DMS) and the part of the DMS which computes the decision starting from the state of the environment of the physical system and possibly also from the state of the physical system itself (hereafter referred to as ES-DMS for Environment State-based DMS).

Let us give some examples of the DMS subsystems to clarify the previous terminology. Consider a storage resource (the physical system in our example) managed by the following rule: 'Order a quantity Q when the inventory level becomes less than q'. In that case, the DMS is restricted to the PSS-DMS because the decision is made starting from the state of the physical system, which is the inventory level.

Let us consider the same physical system obeying the following management rule: 'Order a quantity Q when the difference (inventory level minus demand) is less than q'. Now, we are in the case where the DMS is an ES-DMS, because the decision is based on the state of the environment of the storage resource (i.e. the demand) and on the state of the physical system (i.e. the inventory level).

Again, we consider the storage resource, and we apply the following management rule: 'Order a quantity Q_1 if the inventory level is less than q_1, or a quantity Q_2 if the difference (inventory level minus demand) is less than q_2'. In that case, the DMS is composed of a PSS-DMS which is the following management rule: 'Order a quantity Q if the inventory level is less than q_1', and an ES-DMS which is the following management rule: 'Order a quantity Q_2 if the difference (inventory level minus demand) is less than q_2'.

In turn, the PSS-DMS can be partitioned into two subsystems: the PSS-DMS1 which includes all the decision processes belonging to the PSS-DMS and which can be modeled as a decision-free Petri net **by means of the control places presented hereafter**, and the PSS-DMS2 which groups all the decision processes of the PSS-DMS which do not belong to the PSS-DMS1. In this chapter, we propose a modeling approach of the manufacturing systems in which the set of decision processes belonging either to the ES-DMS, or to the PSS-DMS2 remains outside the Petri net model. Hereafter, we refer to the set (ES-DMS) $\cup$ (PSS-DMS2) as O-DMS for Outside DMS, and to PSS-DMS1 as I-DMS for Inside DMS.

Note that a natural way to specify, model and analyze large-size discrete manufacturing systems is to decompose them into small, thus tractable, subsystems (also called modules), to model each of these modules, and to integrate the module models in order to reach the model of the whole system. The O-DMS design of the whole system is based on the O-DMSs of the module models. If the design is made in an adequate manner, the whole model exposes the classical good properties, namely boundedness, liveness and consistency. The integration problem will be studied in Chapter 4. For the moment, we notice only that the module models have to sometimes be connected to each other and to the O-DMS of the whole system. To make these connections possible, we have to enrich the module models, in the following ways:

(i) **Control places** are introduced in order to allow the operational mode of the module to be influenced by exogenous decisions (controls) made by the O-DMS of the whole system. A control place, usually represented by two concentric circles, contains either zero or one token. A token in a control place means that the output transition of this place is enabled if the other input places of this transition contain at least one token. Firing an output transition of a control place does not result in removing the token contained in this control place. In other words, the state of the control place depends on an exogenous decision which is made outside the part of the module modeled by a decision-free timed Petri net (i.e. in the O-DMS of the whole system),

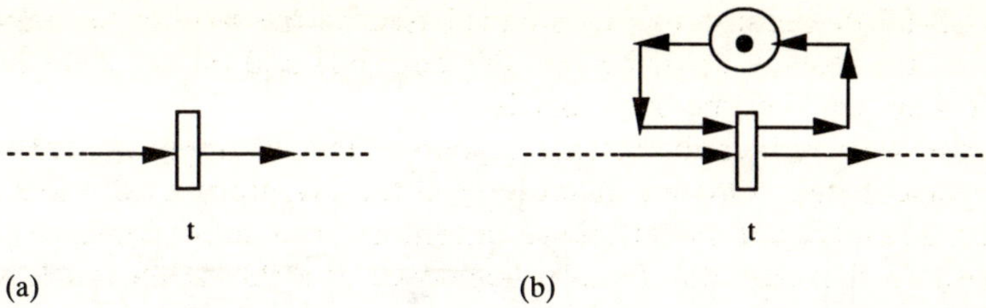

Figure 2.1 (a) Model used; (b) complete model.

and thus can only be changed from outside this part of the module (i.e. by the O-DMS of the whole system). The concept of control places was proposed by Holloway and Krogh (1990).

(ii) **Input** transitions are introduced to model the arrival of parts or raw material from outside the module.

(iii) **Output** transitions are introduced to model parts leaving the module.

In the rest of the chapter, while a transition is firing (which usually represents the utilization of a resource), it is impossible to initiate another firing of the same transition. The initiation of the next firing will be possible only after the firing in progress terminates. In other words, a model like the one represented in Fig. 2.1(a) must be understood as being the model represented in Fig. 2.1(b) (the self-loop is simply neglected in the model represented in Fig. 2.1(a)).

In some of the cases presented in this chapter, the model from which input, output and command places have been removed is a strongly connected timed marked graph (also called strongly connected timed event graph). We know that a strongly connected marked graph has very interesting properties. As a consequence, it will be very easy to analyze the behavior of the related manufacturing systems. This point is studied in Chapter 4.

Let us now present three small examples to illustrate the way to model a manufacturing system using the previous concepts.

The first one concerns a machine M which can manufacture two types of parts denoted by P_1 and P_2 respectively (see Fig. 2.2). Six (resp. four) units of time are required to manufacture one part of type P_1 (resp. P_2). The elementary circuit $\langle Q_1, t_1, Q_2, t_2 \rangle$ contains only one token. Assuming that the initial position of the token is Q_1 (as shown in Fig. 2.2), transition t_1 fires as soon as one token appears in place I_1 (i.e. the manufacturing of a P_1-type part starts as soon as the related raw material is available). Thus, firing input transition T_1 represents the introduction of raw material in the job-shop. The termination of the firing of t_1 represents the completion of a part of type P_1: a token appears in O_1, which means that a part of type P_1 is completed, and a token appears in Q_2, which means that machine M is available to manufacture the next part. Then, assuming that C contains

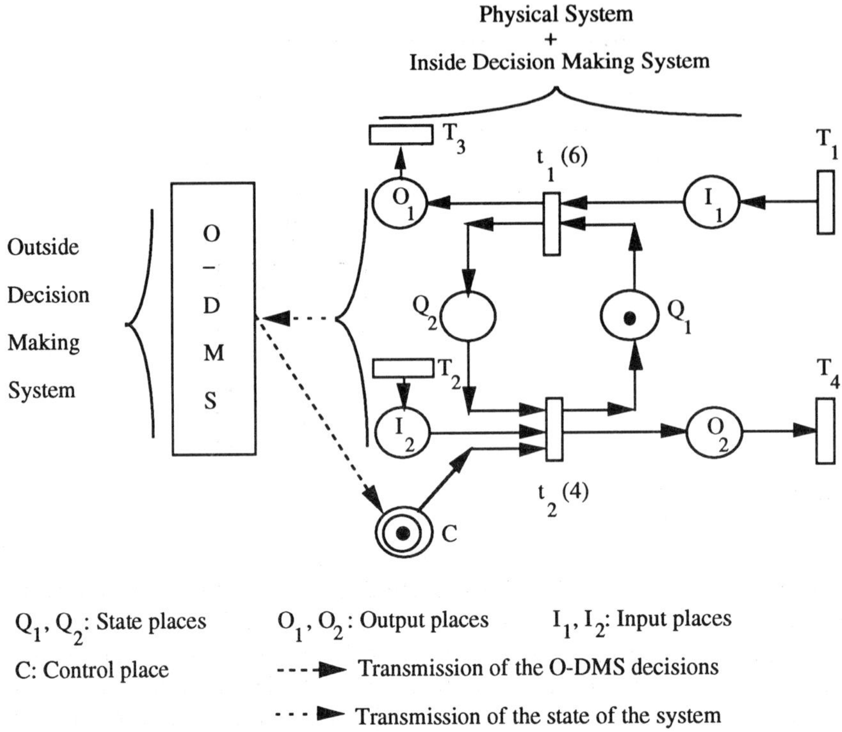

Figure 2.2 Manufacturing system model: first example.

one token, a firing of t_2 initiates as soon as one token appears in I_2, which means that input transition T_2 has been fired at least once, and thus that an external decision has been made to produce parts of type P_2. Firing transition t_2 represents the manufacturing of a part of type P_2. The termination of the firing of t_2 represents the completion of a part of type P_2: a token appears in O_2, which means that a part of type P_2 is completed, and a token appears in Q_1, which means that machine M is available to manufacture the next part of type P_1. The process continues as long as the token remains in C and transitions T_1 and T_2 fire to feed I_1 and I_2 with tokens. Removing the token from C results in stopping the process having the production of a part of type P_1.

Note that: (i) instead of connecting a control place to t_2, we could connect a control place to T_1 and/or T_2, and (ii) we could associate a firing time to T_1 and T_2 (to represent the time needed to introduce raw material from outside the system), and to T_3 and T_4 (to represent the time needed to transfer a part outside the system). If we connect a control place to both T_1 and T_2, we have to put a token in each of these control places alternately to manufacture alternately one P_1 and then one P_2.

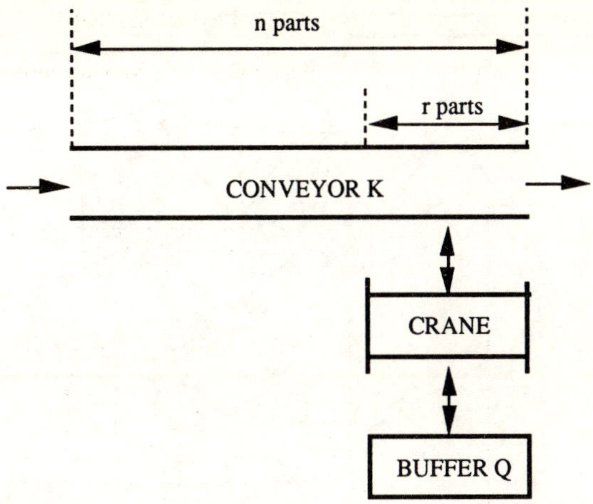

Figure 2.3 The conveyor–buffer system.

In this example, the I-DMS is restricted to the decision to manufacture alternately parts of types P_1 and P_2 or to stop. This I-DMS and the physical system are modeled by drawing the strongly connected event graph (i.e. the elementary circuit $\langle Q_1, t_1, Q_2, t_2\rangle$). The O-DMS remains outside the model. In the system at hand, the O-DMS generates the decision to stop or continue the production. This decision is applied by removing the token located in the control place C, or by putting one token in the empty control place C. In the example at hand, the O-DMS bases its decisions on customer orders (i.e. on the state of the environment of the system).

The second example is a conveyor (denoted by K) able to hold n parts (Fig. 2.3). The parts leave the conveyor on a first-in first-out (FIFO) basis as long as there is a request for these parts. A request for parts is modeled by putting one token in the control place C_2 (see Fig. 2.4). When the conveyor K is full, then r ($r \leq n$) parts are carried to a buffer Q by a crane. The same set of parts is carried back to the conveyor as soon as the conveyor is empty. To reduce the size of the model, we assume that n = 3 and $r = 2$ (i.e. K is able to carry a maximum of three parts, and two parts are moved to buffer Q if K is full, or two parts are carried from Q to K if the conveyor is empty). The size of the buffer Q is not bounded.

In the model of Fig. 2.4, S_1, S_2 and S_3 represent the three places on the conveyor. One token in a place S_i means that the ith position on the conveyor is busy, and one token in the place F_i means that the ith position on the conveyor is free. Thus, at each time, there is exactly one token in each pair of places (S_i, F_i), $i = 1, 2, 3$. Firing *ai* represents one part arriving at the entrance of the system. This part can enter the system only if there is one token in C_1 (which means that the decision made by the O-DMS is to accept new parts in the system) and one token in F_1 (which means that

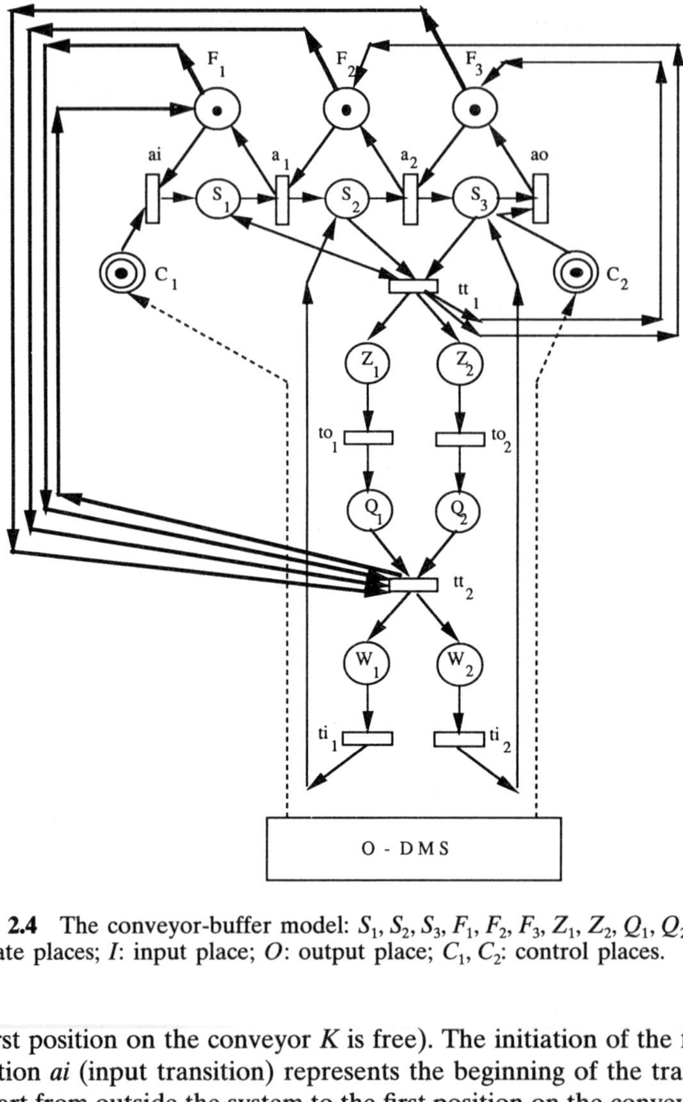

Figure 2.4 The conveyor-buffer model: S_1, S_2, S_3, F_1, F_2, F_3, Z_1, Z_2, Q_1, Q_2, W_1, W_2: state places; I: input place; O: output place; C_1, C_2: control places.

the first position on the conveyor K is free). The initiation of the firing of transition *ai* (input transition) represents the beginning of the transfer of one part from outside the system to the first position on the conveyor. The time associated with transition *ai* represents the time required to perform the previous transfer. The token in F_1 disappears when the firing of transition *ai* initiates, which means that the first position on conveyor K is reserved for the part whose transfer is in progress. The termination of the firing of transition *ai* results in the marking of place S_1 with one token. If place F_2 contains one token, the firing of transition a_1 initiates, which results in removing the token contained in place S_1 and the token contained in place F_2. The time associated to transition a_1 corresponds to the time required to transfer one part from the first position to the second position of conveyor K. When the firing of transition a_1 is completed, one token appears in both places F_1 and S_2: the first position of conveyor K is free for another part, and the part is in the second position of K. Again, if F_3 contains one token (which means that the third position of conveyor K is free), the firing of transition a_2 initiates, and thus tokens in S_2 and F_3 disappear. The firing of transition a_2 takes as much time as the time required to transfer a part from the second to the third position of conveyor K. At the termination of the firing of a_2, one token appears in both S_3 (which means that the part is located in the third position of the conveyor) and F_2 (which shows that the second position of K is free). Finally, if there is one token in C_2 (which means that O-DMS indicates that parts are required from outside the system), the firing of transition *ao* (out-put transition) initiates, which corresponds to the beginning of the transfer of the part outside the system. At the end of the transfer (i.e. when the firing of transition *ao* terminates), F_3 recovers one token (i.e. the third position of conveyor K is free). The same process holds for any part arriving at the entrance of the system.

If places S_1, S_2 and S_3 are marked (i.e. if F_1, F_2 and F_3 are empty), then tt_1, which is timed at 0, fires. Place S_1 recovers one token. Tokens located in S_2 and S_3 disappear as one token appears in each of the places F_2, F_3, Z_1 and Z_2. It means that the second and third positions on the conveyor will be kept free while the two parts previously located at these positions are placed at the disposal of the crane. Firing (simultaneously) transitions to_1 and to_2 represents the transfer from the conveyor to the buffer, which is modeled by places Q_1 and Q_2.

If places S_1, S_2 and S_3 are empty (i.e. if F_1, F_2 and F_3 are marked), and if Q_1 and Q_2 contain at least one token each (i.e. if there are at least two parts in the buffer), then transition tt_2 (timed at 0) fires. As a result, F_1 recovers one token, one token disappears from each F_2 and F_3, one token disappears from Q_1 and Q_2, and one token appears in W_1 and W_2. It means that two tokens are placed at the disposal of the crane to be transported back to the two last positions of conveyor K. Note that these positions are kept free until the arrival of the two parts because tokens in F_2 and F_3 have been removed.

The model provided in Fig. 2.4 is a simplified model of the conveyor–buffer system: the crane is always supposed to be available to move from the conveyor to the buffer, or from the buffer to the conveyor. In fact, to model the system in detail, we should also model the motion of the crane as well as its management. The model would be more complicated, but it would also be based on a decision-free timed Petri net.

The last example is a three-stage manufacturing system in which stages 1 and 3 are each performed by a unique machines, while stage 2 can be performed by one or the other among two machines which may have different performances. This manufacturing system can perform three types of parts denoted by P_1, P_2 and P_3. The decision to launch in production one type of part is made by the O-DMS, according to customer requirements.

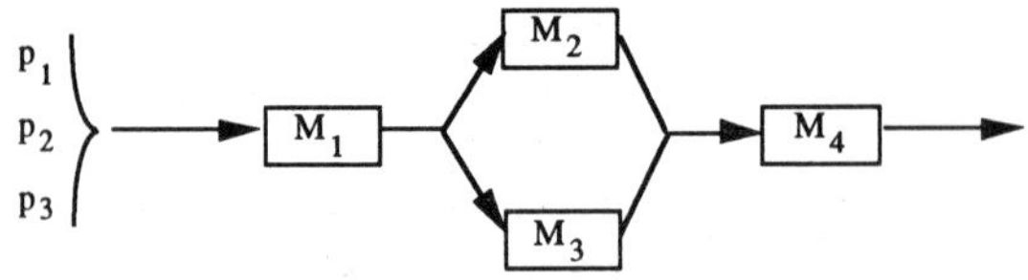

Figure 2.5 A three-stage manufacturing system.

The system is presented in Fig. 2.5.

The model of the three-stage manufacturing system is presented in Fig. 2.6. All the control places are connected to the O-DMS, which is not represented in the figure for simplicity. The main difference between this model and the models of the two previous examples is that the Petri net obtained by removing the control places, the input transition and the output transition is no more a decision-free (i.e. conflict-free) net, because places p_1^1, p_2^1 and p_3^1 have two output transitions each.

A control place is associated to each transition, and there is at most one token in each set $\{C_1^i, C_2^i, C_3^i\}$, $i = 1, 2, 3, 4$ at each point in time. It means that machine M_i, $i = 1, 2, 3, 4$, can be used to manufacture at most one part type at a time. Recall that a self-loop (which is not represented in the figure) is associated to each transition. It means that the same machine cannot be used to manufacture more than one part of the same type at a time. Finally, by combining the two previous conditions, we see that the model prevents a machine being used to manufacture more than one part at a time, whatever the type.

Furthermore, there is at most one token at each point in time in each pair (C_j^2, C_j^3), $j = 1, 2, 3$. It indicates the choice made at the O-DMS level among M_2 and M_3 to manufacture the net part of type P_j.

Note that if we remove from the net of Fig. 2.6 the transitions which are output transitions of the control places used to suppress the structural conflicts and which do not contain a token, as well as the arcs connected to these transitions, we obtain a decision-free (i.e. conflict-free) net. It means that, when a decision is well made at the O-DMS level, then all the uncertainties concerning the behavior of the system are removed.

For instance, assume that the marking of the control places in the model represented in Fig. 2.6 is as follows at a given point in time:

$$C_1^1 = 1;\ C_2^1 = 1;\ C_3^1 = 0$$
$$C_1^2 = 0;\ C_1^3 = 0;\ C_2^2 = 1;\ C_2^3 = 0;\ C_3^2 = 0;\ C_3^3 = 1$$
$$C_1^4 = 1;\ C_2^4 = 1;\ C_3^4 = 0$$

Then the corresponding conflict-free Petri net is given in Fig. 2.7 (control places have been removed). Thus, the dynamics of the Petri net model can be viewed as a sequence of conflict-free subsets as the one represented in Fig. 2.7.

To conclude the first section, the physical part and the I-DMS of the manufacturing module can be modeled either by decision-free timed Petri

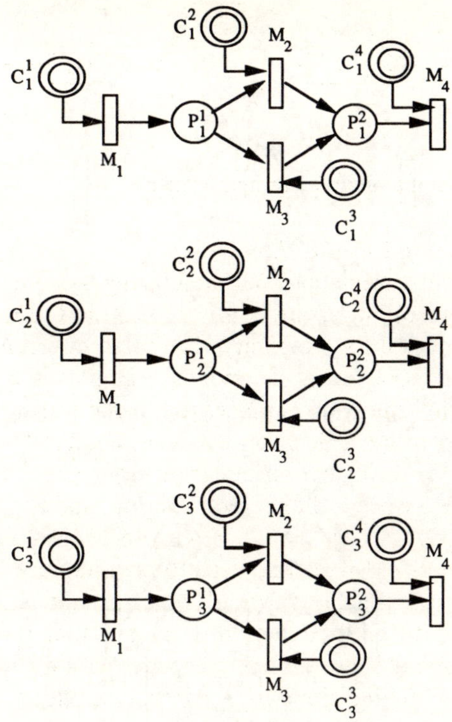

Figure 2.6 The model of the three-stage manufacturing system.

nets (which is sometimes a strongly connected timed event graph) or by no conflict-free nets, enriched by control places, input transitions and output transitions. In the second case, the models should turn into conflict-free nets by means of the marking of the control places.

The O-DMS remains outside the Petri net model. It could be computer software, or a person, or both (in the case when a computer-aided management system (CAMS) is used). The decisions made by the O-DMS are transferred to the physical system using control places.

In the next sections of this chapter, we present the models of the most common modules existing in the manufacturing systems.

2.2 TRANSPORTATION SYSTEM BETWEEN TWO POINTS ON A SHOP-FLOOR

Let us consider a shop-floor on which two positions are connected by a given path. At these positions for instance are located either machines, or storage devices, or manufacturing cells. The goal is to model the path taking into account the fact that a given position on the path can never be taken by more than one transportation resource at the same time. To

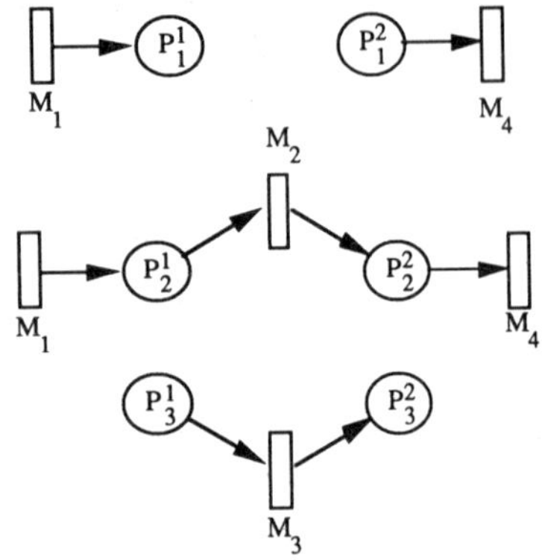

Figure 2.7 The active part of the set at a given point in time.

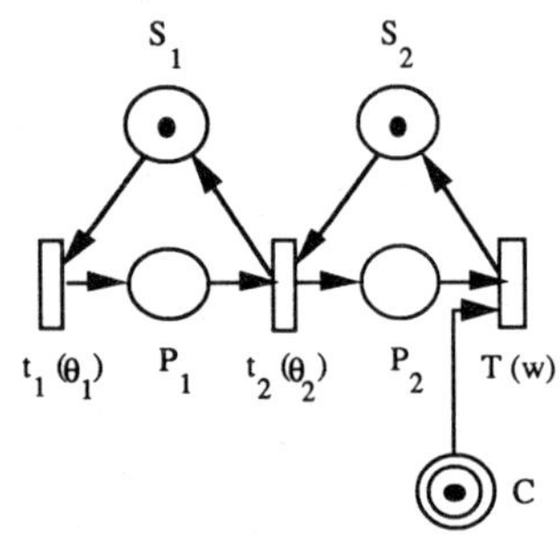

Figure 2.8 Model of a two-section path: P_1, P_2, S_1, S_2: state places; C: command place.

reach this goal, we divide the path into **sections**. The length of a section is such that a section can take at most one transportation resource. The model of such a path is given in Fig. 2.8 in the case of two sections.

A token in S_1 (resp. S_2) shows that the first (i.e. the second) section is free. A time θ_1 (resp. θ_2) is associated to transition t_1 (resp. t_2). It represents the time a transportation resource coming from outside the system (resp. from the first section) needs to move to the first (resp. the second) section. If S_1 contains one token (i.e. if the first section is free), then the firing of input transition t_1 initiates, which means that one transportation resource is moving from the beginning of the path to the first section, and the token contained in place S_1 is removed, which models the fact that the first section is reserved for the transportation resource. The firing of transition t_1 (i.e. the motion of the transportation resource) requires θ_1 units of time. One token appears in P_1 when the firing of transition t_1 terminates: meaning that the transportation resource has arrived at the first section. Then, if S_2 contains one token, the firing of transition t_2 initiates. Tokens in P_1 and S_2 disappear, which models respectively the facts that the transportation resource is moving from the first to the second section, and that the second section is reserved for the transportation resource. Note that, at this point of the process, both S_1 and S_2 are empty. In other words, both the first and the second section are inaccessible by another transportation system, because the transportation system considered is still located partially at the first section and already partially moved to the second section. The firing of t_2 terminates after θ_2 units of time, which results in putting one token in S_1 (i.e. the first section is free), and one token in P_2 (i.e. the transportation resource is arrived on the second section). One token in the control place C means that the part(s) on the transportation resource is (are) required by the resource located at the end of the path. In that case, the firing of output transition T initiates and the token in P_2 disappears. Note that the token located in C remains in C, according to the firing rule related to control places. After w units of time, the firing of T terminates (which means that the part carried by the transportation resource is available at the end of the path) and one token appears in place S_2 (which means that the second section of the conveyor is free).

The same process applies for any transportation resource arriving at the beginning of the path. In the example at hand, at most two transportation resources can be located simultaneously on the path. The model presented in Fig. 2.8 can be easily extended to any number of sections (i.e. to any path). Furthermore, the flow of transportation resources can be controlled by blocking temporarily a transportation resource on a section. This can be modeled by connecting a control place to the transition following the place which represents the section in question.

An important property of the model presented in this section is the following: if we remove the control place(s), as well as the arcs starting from these places, we obtain a strongly connected event graph. This property will be used in Chapter 4 to analyze the behavior of a complex manufacturing system.

2.3 STORAGE FACILITIES

In this section, we consider storage facilities managed in three different ways. The first subsection concerns storage facilities managed on a first-in first-out (FIFO) basis. In the second subsection, we consider the case when there is no connection between the order of the parts at the entrance and at the exit of a storage facility. The last subsection is devoted to the analysis of storage facilities managed on a last-in first-out (LIFO) basis. The capacities of the storage facilities are supposed to be known in all three cases.

2.3.1 Storage facilities managed on a FIFO basis

We consider a storage facility managed on a first-in first-out (FIFO) basis. In such a system, parts leave the storage facility in the same order as they

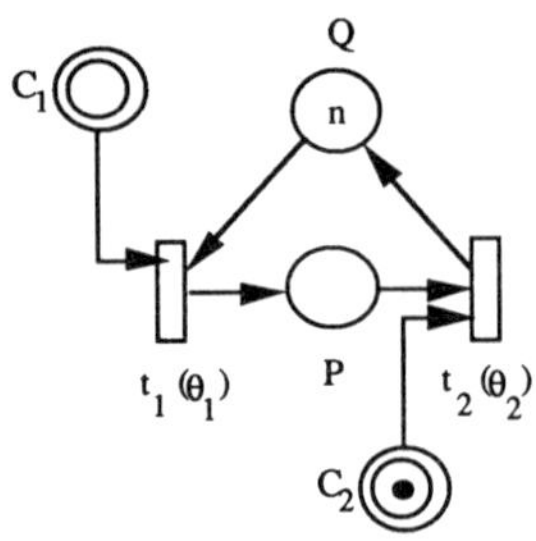

Figure 2.9 Storage facility where input and output orders of the parts are independent of each other.

entered the storage facility. The model is exactly the same as the transportation system model (see Fig. 2.8). Each section of the transportation system becomes a position of the storage facility. Thus the number of sections becomes the storage capacity. The times associated to the transitions represent the times needed to move from one position to the next one.

2.3.2 Case where output and input part orders are independent of each other

The model related to this case is the simplest among the models of storage facilities. It is represented in Fig. 2.9. Transitions t_1 and t_2 are respectively an input and an output transition. Transition t_1 represents the entrance of a part to the storage facility. When the storage facility is empty, place Q contains n tokens, n being the storage capacity. When C_1 contains one token, which means that parts are waiting at the entrance of the system, the firing of transition t_1 initiates if Q contains at least one token (i.e. if there is at least one place free in the storage facility). The initiation of the firing of transition t_1 results in removing one token from place Q, which means that one of the positions available in the storage facility is reserved for the part in process. The firing of transition t_1 takes θ_1 units of time (i.e. the transfer of a part from outside to the storage facility takes θ_1 units of time). The firing of transition t_1 terminates by putting one more token in place P.

If the control place C_2 contains one token (which means that one part is required outside the storage facility), and if the state place P contains at least one token (which means that at least one part is kept in stock), then the firing of transition t_2 initiates (i.e. one part starts moving outside the storage facility). This motion takes θ_2 units of time. When the firing of t_2 terminates, one more token appears in the state place Q, which indicates that one more storage place is available in the storage facility.

Note that if we remove places C_1 and C_2, as well as the arcs starting from these places, we obtain a strongly connected event graph (restricted to the elementary circuits $\langle t_1, P, t_2, Q\rangle$ in the case at hand).

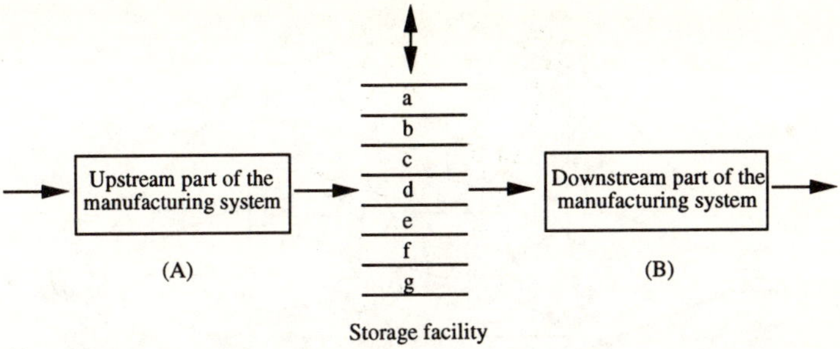

Figure 2.10 A LIFO-type storage facility.

2.3.3 Storage facilities managed on a LIFO basis

We consider a storage facility managed on a last-in first-out basis. This kind of facility can be found in some flow-shop systems, and in particular in the glass industry. Such a system is represented in Fig. 2.10.

Let us assume that the production starts and that the storage facility is empty. In that case, place **a** of the storage is at the same level as the output of the upstream part of the manufacturing system (hereafter denoted by A), which is also the level of the input of the downstream part of the manufacturing system (hereafter denoted by B). When B breaks down, the next part manufactured by A is stored in place **a**. The following part manufactured by A is stored in place **b**, after moving up the storage place **b** at the same level as the output of A and the input of B, and so on until either B is repaired or the storage facility is full. In the last case, A blocks (i.e. the whole manufacturing system stops).

Assume now that A and B are in working order and that the least storage place filled up is place **e**. The storage place **e** is then at the same level as the output of A and the input of B. If A breaks down, the part contained in **e** feeds B. Then the part contained in **d** feeds B after moving down the storage places in order to locate **d** at the same level as the output of A and the input of B, and so on until either A is repaired or the storage facility is empty. In the last case, B stops.

The model of this system is given in Fig. 2.11. In this model, we assume that the storage facility consists of four places denoted by **a**, **b**, **c** and **d** respectively. When the system is empty, P_1 is the only marked state place and contains exactly one token. Assume that C_1 contains one token, which means that the storage facility is opened, at least one part is waiting at the entrance of the system, and that C_2 is empty, which means that no part is required from outside the system (i.e. **b** is out of order). In that case, the firing of input transition t_1 initiates. It represents the beginning of the transfer of the part from outside the system to the place **a** of the storage facility. The token in P_1 disappears, which means that place **a** is reserved. After a time θ_1, one token appears in place **a** (i.e. one part arrives in place

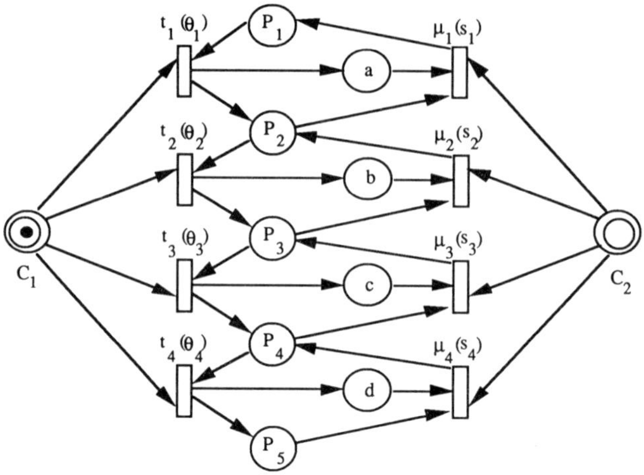

Figure 2.11 A LIFO-type storage facility model.

a), and one token appears in place P_2 (i.e. the next place available is **b**). A firing of t_2 cannot start before the firing of t_1 terminates and a token appears in P_2. At this moment, the firing of t_2 can initiate (i.e. the transfer of a second part from outside the system to place **b** can start), which results in removing the token in P_2 (i.e. place **b** is reserved). After a time θ_2, one token appears in place **b** (i.e. one part is located in place **b**), and one token arrives in place P_3, which indicates that the next position available in the storage facility is place **c**.

Assume now that the token in C_1 is removed and that one token is placed in C_2. It means that no more parts are allowed to enter the system and that parts are required from outside the system. Because places **b** and P_3 contain one token each, the firing of transition μ_2 initiates, which results in removing the tokens in **b** and P_3: it represents the beginning of the transfer of the part in **b** to B. After a time s_2, the firing of μ_2 terminates: one more part is available outside the system and one token appears in P_2, which indicates that the next part available is the part stored in place **a**, and so on.

Note that a t_i ($i \in \{1, 2, 3, 4\}$) transition cannot fire at the same time as a μ_i ($i \in \{1, 2, 3, 4\}$) transition fires, because firing a t_i transition results in removing the token in place P_i, which blocks any other transition firing, and in particular the firing of a μ_i transition. Otherwise, we could reach situations in which a vacant place is surrounded by two occupied places.

Note that if control places C_1 and C_2 never contain a token at the same time, the Petri net model is decision-free: it is one of the constraints applied to the O-DMS. The case where C_1 and C_2 contain both one token is not modeled here: in that case, parts are transferred directly from part A to part B of the system, without passing through the storage system.

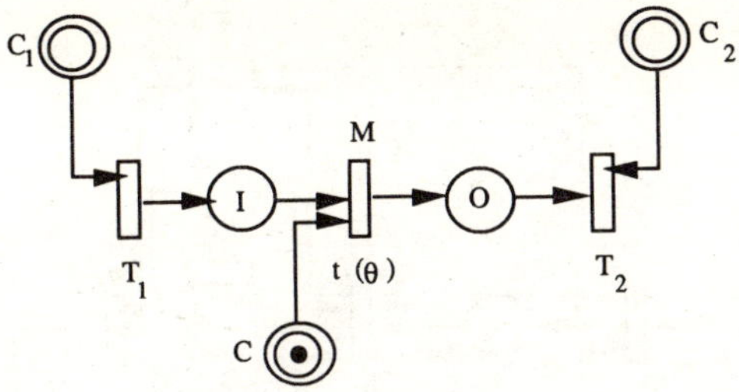

Figure 2.12 Mono-operation machine model.

2.4 OPERATION MODELING

In this chapter, we consider manufacturing module models frequently encountered in real-life situations.

2.4.1 Mono-operation machine

This is the case when a machine always performs the same operation, i.e. always operates on the same type of product. Figure 2.12 presents a model of this type of system. The machine is in working order if there is one token in the control place C. In that case, when I contains at least one token (which means that at least one part is waiting in front of the machine), the transition t starts firing. The firing terminates θ units of time later, which means that the part is manufactured, and one more token appears in place O. If the token in C is removed, then no other firing arises, which means that machine M is stopped. Input transition T_1 fires when there is one token in C_1. This models the arrival of a new part in front of machine M. Output transition T_2 fires when control place C_2 contains one token, which means that a part is required from outside the system. The time associated to T_1 (resp. T_2) represents the time needed by a part to take position in front of the machine (resp. to leave the machine).

2.4.2 Multi-operation machine

We still consider one machine M, but this machine can produce parts of types $P_1, P_2, \ldots, P_n$. We denote by θ_i the time required to produce one part of type P_i $(i = 1, 2, \ldots, n)$.

(a) Assume first that the production is cyclic and that the sequence of parts to produce is known. Consider, for instance, the case $n = 2$ and assume that the production sequence is P_1, P_2, P_2 (i.e. the production ratios are $1:3$ and $2:3$ respectively). The related model is given in Fig. 2.13. Firing t_1 represents the manufacturing of one part of type P_1 (which requires θ_1 units of time), while firing t_2^1 or t_2^2 represents the manufacturing of one part of type P_2 (which requires θ_2 units of time).

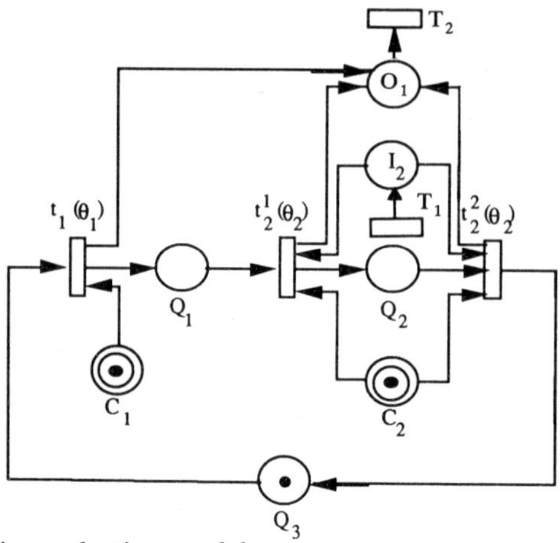

Figure 2.13 Cyclic production model.

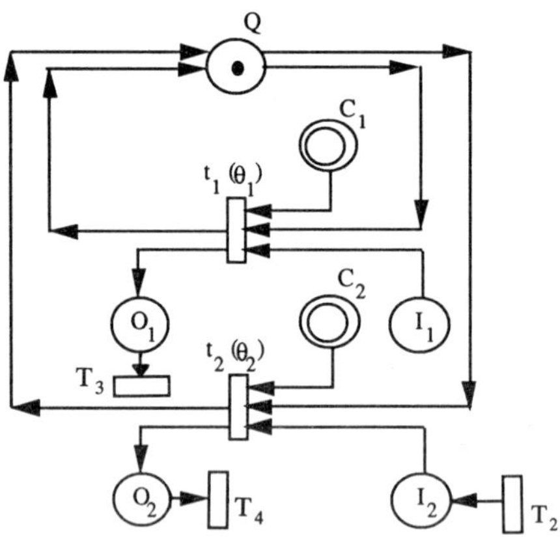

Figure 2.14 Non-cyclic production model.

Note that the elementary circuit $\langle t_1, Q_1, t_2^1, Q_2, t_2^2, Q_3\rangle$ (which is the basic decision-free timed Petri net of the model) is a strongly connected timed event graph and contains only one token, which guarantees that machine M manufactures at most one part at a time, and that parts are produced according to sequence P_1, P_2, P_2.

(b) Assume now that production is not cyclic. In that case, we allow the O-DMS the responsibility of deciding which type of part to manufacture next when a part is completed. The model is given in Fig. 2.14 for the case of two product types. Product types are denoted by P_1 and P_2 respectively.

If control place C_1 contains one token (which means that the manufacturing of a part of type P_1 is required), then the firing of transition t_1 initiates, which represents the beginning of the manufacturing of a part of type P_1. The token in place Q_3 disappears (which means that the machine

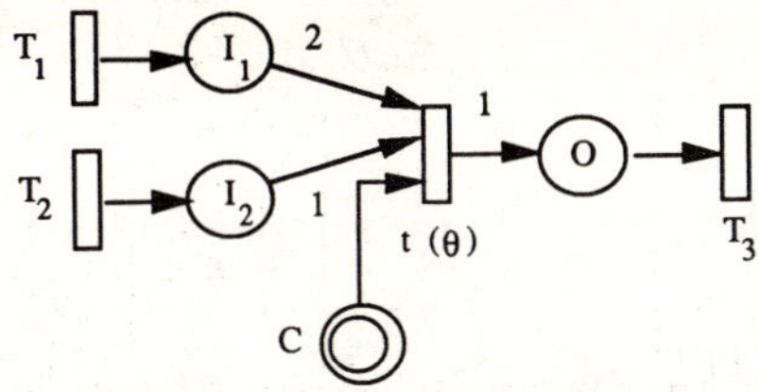

Figure 2.15 A mono-operation assembly machine model.

is busy and cannot accept another task). After a time θ_1, the firing of t_1 terminates and:

- a token appears in place Q_1, which means that the machine is ready for the next task;
- a token appears in place O_1, which means that a new part P_1 is available.

The same process holds for parts of type P_2 if we put one token in the control place C_2. Note that putting one token in both C_1 and C_2 simultaneously is not allowed, otherwise the Petri net would not be decision-free. The O-DMS must comply with this constraint.

2.4.3 Assembly machine

In this case, we consider a machine M which performs an assembly operation. This assembly operation produces one part P out of n_1 components of type K_1, n_2 components of type $K_2, \ldots, n_r$ components of type K_r. In Fig. 2.15, we consider the case of an assembly machine producing one unit of part type P out of two components K_1 and one component K_2. Machine M is thus a mono-assembly operation machine.

If there are at least two tokens in place I_1 (which represents components K_1) and one token in the input place I_2 (which represents component K_2), and if there is one token in the control place C, then transition t fires and produces one token in place O after θ units of time. It corresponds to the production of one part of type P. The process restarts immediately if transition t is enabled (i.e. if there is one token in C and at least two tokens in I_1 and one token in I_2). Firing input transitions T_1 or T_2 models components arriving in the system, while firing output transition T_3 models a part leaving the system.

A model of a multi-assembly operation machine is given in Fig. 2.16. It concerns a non-cyclic production. It has to be compared with the model given in Fig. 2.14. We restrict ourselves to a two-operation model. The first assembly operation needs one component of type K_1 and two components of type K_2, while the second assembly operation requires one component of type K_2, one component of type K_3 and three components of type K_4.

In this model, one token in I_k ($k = 1, 2, 3, 4$) represents the availability

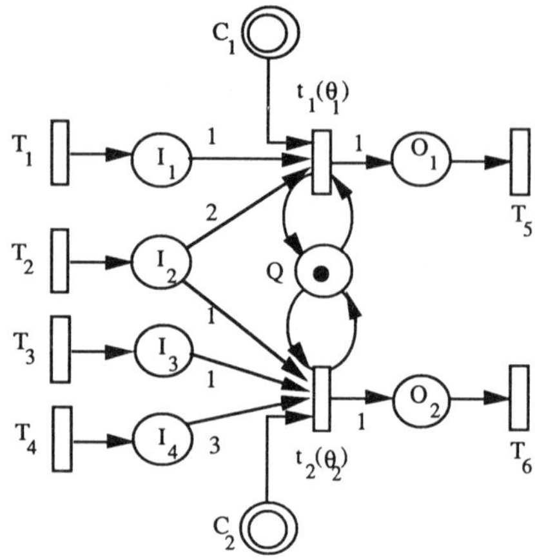

Figure 2.16 A two-operation assembly machine.

of one component of type K_k. Transition t_1 (resp. t_2) represents the first (resp. the second) operation performed on the machine. The token in place Q is to make sure that at most one operation is performed on the machine at any time. If the control system, which is in charge of putting in or removing the tokens from the command places, is such that there is at most one token in the control places (i.e. either one token in C_1 and no token in C_2, or no token in C_1 and one token in C_2), then the model is decision-free. In other words, when the choice of the next operation to be performed (when the machine is idle) is made by a control system external to the system we want to model, then the system is decision-free.

Assuming that there is one token in the control place C_1, and no token in the control place C_2, then transition t_1 is enabled (i.e. the machine is ready to perform the first assembly operation) if there is at least one token in I_1 (i.e. at least one component of type K_1 is available), two tokens in I_2 (i.e. at least two components of type K_2 are available) and one token in place Q (which means that the machine is idle). Firing transition t_1 means performing the first assembly operation. It needs θ_1 units of time, after which one token appears in O_1 (which means that a first type part is completed). If t_1 is still enabled, t_1 starts to fire for the second time, which represents the production of the second part of the first type, and so on.

Similarly, t_2 is enabled (i.e. the system is ready to start assembling a second type part) if there is at least one token in I_2 (i.e. one component of type K_2 is available), one token in I_3 (i.e. one component of type K_3 is available), and three tokens in I_4 (i.e. three components of type K_4 are available), and if C_2 and Q contain one token each, which means that the external decision is to perform the second assembly operation and that the machine is idle, respectively. Firing transition t_2 requires θ_2 units of time,

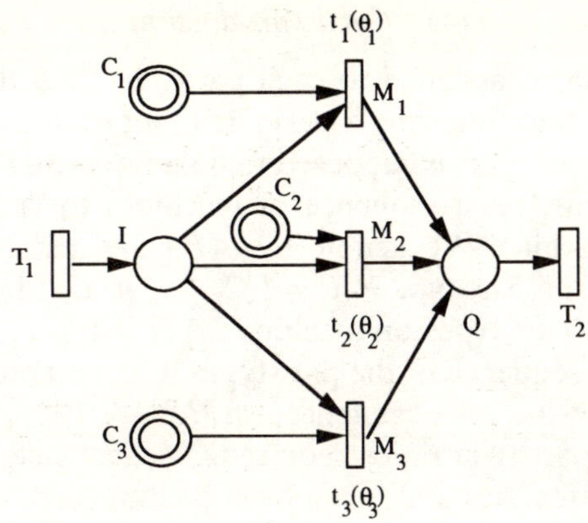

Figure 2.17 Three machines working in parallel.

which is the time needed to perform the second operation, after which one token appears in O_2, which represents the fact that one part of the second type is completed. Input transitions, whose firings model the arrival of components in the systems, are T_1, T_2, T_3 and T_4. Output transitions are T_5 and T_6. The firing of these transitions allows us to model parts leaving the system.

2.4.4 Machines working in parallel

This model is presented in Fig. 2.17 in the case of three machines working in parallel. These machines, denoted by M_1, M_2 and M_3 respectively, perform the same operation, in different amounts of time (θ_1, θ_2, θ_3 respectively). One token in the command place C_i ($i = 1, 2, 3$) means that machine M_i is ready to accept a new part. If, moreover, transition t_1 is not firing and at least one token is in place I (i.e. raw material for manufacturing at least one part is available), then t_1 is enabled and can fire, which means that one part starts to be manufactured on machine M_i. A conflict arises if, for instance, there are less tokens in I than the number of transitions which can start firing. We assume that this problem is solved by the control system which manages the tokens in the control places in the following way: there are never more idle machines in working order (i.e. more control places activated) than the number of tokens in I. T_1 (resp. T_2) is the input (resp. output) transition.

2.4.5 Job-shop modeling

This section is devoted to job-shop modeling. We first consider the case where production is cyclic, i.e. when the parts are manufactured according to given ratios. We then model the most general case.

(a) *Cyclic production*

In this case, a ratio is assigned to each part-type, and the part-types are released in the system following a periodic input sequence. The manufacturing process of each part is supposed to be uniquely defined as a routing through the system (i.e. a sequence of machines to visit) with the time spent on each machine. We denote by $\sigma(P_i)$ the manufacturing process corresponding to the part-type P_i ($i = 1, 2, \ldots, n$) and by $\sigma(M_j)$ the fixed sequencing of the part-types on machine M_j ($j = 1, 2, \ldots, m$). Finally, σ_0 denotes the input sequence of the part-types in the manufacturing system. Sequence σ_0, as well as the sequences $\sigma(M_j)$, $j = 1, 2, \ldots, m$, are given according to the part-type ratios required. For instance, assume that the manufacturing system at hand is a job-shop composed of three machines M_1, M_2 and M_3 ($m = 3$) which can produce three types of parts P_1, P_2 and P_3.

The manufacturing processes are given hereafter (the values in parentheses are the times the parts spend on the machines):

$$\begin{aligned}\sigma(P_1) &= \langle M_1(3), M_2(1), M_3(4)\rangle\\ \sigma(P_2) &= \langle M_3(2), M_1(1)\rangle\\ \sigma(P_3) &= \langle M_2(6), M_1(4), M_3(3)\rangle\end{aligned}$$

We assume that the ratios assigned to P_1, P_2 and P_3 are 25%, 50% and 25% respectively. We thus can choose $\sigma_0 = \langle P_1, P_2, P_2, P_3\rangle$ as the input sequence, because it contains P_2 twice and P_1 and P_3 once each, which corresponds to the production ratios required. Furthermore, the following sequencing of the part-types on the machines fit with the production ratios:

$$\begin{aligned}\sigma(M_1) &= \langle P_1, P_2, P_2, P_3\rangle\\ \sigma(M_2) &= \langle P_1, P_3\rangle\\ \sigma(M_3) &= \langle P_1, P_2, P_3, P_2\rangle\end{aligned}$$

Note that, because M_2 is not used to produce P_2-type parts, $\sigma(M_2)$ does not contain P_2, but only P_1 and P_3 in the same proportions. We propose the following three-step modeling process.

(i) First step

The first step consists of modeling the manufacturing process of each part-type of the input sequence σ_0, as shown in Fig. 2.18 for the example above. The first transition of each model represents the entrance into the manufacturing system (loading–unloading station if the job-shop considered is an FMS, for instance). Each of the other transitions models an operation. They are arranged in the same order as the operations in the manufacturing processes. The firing times are the times required to perform the operations. A token in such a model represents a part, as it flows through the system, and the places correspond to storage buffers. Thus, places are referred to as buffer places.

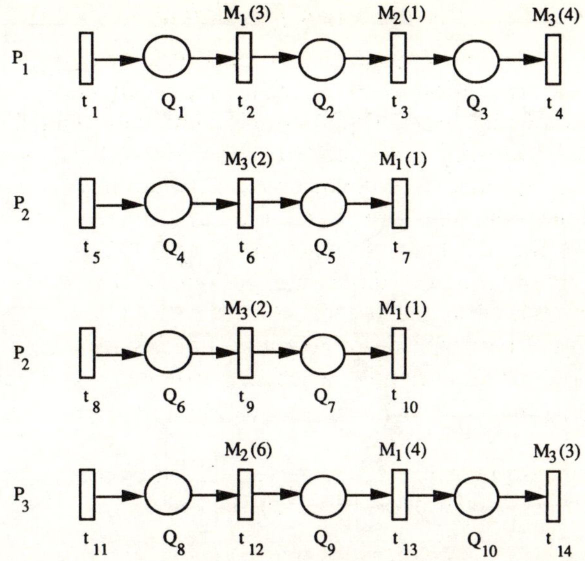

Figure 2.18 Job-shop cyclic production: first step of the modeling process.

(ii) Second step

The second step is to model the cyclic operating mode of the system, assuming that a new part is released to the system as soon as a part is completed. The model represented in Fig. 2.19 is self-explanatory. The circuit which composes this model will be referred to as **processing circuits**, and the loop-closing places will be called **resource places**. A token in such a place can model a free transportation resource (pallet, cart, etc.) for loading a new part of the same type into the system. Note that, in this model, we assume that no time is required for a transportation resource to go from the output to the input of the job-shop. This hypothesis is made only to simplify the model. Introducing the recycling time would need one more transition and one more place in each processing circuit, as shown in Fig. 2.20 for the processing circuit related to part-type P_1. In this model, the time assigned to transition T represents the time needed by a transportation resource to move from the output to the input of the job-shop. We could similarly introduce the time needed to move from one machine to the next one.

(iii) Third step

The final step consists of modeling the sequencing of the part-types on the machines as well as the input sequence. This is done by connecting all the transitions that correspond to operations performed by the **same** machine in a processing circuit. The order of the transitions in these circuits is determined by the sequencing of the part-types on the corresponding

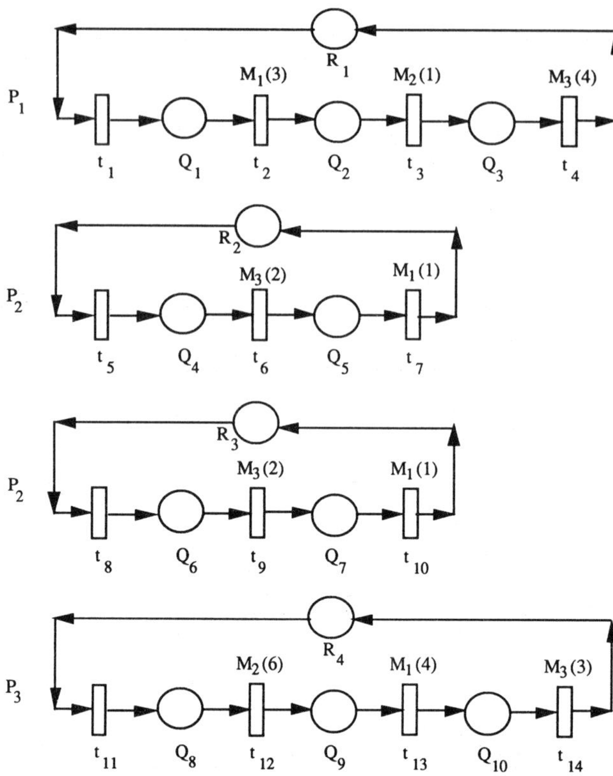

Figure 2.19 Job-shop cyclic production: second step of the modeling process.

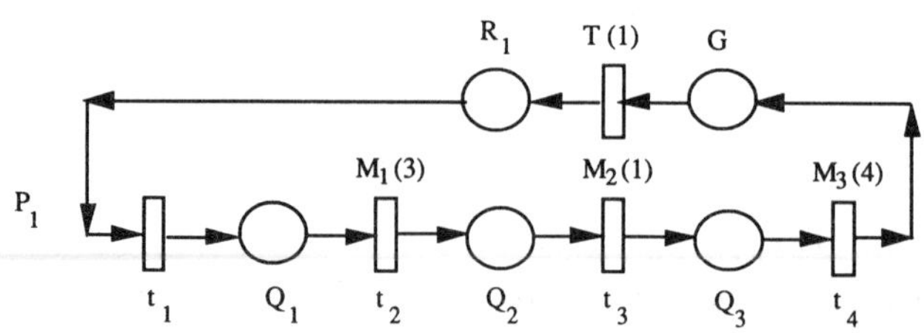

Figure 2.20 A processing circuit taking into account the recycling time.

machine. These circuits are referred to as **command circuits**. A token in a **command circuit** represents the availability of the machine to process a specific job. As a consequence, there can be only one token in each command circuit. The initial position of the token in a command circuit is given by the first part-type in the input sequence or in the machine sequencing. The model is given in Fig. 2.21. It should be completed by adding:

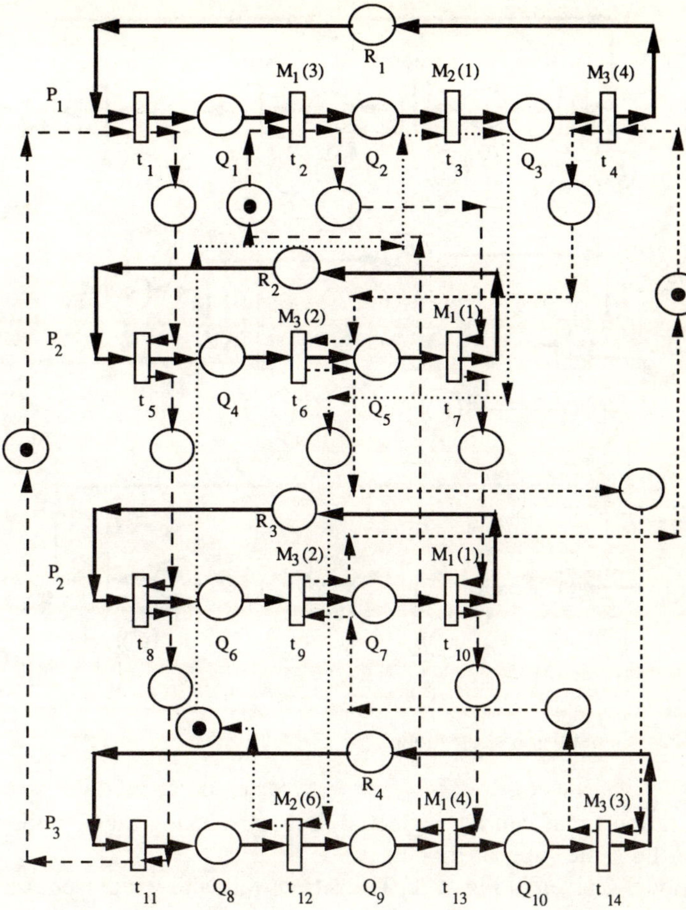

Figure 2.21 Job-shop cyclic production: last step of the modeling process.

- input transitions connected through places to t_1, t_5, t_8 and t_{11}
- output transitions connected through places to t_4, t_7, t_{10} and t_{14}
- control places to some of the transitions (for instance t_1 and/or t_5 and/or t_8 and/or t_{11}).

Note that the model as shown in Fig. 2.21 (i.e. without input transitions, output transitions and control places) is a strongly connected timed marked graph.

(b) The general case

We now propose a job-shop model for the general case (i.e. the case when the types of parts to manufacture are decided at the O-DMS level). The explanations are based on the previous example. We still consider that

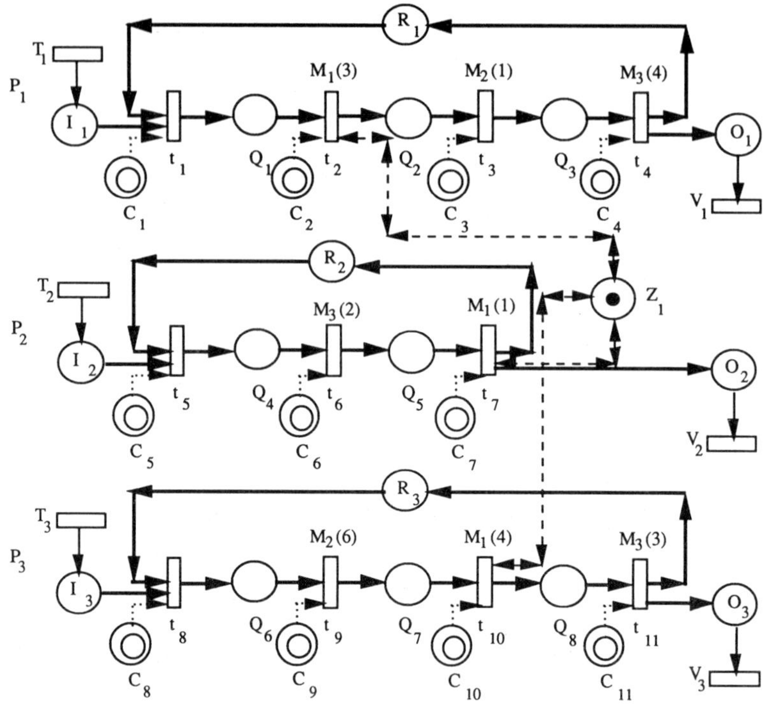

Figure 2.22 General job-shop model.

transportation times can be neglected, but introducing these times would result in the same type of model.

The model is given in Fig. 2.22. Transitions t_1, t_5 and t_8 represent the tasks of releasing part-types P_1, P_2 and P_3 (respectively) to the manufacturing system. Command circuits, as introduced in the case of a cyclic production, disappear and are replaced by a system which guarantees that a given machine cannot be utilized to produce more than one part at a time. In Fig. 2.22, place Z_1 contains exactly one token if machine M_1 is idle. This token disappears if one of the transitions t_2, t_7 or t_{10} is firing, which means that machine M_1 is used to manufacture a part of either type P_1, or P_2, or P_3. This guarantees that none of the two other transitions can start firing (i.e. that M_1 cannot start manufacturing another part). When the firing in progress terminates, a token reappears in Z_1, which indicates that machine M_1 is idle and ready to manufacture another part. The same approach applies to machines M_2 and M_3. We chose not to reflect this in order to keep the figure readable. This model is decision-free if and only if:

- there is at most one token in (C_2, C_7, C_{10}) (a)
- there is at most one token in (C_3, C_9) (b)
- there is at most one token in (C_4, C_6, C_{11}) (c)

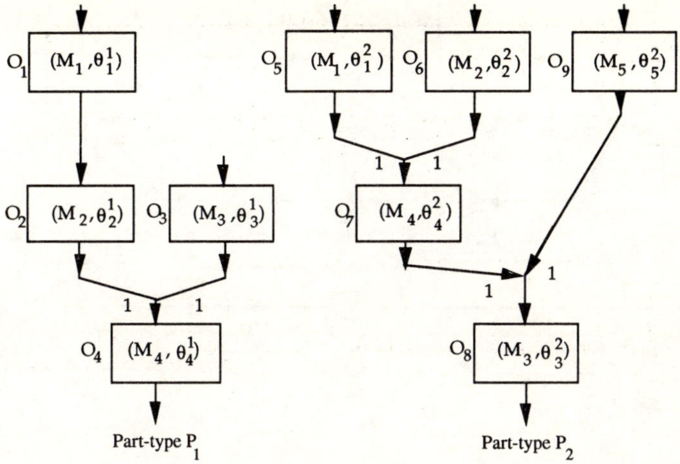

Figure 2.23 Manufacturing processes of two assembled parts.

One token in C_1 (resp. C_5, C_8) means that the decision has been made to release a part-type P_1 (resp. P_2, P_3).

The state of the control places at each time is the result of the scheduling process which belongs to the O-DMS and which leads to a schedule verifying conditions (a), (b) and (c).

The previous approaches can be easily extended to any job-shop system.

2.4.6 Assembly system modeling

Similar to the presentation of the job-shop system, we first consider assembly systems in the case of cyclic production. We then extend the study to the general case.

(a) Cyclic production

As for the job-shop case, a ratio is assigned to each part-type, and the part-types are released to the assembly system following a periodic input sequence which reflects these ratios. Each part-type is assumed to be attached to a unique manufacturing process. Again we use a short example to explain how to model such an assembly system. This example concerns two part-types P_1 and P_2 whose manufacturing processes are presented in Fig. 2.23.

In Fig. 2.23, M_i $(i = 1, 2, 3, 4, 5)$ are the machines, O_i $(i = 1, \ldots, 8)$ represent the operations and θ_i^j represent the manufacturing times. Two types of operations are involved in the manufacturing process. Operations such as O_1, O_2 just transform a component of the final product. Below, we refer to these operations an **regular operations**. Other types of operations

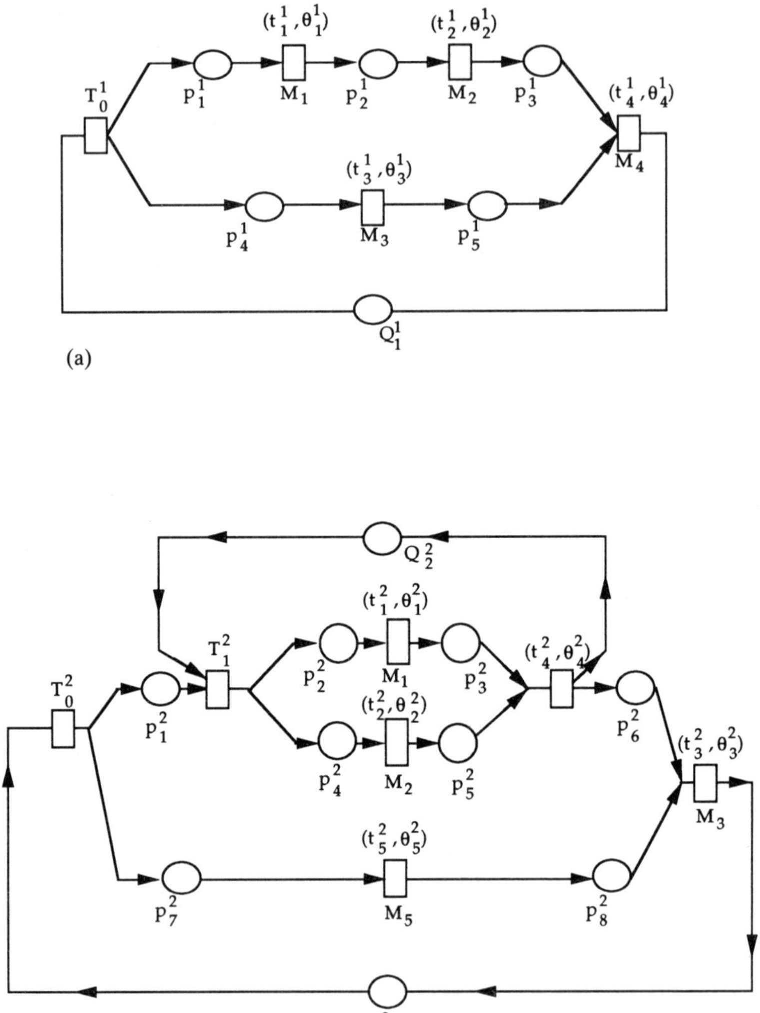

Figure 2.24 (a) Model related to manufacturing process P_1; (b) model related to manufacturing process P_2.

are the **assembly operations**. For instance, O_4, O_7 and O_8 are assembly operations. An assembly operation puts components together to obtain a more complex component of a final product or the final product itself. In Fig. 2.23, the integer values assigned to the arrows represent the number of components of each type of product required to obtain one unit of the next component.

In Figs 2.24(a) and 2.24(b), we present the models related to the above manufacturing processes. Note that these models make use of transitions which do not represent operations but the beginning of the manufacturing of components which are required to assemble a more complex component or the final product. In Fig. 2.24(a), such a transition is T_0^1. This is also the case with T_0^2 and T_1^2 in Fig. 2.24(b). We call these transitions **releasing transitions**. A directed path connects each transition representing an assembly operation to the related releasing transition. In practice, this means that the transportation resources used to carry the components at this assembly level recirculate as soon as the assembly operation is completed. Thus, transportation resources are assigned to each manufacturing level.

The manufacturing process models correspond to the process circuits in the case of job-shops. These models are completed by command circuits which model the sequencing of the part-types in the machines. In Fig. 2.25, we represent the command circuit related to machine M_1, assuming that the sequencing related to M_1 is $\langle P_1, P_2, P_2 \rangle$.

Each command circuit contains only one token as in the job-shop case. Using the modeling process just described in the previous brief example, we obtain a strongly connected event graph, as in the case of the job-shop. This model has to be completed by adding:

- input transitions connected to transitions $t_1^1, t_3^1, t_1^2, t_2^2$ and t_5^2, through places;
- output transitions connected to transitions t_4^1 and t_3^2, through places.

Control places can be added, for instance, to T_0^1, T_0^2 and T_1^2.

(b) The general case

The extension of the cyclic assembly to the general case is similar to the extension of the job-shop cyclic production to the job-shop general case. We associate a model to each part-type, similar to the models presented in Figs 2.24(a) and 2.24(b), enhanced by adding:

- input transitions T_1, T_2, T_3, connected through places to the first transition of each sequence of transitions related to an elementary component;
- an output transition V connected through a place to the transition related to the last operation performed on a product;
- a control place to at least each transition related to an operation.

Figure 2.26 is the model presented in Fig. 2.24(b) after it has been enhanced. The model of the manufacturing system is the set of the enhanced part-type models, completed by a system which guarantees that no more than one transition (among a set of transitions representing the operations which can be performed on the same machine) can fire simultaneously.

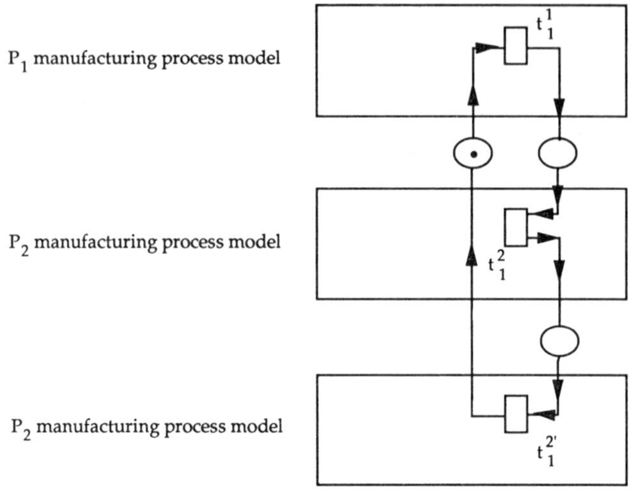

Figure 2.25 The command circuit related to M_1.

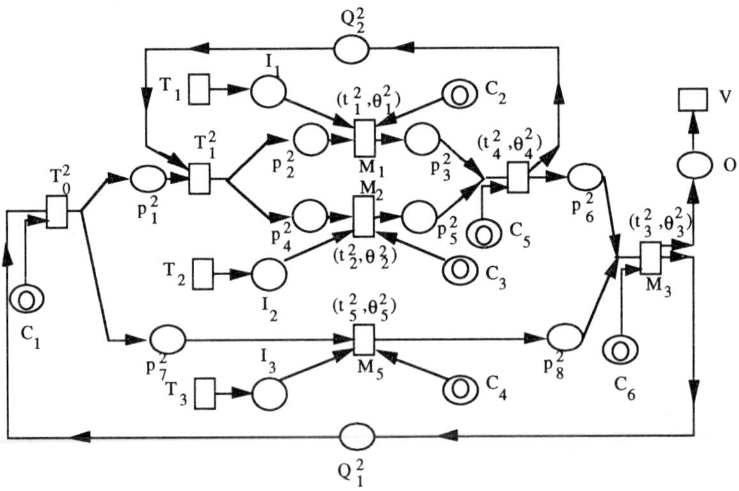

Figure 2.26 An enhanced model for a part type: C_i, $i = 1, 2, 3, 4, 5, 6$: control places; I_i, $i = 1, 2, 3$: input places; O: output place.

This system is the same as the one represented in Fig. 2.22, to ensure that transitions t_2, t_7 and t_{10} cannot fire simultaneously.

The following constraint applies to the set of control places: the set of control places related to a given machine never contains more than one token. This is the condition for the whole Petri net model to be decision-free. The scheduling of the system is computed at the O-DMS level, that takes into account the previous constraint.

2.5 TOOL SYSTEMS

Several types of tools are used in manufacturing systems. In this section, we consider two types of tools: the fixture devices, which are used to fix a part on machines, and the cutting tools, which are used to manufacture parts.

The fixture devices can be reutilized indefinitely, and are usually related to a few part-types. The cutting tools are subject to wear and breakdown. Thus, they can be utilized a finite number of times. In the following subsections, we propose some ways to model tool systems which concern fixture devices and cutting tools.

2.5.1 Fixture systems

(a) The cyclic production case

We first consider the cyclic production case, i.e. the case where part-types are manufactured according to given ratios and where these ratios are reached by assigning an input sequence to each machine; this input sequence complies with the desired ratios.

To illustrate the modeling of such a system, we restrict ourselves to a case of one machine which manufactures two parts denoted by P_1 and P_2, respectively. Assume that manufacturing of a P_1 requires a fixture device of type F_1 and that manufacturing of P_2 requires a fixture device of type F_2. We aim at manufacturing two units of P_1 and one unit of P_2. To reach this goal, we fix the sequence $\sigma = \langle P_1, P_1, P_2 \rangle$ on the machine. This means that the machine will manufacture two units of type P_1, then one unit of type P_2, and again two units of type P_1 and one unit of type P_2, and so on. As a consequence, we have to first set up a fixture device of type F_1, and then replace it by a fixture device of type F_2 (after having manufactured two units of P_1), then to come back to a fixture device of type F_1 (after having manufactured one unit of P_2), and so on.

The model of this manufacturing system is given in Fig. 2.27. Firing transition m_1 represents the setting up of an F_1-type fixture device, which requires a time w_1. Firing transition t_1 and transition t_2 represents the manufacturing of two units of P_1. The F_1-type fixture device is then removed, which is represented by firing transition d_1 (with firing time s_1). The firing of d_1 ends by replacing one token in Q_1 (which means that an F_1-type fixture device is ready to be used again) and by putting one token in S_1 (which indicates the start of putting an F_2-type fixture device on the machine, i.e. m_2 will start firing). When the firing of m_2 ends, one token appears in R_4 and t_3 can start firing (i.e. a part of type P_2 starts its processing on the machine). The firing of t_3 ends after σ_2 units of time. θ_2 is the processing time of a P_2-type part. Then, one token appears in R_5, and d_2 can start firing, which represents the beginning of dismounting the F_2-type

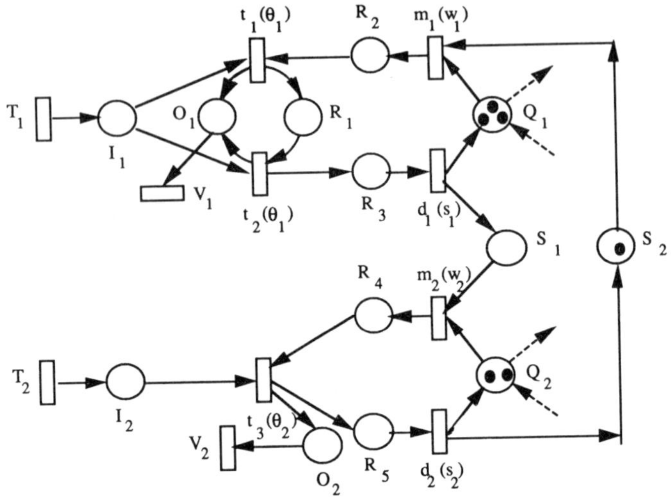

Figure 2.27 Manufacturing system with fixture devices: the cyclic production case.

fixture device. The firing of d_2 ends by replacing one token in Q_2 (which means that an F_2-type fixture device is ready to be used again), and by one token in S_2 (which means that the system is ready to manufacture a new pair of P_1-type parts).

Note that the elementary circuit $\langle S_2, m_1, R_2, t_1, R_1, t_2, R_3, d_1, S_1, m_2, R_4, t_3, R_5, d_2 \rangle$ contains only one token, which guarantees that the machine is reserved by only one part at a time.

This model, presented in Fig. 2.27, is a strongly connected event graph if we remove the dotted arcs as well as the input transitions T_1 and T_2 and the output transitions V_1 and V_2 and the related places. The tokens in Q_1 (resp. Q_2) represent the availability of F_1 (resp. F_2)-type fixture devices. These devices can also be used by another machine belonging to the manufacturing system considered: this is represented by the dotted arcs in Fig. 2.27.

(b) The general case

We still consider one machine which can manufacture two types of parts. The model is represented in Fig. 2.28 where:

- transition t_1 represents the setting up of an F_1-type fixture device;
- transition μ_1 represents the manufacturing of a P_1-type part;
- transition d_1 represents the dismounting of the F_1-type fixture device;
- transition t_2 represents the setting up of an F_2-type fixture device;
- transition μ_2 represents the manufacturing of a P_2-type part;
- transition d_2 represents the dismounting of the F_2-type fixture device.

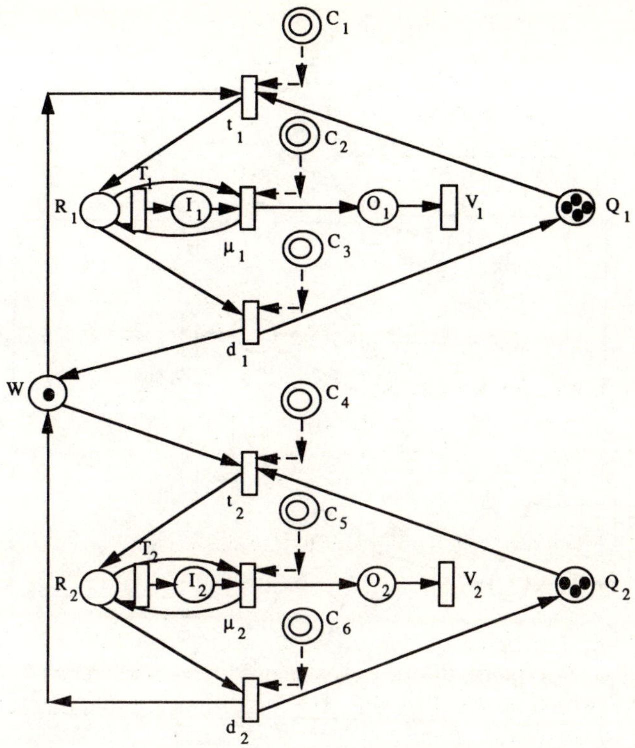

Figure 2.28 Manufacturing system with fixture devices: the general case.

A token in Q_1 (resp. Q_2) represents one F_1 (resp. F_2)-type fixture device available. At the beginning of the process, W contains one token: this is to make sure that transition t_1 (or μ_1, or d_1) does not fire concurrently with transition t_2 (or μ_2, or d_2), which would mean that the machine is used simultaneously to manufacture a part of type P_1 and a part of type P_2.

C_1, C_2, C_3, C_4, C_5 and C_6 are the control places. At each point in time, this set of control places must not contain more than one token, otherwise the model could be subject to conflicts. For instance, one token in C_2 and C_3 simultaneously, or one token in C_1 and C_4 simultaneously results in a conflict.

Note also that the circuit $\langle W, t_1, R_1, d_1, W\rangle$ guarantees the sequence: setting up of an F_1-type fixture device, manufacture of one or more P_1-type parts, and dismounting of the F_1-type fixture device. The elementary circuit $\langle W, t_2, R_2, d_2, W\rangle$ plays the same role for part-type P_2. T_1 (resp. T_2) is the input transition for P_1 (resp. P_2), while V_1 (resp. V_2) is the output transition for P_1 (resp. P_2).

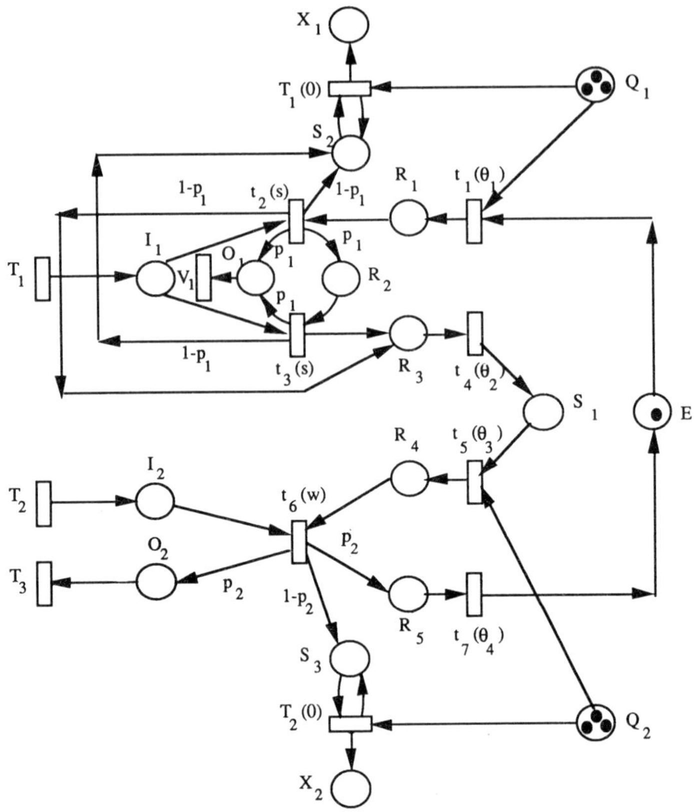

Figure 2.29 Manufacturing system with cutting tools: the cyclic production case.

2.5.2 Cutting-tool systems

The cutting-tool case differs from the fixture device case because (i) there is wear on a cutting tool, while a fixture device can (theoretically) be used indefinitely, and (ii) a cutting tool may break down, while it is assumed that a fixture device does not.

(a) The cyclic production case

To illustrate this case, we still consider two part-types, denoted by P_1 and P_2, and one machine. We assume that we want to manufacture two P_1-type parts while manufacturing one P_2-type part, which is obtained using the sequencing $\sigma = \langle P_1, P_1, P_2 \rangle$ on the machine. The related model is given in Fig. 2.29. Manufacturing a P_1 (resp. P_2)-type part requires cutting-tool A_1 (resp. A_2). The cutting-tool A_1 (resp. A_2) can be used to manufacture n_1 (resp. n_2) units of P_1 (resp. P_2)-type parts. Place Q_1 (resp. Q_2) contains n_1

(resp. n_2) tokens at the beginning of the process. Transition t_1 (resp. t_5) represents the set-up of the tool A_1 (resp. A_2). Transition t_2 and t_3 (resp. t_6) represents the manufacturing of a P_1 (resp. P_2)-type part. When the manufacturing is successful (probability p_1 for P_1 and p_2 for P_2), the process continues. The cutting-tool A_1 is dismounted after two P_1-type parts are manufactured: this is represented by transition t_4. Similarly, transition t_7 represents the dismounting of A_2.

A breaking down of A_1 (resp. A_2) results in firing T_1 (resp. T_2) as many times as the number of tokens contained in Q_1 (resp. Q_2), after which the system blocks. Thus, the model at hand is a stochastic Petri net. If place E contains one token and Q_1 contains at least one token, then transition t_1 fires, which represents the setting up of the cutting-tool A_1 on the machine.

One token disappears from Q_1, which represents the fact that A_1 wears, and E becomes empty. The firing of transition t_1 ends after θ_1 units of time, by placing one token in R_1. If the place I_1 contains at least one token, which means that enough raw material is available to manufacture at least one unit P_1, transition t_2 starts firing, which represents the beginning of manufacturing one unit of P_1. One token disappears from I_1, and R_1 becomes empty. After s units of time, which is the time needed to manufacture one unit of P_1, transition t_2 firing ends, which results in:

(i) either placing one token in R_2, which means that the machine is ready to manufacture the next P_1-type part, and one token in O_1, which means that one more P_1-type part is available: this occurs with the probability p_1;
(ii) or placing one token in S_2 (probability $1-p_1$), which means that tool A_1 is broken down, and one token in R_3, which means that tool A_1 is ready to be dismounted; in this case, T_1 is fired as many times as the number of tokens remaining in Q_1, which results in emptying Q_1, and the P_1-type part is lost.

If there is one token in R_2 and still at least one token in I_1, then transition t_3 starts firing, which means that the next P_1-type part starts its manufacturing process on the machine. After s units of time, the firing of transition t_3 ends, again with two possibilities:

(i) either one token appears in R_3, which means that A_1 is ready to be dismounted, and one more token appears in O_1, which means that one more P_1-type part is available; this occurs with the probability p_1; or
(ii) one token appears in S_2 (probability $1-p_1$), which results in emptying Q_1, and one token in R_3, which means that tool A_1 is ready to be dismounted.

Transition t_4 starts firing as soon as there is one token in R_3. The firing process requires θ_2 units of time, which represents the time required to dismount A_1, after which one token appears in S_1: the machine is idle and can manufacture one unit of P_2. At this point, if there is at least one token in Q_2 (which means that A_2 has manufactured less than n_2 parts), transition t_5 starts firing, which models the beginning of the setting up of A_2. Place S_1 becomes empty, and one token disappears from place Q_2. After θ_3 units of time, which is the time required to set up A_2, one token appears in R_4. If I_2 contains at least one token, which means that there is enough raw material to manufacture at least one P_2-type part, transition t_6 starts firing. Transition t_6 ends firing after w units of time, which is the time required to manufacture one P_2-type part. Then, two possibilities may occur:

(i) either one token appears in R_5 and one token in O_2, which means that the manufacturing process ended (probability p_2);
(ii) or one token appears in R_5 and one token in S_3 (probability $1-p_2$).

In both cases, tool A_2 starts being dismounted (i.e. transition t_7 starts firing). In case (i), one more unit of P_2 is available. In case (ii), transition T_2 (which is timed at 0) fires as many times as the number of tokens in Q_2, and A_2 is no longer available.

In this model, the elementary circuit $\langle E, t_1, R_1, t_2, R_2, t_3, R_3, t_4, S_1, t_5, R_4, t_6, R_5, t_7, E\rangle$ guarantees that the machine cannot be used to manufacture more than one part at a time.

Note that this model is decision-free, but is not live: it blocks after manufacturing n_1 units of P_1 (or n_2 units of P_2), except if other P_1- and P_2-type tools are available.

(b) The general case

We now consider the general case, i.e. the case when the production ratios can change in the course of the manufacturing period. This model is represented in Fig. 2.30 and concerns a machine which can manufacture two types of parts, denoted by P_1 and P_2, using cutting-tools A_1 and A_2 respectively.

Transition t_1 (resp. t_2) represents the set-up of cutting-tool A_1 (resp. A_2) and is timed at the associated set-up time. Transition t_1 (resp. t_2) fires if there is one token in the control place C_1 (resp. C_2). This results in putting one token in S_2 (resp. S_3). If the token remains in the control place C_1 (resp. C_2), if there is at least one token in I_1 and Q_1 (resp. I_2 an Q_2), and if C_5 (resp. C_6) is empty, transition t_3 (resp. t_4) fires, which represents the manufacturing of one unit of P_1 (resp. P_2)-type part. Transition t_3 (resp. t_4) firing ends by:

(i) either putting one token in S_2 (resp. S_3), which means that the machine is ready to restart the production (the probability of this situation is p_1 (resp. p_2));
(ii) or putting one token in S_1 (resp. S_4), which results in emptying Q_1 (resp. Q_2) and one token in R_1 (resp. R_2); this case means that A_1 (resp. A_2) is broken down and that it has to be dismounted (probability $1-p_1$ (resp. $1-p_2$)).

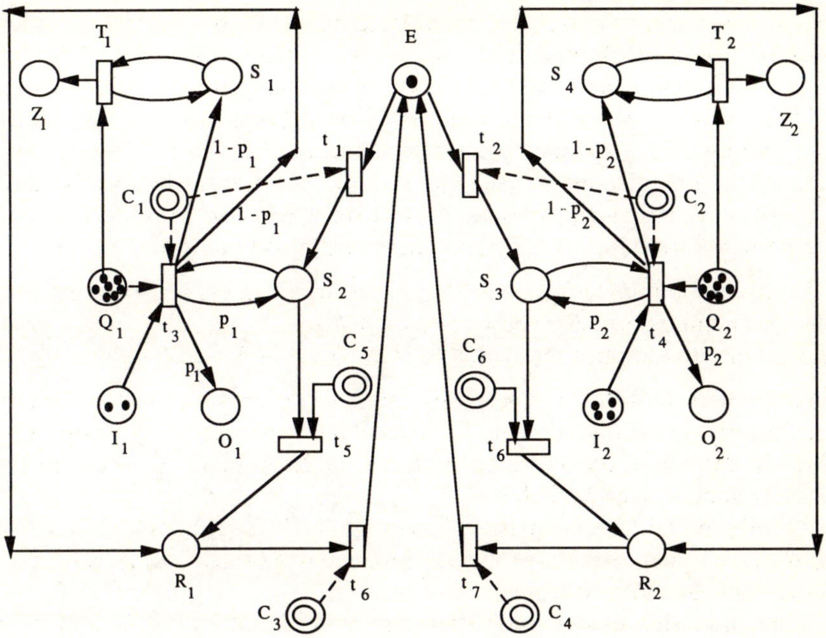

Figure 2.30 Manufacturing system with cutting tools: the general case.

In case (i), manufacturing of parts continues until either the system switches to case (ii), or Q_1 (resp. Q_2) is empty; which means that tool A_1 (resp. A_2) is worn out, or I_1 (resp. I_2) is empty; which means that there is no more raw material available, or the token is removed from control place C_1 (resp. C_2); which means that the decision has been made to stop manufacturing P_1 (resp. P_2). In all cases, a token appears finally in R_1 (resp. R_2) if one token is put in C_5 (resp. C_6). If the decision is made to remove the tool, which is modeled by putting one token in the control place C_3 (resp. C_4), then transition t_7 (resp. t_8) fires, and the firing time is the time required to dismount the cutting tool. The transition firing ends by placing one token in place E: the system can restart, i.e. the system is ready to manufacture either a unit of P_1 or a unit of P_2. Note that the O-DMS (Outside Decision-Making System) has to comply with the constraint of not placing a token simultaneously in C_1 and C_2, or in C_1 and C_5, or in C_2 and C_6.

Remarks

1. The number of tokens in Q_1 (resp. Q_2) represents the number of times tool A_1 (resp. A_2) can be used to manufacture a part, assuming that it does not break down.
2. The model concerns only two tools, namely A_1 and A_2. The replacement of these tools is not taken into account in the model.

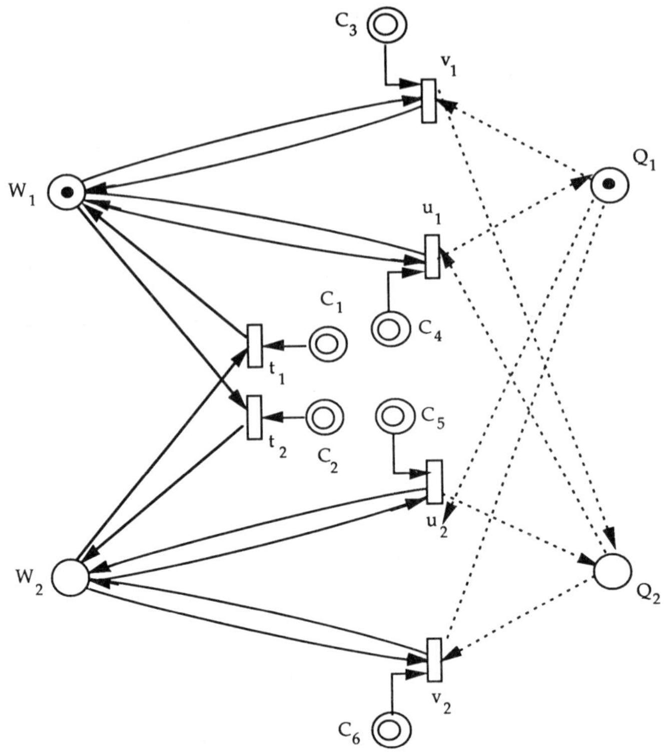

Figure 2.31 Crane model.

2.6 MATERIAL-HANDLING SYSTEMS

In this section, we present two more modeling examples. The first one concerns a crane which is in charge of transporting parts from a position of the storage facility to another one. The second one presents a robot moving parts from a machine M_1 to a storage facility, and from the storage facility to a machine M_2.

2.6.1 A crane model

Let us consider a crane which is in charge of moving heavy parts into a storage facility. Moving a part is a three-step operation which consists of picking up the part, transporting the part to its next location and, finally, laying down the part. The model of such a module is given in Fig. 2.31.

To simplify the model, we restrict ourselves to a two-position storage resource, represented by places W_1 and W_2. Transition v_1 (resp. v_2) represents the operation of picking up a part located in position W_1 (resp. W_2). Transition u_1 (resp. u_2) represents the operation of laying down a part in position W_1 (resp. W_2). Transition t_1 (resp. t_2) represents the transportation of a part from W_2 to W_1 (resp. from W_1 to W_2). The firing times of these transitions are the times needed to perform the related operations. Places Q_1 and Q_2 are introduced to make sure that neither two pick-up operations can be performed without inserting one lay-down operation in between, nor two lay-down operations can be performed without inserting a pick-up operation in between.

This model has neither an input place nor an output place. It is easy to see that the control place states must verify the following constraints:

(i) the O-DMS system cannot put more than one token simultaneously in (C_1, C_3, C_4) to avoid a conflict between the move of the crane from W_1 to W_2 and the pick-up or the lay-down operations (depending on the position of the token in Q_1 or Q_2);
(ii) similarly, the O-DMS system cannot put more than one token simultaneously in (C_2, C_5, C_6).

Furthermore, as we can see, the control place state has to fit with the state place marking to be able to influence the behavior of the system. For instance:

(iii) a token in C_1 needs a token in W_2 to be effective, i.e. the decision to move the crane from W_2 to W_1 requires that the crane be in position W_2;
(iv) similarly, a token in C_2 needs a token in W_1;
(v) a token in C_3 needs a token in W_1 and in Q_1, i.e. the decision to pick up a part from position W_1 requires that the crane be in position W_1 and that it is empty;
(vi) similarly, a token in C_6 requires a token in W_1 and in Q_2;
(vii) a token in C_4 needs a token in W_1 and in Q_2, i.e. the decision to lay down a part in position W_1 requires that the crane be in position W_1 and that it carries a part.

2.6.2 A robot used as a transportation system

We consider a robot moving parts from machine M_1 to a storage facility S, and from S to a machine M_2. This system is depicted in Fig. 2.32.

It is assumed that there are always parts available at the input of the system and that there is no storage limitation at the output of the system. The corresponding model is given by Fig. 2.33. Transitions t_1 and t_4 correspond to operations executed on M_1 and M_2, respectively, and are timed by the duration of these operations.

Transition t_2 represents the transfer operation from machine M_1 to buffer S while transition t_3 represents the transfer operation from buffer S to machine M_2. The tokens indicated in P_1 and P_2 are used to regulate operations on M_1 and M_2. The n tokens initially appearing in Q indicate that

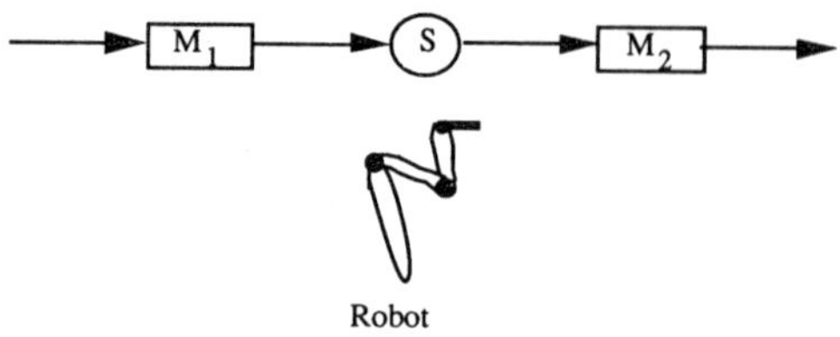

Figure 2.32 Robot and machines on a line.

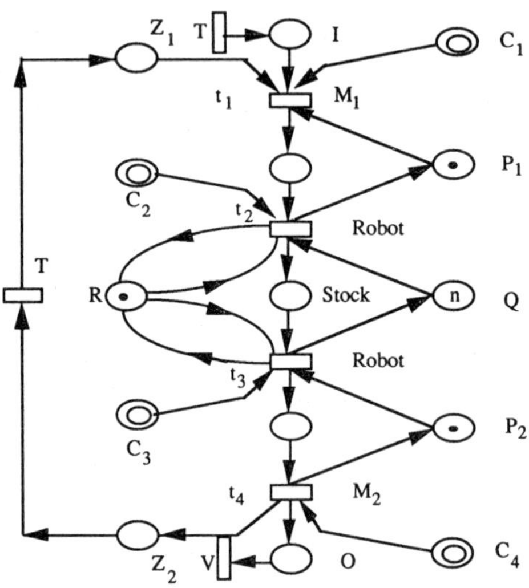

Figure 2.33 Robot moving parts from M_1 to a stock, and from the stock to M_2.

the size of the part buffer is limited to n units. Finally, the token in R makes sure that the robot is only used for one transfer at a time.

The control place C_1 (resp. C_4) makes possible the use of machine M_1 (resp. M_2). The control places C_2 and C_3 are used to manage the utilization of the robot: a token in C_2 allows the use of the robot to unload M_1, while a token in C_3 allows the use of the robot to load M_2. Consequently, a constraint to be applied to the management policy of control places is that both C_2 and C_3 cannot hold tokens simultaneously.

Any control place marking which does not contain one token simultaneously in both C_2 and C_3 is admissible (i.e. guarantees that the model is decision-free). Input transition T is used to model the arrival of a part in front of the system, while output transition V is used to model a part leaving the system. Control places can be connected to these transitions to control the flow of the system.

2.7 CONCLUSION

In this chapter, we chose to use the less sophisticated Petri nets to model the manufacturing modules. Of course, this is certainly not the best choice if we aim at obtaining very compact models. There are mainly two reasons for this choice. First, we obtain simple, thus readable models, which are usually accepted by engineers. Second, the less sophisticated a Petri net, the greater the number of its properties. In other words, it is easier to find analytical properties for the manufacturing system when using simple Petri nets (like events graphs) than sophisticated nets like colored Petri nets.

Furthermore, as we will see in Chapter 4, a natural way to specify and model large size manufacturing systems is to decompose them into small, thus tractable, modules with 'good' qualitative properties, and then to integrate the module models in an adequate manner (i.e. in a manner which optimizes a criterion or provides a good compromise between several criteria and, in addition, preserves the qualitative properties). As a consequence, the size of the whole model is of little importance: it is enough to define modules whose models are of a reasonable size or which are the generalization of small-size models (for instance, the crane model given in Fig. 2.31 can be easily generalized to a storage facility having more than two positions).

The reader will find several other modeling examples in the following chapter.

BIBLIOGRAPHY

Agerwala, T. and Choed-Amphai, Y. A synthesis rule for concurrent systems. *Proceedings of the 15th Design Automation Conference*. Las Vegas, NV, pp. 305–311, June 1978.

Alaiwan, H. and Toudic, J. M. Recherche des semi-flots, des verrous et des trappes dans les réseaux de Petri. *Techniques et Science Informatiques*, vol. 4, no. 1, pp. 103–112, 1985.

Baccelli, F. Ergodic theory of stochastic Petri networks. *INRIA Research Report No. 1037*, May 1989.

Benson, H. P. A finite algorithm for concave minimization over a polyhedron. *Navicular Respect Logis.*, vol. 32, pp. 165–177, 1985.

Chretienne, P. Les réseaux de Petri temporisés. Université de Paris VI, Paris, France, Thèse d'Etat, 1983.

Claver, J. F., Harhalakis, G., Proth, J. M., Savi, V. M. and Xie, X. L. A step-wise specification of a manufacturing system using Petri nets. *Conference IEEE Systems, Man and Cybernetics*, Charlotte, Virginia, October 1991.

Commoner, F., Holt, A., Even, S. and Pnueli, A. Marked directed graphs. *Journal of Computer and System Science*, vol. 5, no. 5, 1971.

Di Mascolo, M., Frein, Y., Dallery, Y. and David, R. *A Unified Modeling of Kanban Systems using Petri Nets*. Technical Report No. 89-06, LAG. Grenoble, France, September 1989.

Hillion, H. P. and Proth, J. M. Performance evaluation of job-shop systems using timed event-graphs. *IEEE Transactions on Automatic Control*, vol. 34, no. 1, pp. 3–9, 1989.

Holloway, L. E. and Krogh, B. H. Synthesis of feedback control logic for a class of controlled Petri nets. *IEEE Transactions on Automatic Control*, vol. 35, no. 5, May 1990.

Koh, I. and DiCesare, F. Transformation methods for generalized Petri nets and their application in flexible manufacturing systems. *Proceedings of the 2nd International Conference of the CIM*, Reusselaer Polytechnic Institute, Troy, NY, May 1990.

Krogh, B. H. and Beck, C. L. Synthesis of place/transition nets for simulation and control of manufacturing systems. *Proceedings of the IFIP Symposium on Large Scale Systems*, Zurich, August 1986.

Laftit, S., Proth, J. M. and Xie, X. Optimization of invariant criteria for event graphs. *IEEE Transactions on Automatic Control*, May 1992.

Mu Der Jeng and DiCesare, F. A review of synthesis techniques for Petri nets. *Proceedings of the International Conference on Computer Integrated Manufacturing*, Troy, NY, pp. 348–355, 1990.

Murata, T. Petri nets: properties, analysis and applications. *Proceedings of the IEEE*, vol. 77, no. 4, April 1989.

Narahari, Y. and Visvanadham, N. A Petri net approach to the modeling and analysis of flexible manufacturing systems. *Annals of O.R.*, vol. 3, pp. 449–472, 1985.

Peterson, J. L. *Petri Nets Theory and the Modeling of Systems*. Prentice-Hall, Englewood Cliffs, NJ, 1981.

Ramamoorthy, C. V. and Ho, G. S. Performance evaluation of asynchronous concurrent systems using Petri nets. *IEEE Trans. Software Eng.*, vol. SE-6, no. 5, pp. 440–449, 1980.

Ramchandani, C. Analysis of asynchronous concurrent systems by timed Petri nets. *Laboratory of Computer Science, Massachusetts Institute of Technology*, Cambridge, MA, Tech. Rep. 120, 1974.

Sifakis, J. A unified approach for studying the properties of transition systems. *Theoretical Computer Science*, vol. 18, pp. 227–258, 1982.

Van Thoai, N. and Tuy, H. Convergent algorithms for minimizing a concave function. *Mathematics of Operations Research*, vol. 5, no. 4, pp. 556–566, 1980.

Warland, J. *An Introduction to Queueing Networks*. Prentice-Hall, Englewood Cliffs, NJ, 1988.

3

Synthesis for manufacturing systems integration

F. DiCesare, with Mu Der Jeng

The purpose of this chapter is to present and give examples for Petri net synthesis methods including bottom-up, top-down and hybrid techniques. Emphasis is placed on application of synthesis methods for the design and control of concurrent systems. Each section will present the techniques, illustrate them with a manufacturing systems example, and discuss the effect of the synthesis method on net properties such as liveness, boundedness and reversibility.

Section 3.1 provides an introduction to Petri net modeling for manufacturing systems and motivates the need for synthesis methods. Section 3.2 provides a review of bottom-up Petri net synthesis techniques. Section 3.3 investigates top-down Petri net synthesis techniques. Section 3.4 reviews recent work combining bottom-up and top-down methods into a hybrid methodology. Section 3.5 summarizes the chapter.

3.1 INTRODUCTION TO SYNTHESIS

Since their introduction Petri nets have been widely studied and applied for modeling concurrent systems, such as computer systems (Ayache *et al.*, 1982; Baer and Ellis, 1977; Kluge and Lautenbach, 1982; Miller, 1973; Ramamoorthy and Ho, 1980; Voss, 1980); communication protocols (Berthelot and Terrat, 1982; Diaz, 1982; Merlin, 1979); manufacturing systems (Martínez *et al.*, 1986; Martínez *et al.*, 1988; Narahari and Viswanadham, 1985; Velilla and Silva, 1988; Villarroel *et al.*, 1988). Using Petri nets to represent concurrency, conflict and mutual exclusion in a system is convenient and straightforward, but problems arise when the system to be modeled is very complex. In these cases the final models will be large and difficult to analyze. Thus, a number of approaches have been proposed, one of which is the systematic synthesis of models such that important system properties are preserved or can be easily analyzed.

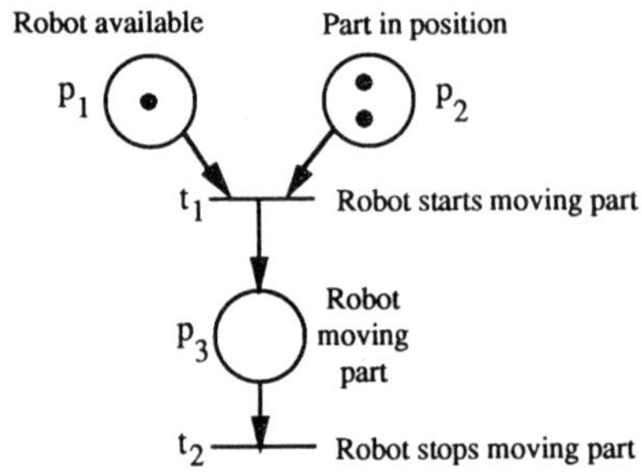

Figure 3.1 Illustration of Petri net modeling of conditions (status) and events.

3.1.1 Net interpretations for manufacturing systems

A very straightforward view of manufacturing is one of sets of activities and resources which will result in a product. The activities are the manufacturing processes including materials handling and information processing that must occur to make products. The resources are the humans, machines, raw materials and so on, that are required to perform these activities. The product process plan specifies the activities and resources in great detail. This includes the precedence relations between activities, i.e., some activities must occur before others, otherwise they could occur concurrently given the necessary resources. With regard to the resources required for each activity, the process plan may specify several alternate resources and give some priority to which is preferred.

Petri nets as defined in graphical and mathematical form have no physical meaning. For their use in applications, the modeler must provide the domain interpretation by the labeling of places and transitions. Generally in Petri net modeling, places (circles) represent conditions and transitions (bars) represent events. There are at least two common ways to model manufacturing systems. Using the first method, the places may represent resource (Krogh and Beck, 1986) availability or activity status as illustrated in Fig. 3.1. For example, places p_1 and p_2 in Fig. 3.1 model resource availability status with p_1 indicating that the robot is available when it is marked by a token and not available when p_1 is not marked. Similarly, when marked p_2 indicates the availability of parts to be moved, the number of tokens representing the number of parts and when unmarked indicates that no parts are available. The marking of places p_1 and p_2 represents the conditions that must jointly be true before the activity of the robot moving the part can occur. p_3 models activity status, specifically p_3 being marked indicates the occurrence of the activity 'robot moving part'. t_1 and t_2 represent the starting and stopping events in this system.

Figure. 3.2 illustrates a second common method of modeling manufacturing systems. In this method the places again represent conditions. The activity, however, is considered an event and modeled as a transition. Both

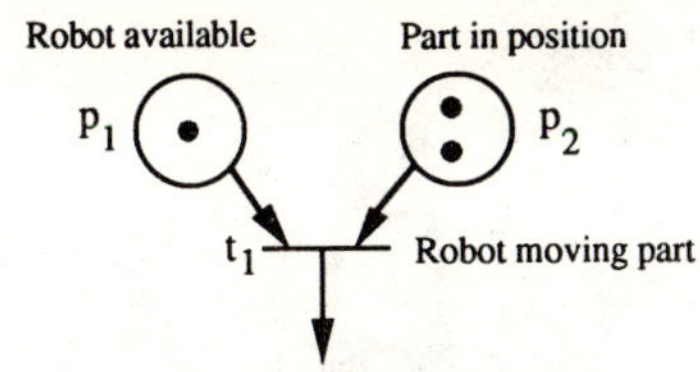

Figure 3.2 Illustration of Petri net modeling of an activity as an event.

of these methods will be used in this chapter and, as will be seen, they are in many ways equivalent.

3.1.2 Modeling methods

Modeling methodology is the key to the application of Petri nets to manufacturing systems. We start with one approach to modeling. Other approaches are possible.

1. Identify the activities and resources required for the production of one item of each product.
2. Order activities by the precedence relations as given in the process plans.
3. For each activity in order: create and label a place to represent the status of that activity; add a transition (start activity) with an output arc(s) to the place(s); add a transition (stop activity) with an input arc(s) from the activity place(s). In general, the stop transition for one activity will be the same as the start transition for the next activity. When the net is executed, a token in an activity place will indicate that the activity is taking place. Multiple tokens will indicate the activity occurring in multiplicity; for example, in a buffer place, two tokens might represent two parts being stored at the same time. The firing of the start transition represents starting the activity or process and the firing of the stop transition represents the completion of the activity and may also represent the start of the next activity.
4. For each activity in order: if such a place has not been already created, create and label a place for each resource which must be available to start the activity. Connect all appropriate resource availability places with arcs such that each inputs to the starting transition for the activity. Create output arcs to connect the stop transition following the activity to any resource places representing resources which become available (are released) upon completion of the activity.
5. Specify the initial marking for the system.

To illustrate the modeling methodology, take a simple manufacturing system as shown in Fig. 3.3. It consists of two machining stations M_1 and M_2, a shared robot R for unloading, and a buffer B for storage of intermediate parts. The process plan for any part is straightforward: it must

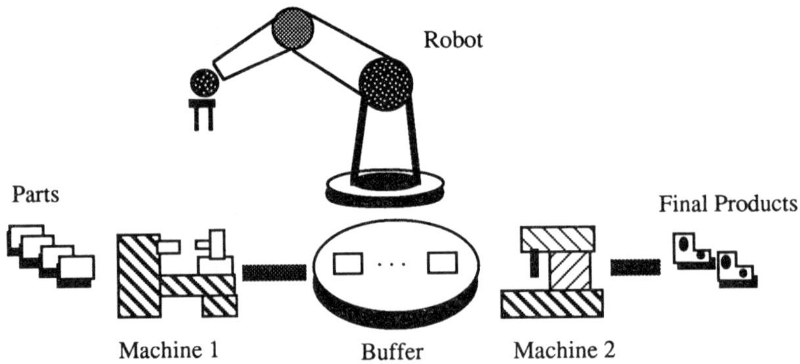

Figure 3.3 A simple manufacturing system (Zhou, 1990).

be machined by M_1, then M_2. Parts enter the systems and at machining station M_1 each is fixtured to a pallet and loaded into the machine. After processing, the robot unloads the intermediate part from M_1 into the buffer B. At machining station M_2 intermediate parts are loaded into M_2 and processed. When M_2 finishes processing a part, the robot R unloads the final product, defixtures it and returns the fixture to M_1. It is assumed that parts are always available to be fixtured. Following the methodology:

(1) The activities required are machining (pallet fixturing, loading and processing), storage (buffering) and unloading. The resources are M_1, M_2, R, B, pallets and parts.

(2) The order of activities is as follows:

- M_1P: M_1 loads, fixtures and processes a palleted raw part;
- RU_1: R unloads an intermediate part to the buffer;
- BS: B stores an intermediate part;
- M_2P: M_2 loads and processes an intermediate part;
- RU_2: R unloads a final product from M_2, defixtures and returns the pallet.

(3) As shown in Fig. 3.4(a), places M_1P, RU_1, BS, M_2P and RU_2 are created to model the activity sequence for one part. Transition t_1 models the start of activity M_1P; t_2 models the stop of activity M_1P and the start of activity RU_1; t_3, the stop of RU_1 and the start of the storage activity BS; t_4, the stop of BS and the start of the M_2P activity. Similarly, t_5 models the stop of activity M_2P and the start of RU_2, and t_6 models the stop of RU_2.

(4) For activity M_1P we require that a pallet and the machine M_1 be available. Figure 3.4(b) shows the Petri net model with place PA representing pallets available, and place M_1A representing machine 1 availability. Each has an input arc to transition t_1. Next consider t_2 as the stop transition for activity M_1P. When the machining activity is complete, no resources can be released since the machine must be unloaded before becoming available again and the pallet will be used throughout manufacture.

Considering t_2 as the start of the robot unloading machine 1 activity, RU_1, the only additional resource required is the robot R. This is modeled

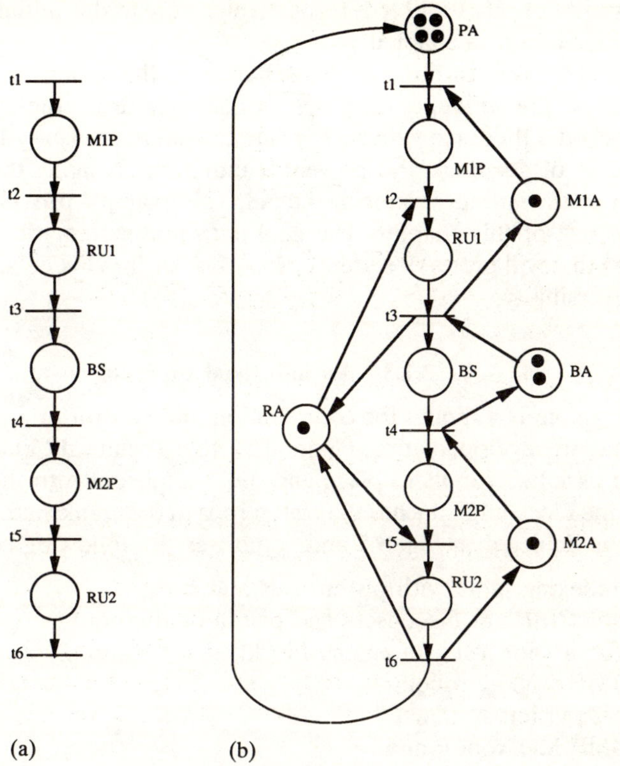

Figure 3.4 (a) Petri net graph of the activity sequence for the manufacturing example; (b) Petri net with the resource places added and connected and the initial marking.

by the place *RA* representing robot available and the input arc from *RA* to t_2. With t_3 as the stop transition of activity RU_1 the release of M_1 can be modeled as an output arc from t_3 to M_1A and the release of *R* as an output arc from t_3 to *RA*.

Next consider the buffer activity, *BS*. In order to store intermediate parts in the buffer, empty spaces in the buffer must exist. This is modeled by place *BA* where the number of tokens indicates the number of empty spaces and the arc from *BA* to t_3, the start transition of *BS*. The completion of the buffer activity for a part and the release of an empty space is represented by transition t_4 and the arc from t_4 to *BA*.

The activities and resources for M_2P and RU_2 are modeled in the same way as for M_1P and Ru_1 except for the output arc from t_6 to place *PA*. This represents the return of an empty pallet after M_2 is unloaded and the product is defixtured.

(5) The initial marking is formulated for system startup. There are four pallets available for parts as indicated by the tokens in *PA* in Fig. 3.4(b); both machining stations and the robot are available as shown by the single

tokens in M_1A, M_2A and RA, respectively. Note in this initial marking that only transition t_1 is enabled.

It is interesting to observe the behavior of the system modeled in this manner. If the sequence $t_1, t_2, t_3, t_1, t_2, t_3, t_1, t_2$ is fired, the system will be deadlocked. Of course with such a simple system, it is easy to identify the possibility of deadlock and prevent it either by changing the structure of the net or the initial marking of the net. This example provides motivation for the rest of this chapter. The goal is to find systematic synthesis procedures that will preserve systems properties such as liveness, boundedness and reversibility.

3.1.3 An industrial example

This example is a model for coordination and control of a piston-stuffing workstation developed at the General Motors Technical Center. The workstation uses two robots to place and pull the pistons into the cylinders of an engine block and attaches the piston rods to the crankshaft. This system's layout is depicted in Fig. 3.5 and comprises the following components:

1. an indexing table with piston rods and caps;
2. a table with nut dispensers and piston-pulling tools;
3. a workstation with an engine block;
4. a GMF S-380R robot;
5. a V-20 vision system;
6. a GMF M-1 robot; and
7. some sensors installed in the robots.

The basic activities involved in the system are as follows:

1. the S-380 robot moves the crankshaft to align the crankshaft position for piston insertion;
2. the vision system checks and verifies the crankshaft position in the engine block;
3. the S-380 robot picks up a piston, with its rod attached, from the indexing table and brings it to an engine-block cylinder;
4. the M-1 robot picks up the piston-pulling tool, two nuts and a cap and moves to the workstation, respectively. It pulls the piston rod into the engine block, installs the cap, and attaches and turns two nuts to secure this cap.

Repeat steps (1) through (4) for the remaining pistons.

The resulting Petri net model (see Fig. 3.6) clearly shows the sequence of activities and the resource requirements for each of the activities.

3.1.4 Properties of Petri nets and their relation to manufacturing

One of the major advantages of Petri nets is the ability to analyze these nets for properties related to manufacturing control. These properties

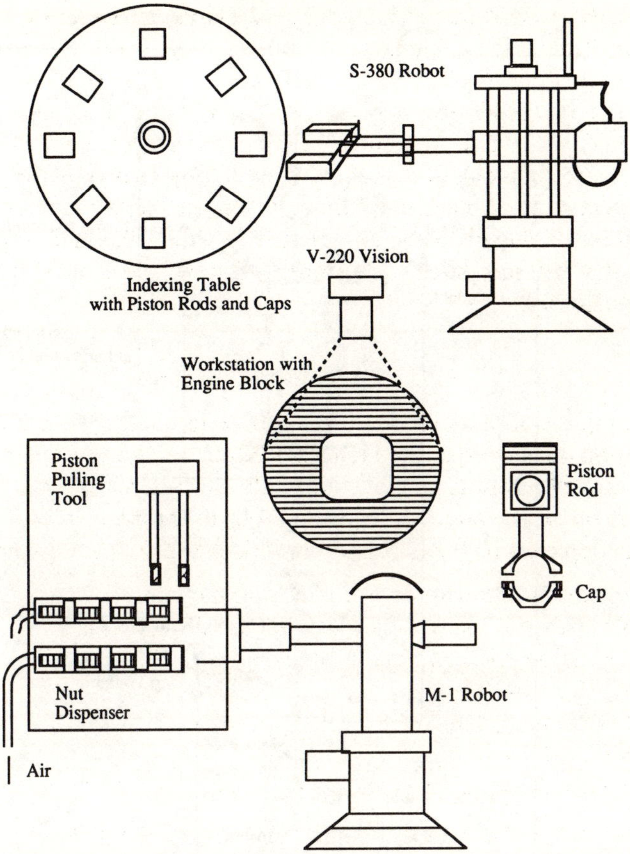

Figure 3.5 Industrial system for motor vehicle engine piston insertion.

include boundedness or safeness, liveness and reversibility. Their significance to manufacturing systems is briefly stated as follows:

1. Boundedness or safeness implies the absence of overflows. Safeness of an operation place guarantees that there is no attempt to request execution of an ongoing process. Safeness of a resource place indicates the availability of only a single resource and is often used to guarantee the safeness of some related operation places.
2. Liveness implies the absence of deadlocks. This property guarantees that a system can successfully produce. Moreover, it ensures that all modeled processes can happen.
3. Reversibility implies the fulfillment of cyclic behavior of a system which will perform its functions repeatedly. This also has implications for error recovery in the manufacturing context.

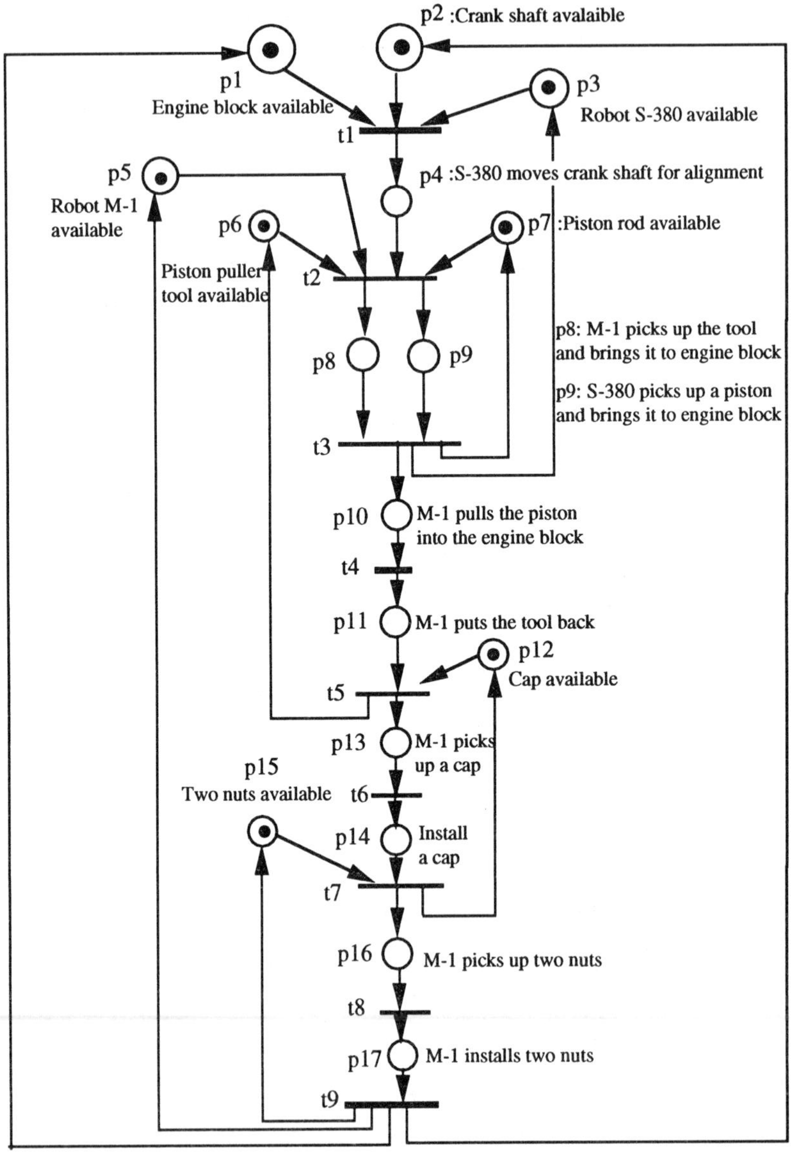

Figure 3.6 Petri net model for the piston insertion example.

3.1.5 Preservation of properties vs. analysis

There are currently two ways to approach Petri net modeling for manufacturing control. One is to model the system with a Petri net using a systematic procedure, as shown earlier in this section, and follow that by analysis of the system for the desired properties. For any system of even moderate size the number of states makes analysis, particularly for liveness, practically impossible. To address this problem, transformation methods which reduce the size (state space) of the net while maintaining properties of interest have been developed (Andre *et al.*, 1976; Berthelot, 1985; 1986; Dong, 1983; Johnsonbaugh and Murata, 1981; Kowalk and Valk, 1979; Kwong, 1977; Lee and Favrel, 1985; Lee *et al.*, 1987; Lipton, 1981; Mayr and Meyer, 1981; Murata and Koh, 1980; Silva, 1981). Analysis can then be effectively performed on the reduced net. These methods are very powerful. However, in a system environment with many shared resources, such as might be found with flexible automation, even these methods may not suffice. An alternate approach is to devise systematic modeling methods which guarantee the desired properties. These synthesis methods obviate the need for analysis and thus may circumvent the state space explosion problem. The remainder of this chapter presents and discusses several approaches to synthesis: bottom-up, top-down and hybrid methods.

3.2 BOTTOM-UP SYNTHESIS

The use of bottom-up or modular composition methods is a commonly used systems design methodolodgy. This involves the specification of subsystems or modules and some systematic procedure for combining these modules with their interactions into an integrated system. This method can be used in Petri net modeling for manufacturing systems. First, incomplete subsystems are specified separately in detail. These subsystems are usually very simple and easy to verify. Some interactions are represented by common places, transitions or paths in the individual subsystems. Then, at each synthesis step, these interactions are considered, and the corresponding subsystems are combined through merging these places and/or transition into a larger subsystem. Analysis of the combined net is usually done immediately after each synthesis step, so when the final stage of synthesis is reached, the analysis can be simplified. At the end of the synthesis steps, the final system and some of its important properties are obtained.

3.2.1 Merging of Places

Methods

Agerwala and Choed-Amphai (1978) have done pioneering work in bottom-up techniques and have proposed a systematic bottom-up approach for synthesizing concurrent systems modeled by Petri nets. They suggest that synthesis can start with basic structures or simple nets, which can be verified easily. A basic structure, as given in Fig. 3.7, is defined as a transition t_1 with a set $X = \{p_1, \ldots, p_r\}$ of input places and a set $Y = \{q_1, \ldots, q_s\}$ of output places where $X \cap Y = \emptyset$.

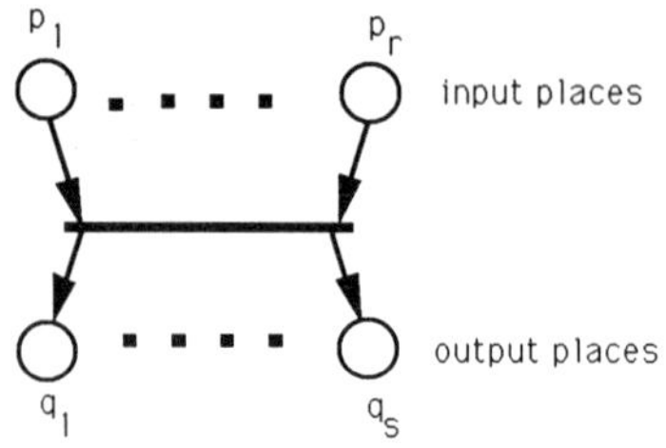

Figure 3.7 Basic structure (Agerwala and Choed-Amphai, 1978).

At each synthesis step, subnets can be merged in such a way that a set of places are merged into a new place. This is called a 1-way merge, as described in the following paragraph.

Given a Petri net $N = (P, T, I, O)$, select a set of places to be merged $P_m \subseteq P$, such that

1. For any $p_i, p_j \in P_m$, if $(p_i, t) \in I$ and $(p_j, t) \in I$, then $i = j$ (no two or more places in P_m are input places for some transition).
2. For any $p_i, p_j \in P_m$, if $(p_i, t) \in O$ and $(p_j, t) \in O$, then $i = j$ (no two or more places in P_m are output places for some transition).

Remember that the places in P_m will be merged. Then, construct the resultant net $N' = (P', T', I', O')$, such that

1. $T' = T$;
2. $P' = (P - P_m) \cup \{p\}$ where $p \notin P$;
3. I' and O' are obtained by replacing every occurrence of each $p_i \in P_m$ in I and O by p.

In other words, all places in P_m are merged to a place named p. For example, Fig. 3.8(b) is obtained by merging p_3 and p_6 of Fig. 3.8(a) into p_3'.

At the end of the synthesis, we obtain the system model, the merged Petri net, which results from all the 1-way merges.

In order to facilitate the analysis, Agerwala and Choed-Amphai (1978) have provided a theorem which states that after every 1-way merge, the P-invariants of the resultant net can be known from the P-invariants of the subnets, as described below.

Theorem 1 (Agerwala and Choed-Amphai, 1978): Consider a net $N = (P, T, I, O)$ on which the 1-way merge operation is applied to construct a net $N' = (P', T', I', O')$. All the P-invariants of N' are obtained from the P-invariants of N as follows: $PI' \subseteq P'$ is a P-invariant of N' if and only if there exists a P-invariant, PI, of N, such that

1. if $P_m \subseteq PI$ then $PI' = (PI - P_m) \cup \{p\}$;
2. if $P_m \cap PI = \varnothing$ then $PI' = PI$.

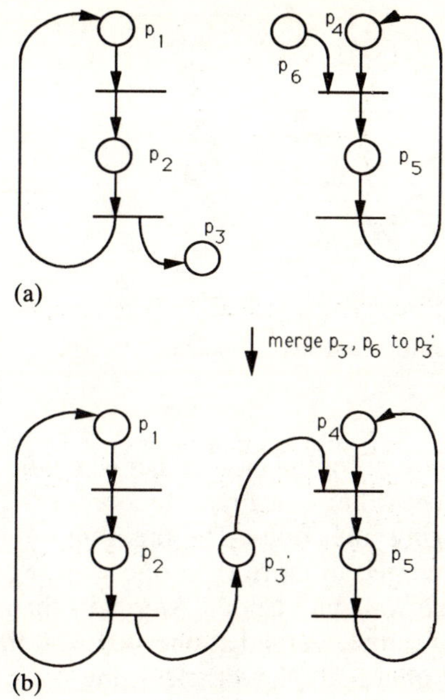

Figure 3.8 An example of 1-way merge (modified from Agerwala and Choed-Amphai, 1978).

Note that if the synthesis rule is restricted in such a way that a merge operation is allowed only if each place in the resultant net is in some *P*-invariant, then the class of nets synthesized in this way is equivalent to the class of bounded Petri nets, because all the places are in some *P*-invariants.

Similar work has been investigated by Narahari and Viswanadham (1985) for modeling and analyzing flexible manufacturing systems. In their approach, a Petri net is first created for every basic operation such as an intermediate machine operation for a product. Because a basic operation is simple, its corresponding Petri net is easy to verify and analyze. A product might need more than one machine operation, and there might be several choices for selecting machines to execute these machine operations. In addition, several different types of products might be manufactured in the system. In this way, the subnets that represent the products can be obtained by merging the places of the subnets that denote the machine operations. Therefore, the system Petri net can be constructed from merging the places of the subnets that represent the products.

Narahari and Viswanadham have developed two theorems for finding invariants in order to verify important qualitative properties of the net, such as existence/absence of deadlock (liveness), conservativeness, and boundedness after each synthesis step. One of the theorems deals with nets obtained by merging places from the subnets. It facilitates the

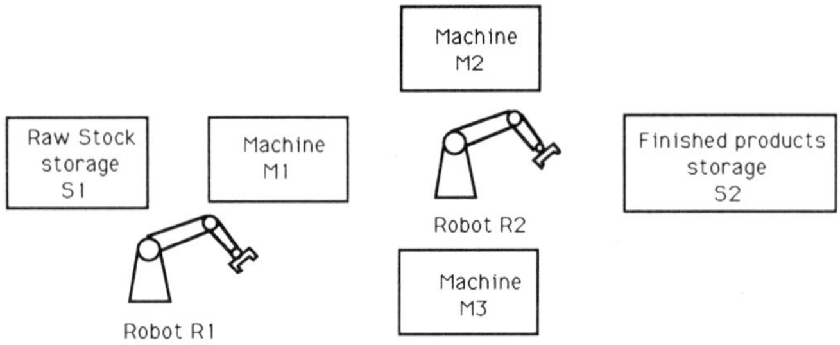

Figure 3.9 A simple automated manufacturing system.

computation of *P*-invariants of the merged net when the *P*-invariants of the subnets are known. The other is a similar theorem for *T*-invariants for the net obtained by merging transitions. The theorem for *P*-invariants, which allows places to be merged in more than one way at each synthesis step, is an extension of Agerwala and Choed-Amphai's theorem. That is, more than one set of places can be merged at one step, and the properties of the resultant net can be obtained. Nevertheless, the basic principles are the same.

3.2.2 Model synthesis example

Example 1

In this illustration, we show how to design a simple automated manufacturing system consisting of two robots R_1, R_2, and three machines M_1, M_2, M_3, as shown in Fig. 3.9, using Agerwala and Choed-Amphai's method. This example could also be accomplished by using Narahari and Viswanadham's method. The system that we are going to model produces only one type of product, and requires two machine operations: a product is produced by a machine operation 1 on M_1 followed by a machine operation 2 on M_2 or M_3. There are two assumptions in this example:

1. The supply of raw stock is never exhausted; that is, there is always raw stock in its storage place ready for use.
2. The finished product will be taken away so that there is no overflow problem with its storage place.

The procedure to manufacture a product is described as follows:

1. R_1 takes raw stock from storage S_1 and loads M_1;
2. M_1 starts machining;
3. after M_1 finishes its operation, R_2 takes the intermediate product from M_1 to M_2 or M_3 for further machining;
4. Any finished product on M_2 or M_3 will be moved by R_2 to storage S_2.

Although simple in its configuration this example demonstrates some important characteristics of concurrent systems:

1. concurrency – any robot or machine is an autonomous system which can operate in parallel with others;
2. conflict – a second machine operation can be done on either M_2 or M_3;
3. resource sharing – robot R_2 is shared.

In addition, it demonstrates some flexibility in the context of manufacturing since the system produces one product type but employs two possible sequences for making it.

Before using one-way merges, we specify the individual subsystems separately. Remember that at this stage any interaction among the subsystems is neglected. Figure 3.10(a) shows operations on M_1 and an initial merge of p_{12} and p_{15} into p'_{12}. Using the state equation of the model in (a) it is easy to find that the P-invariants of the model after the initial merges in Fig. 3.10(a) are $\{p_{11}, p_{13}, p_{14}, p_{16}\}$ and $\{p'_{12}, p_{13}\}$. Figure 3.10(b) denotes operations on M_2, initial merges of p_{23}, p_{27}, p_{28} and p_{211} into p'_{23}, and an initial merge of p_{22} and p_{210} into p'_{22}. Its P-invariants are $\{p_{21}, p_{24}, p_{25}\}$, $\{p'_{22}, p_{24}, p_{26}, p_{29}\}$ and $\{p_{23}, p_{24}, p_{29}\}$. Figure 3.10(c) describes operations on M_3, initial merges of p_{33}, p_{37}, p_{38} and p_{311} into p'_{33}, and an initial merge of p_{32} and p_{310} into p'_{32}. Its P-invariants are $\{p_{31}, p_{34}, p_{35}\}$, $\{p'_{32}, p_{34}, p_{36}, p_{39}\}$ and $\{p'_{33}, p_{34}, p_{39}\}$. Note that places with the same description will be merged to a single place after a 1-way merge.

After several 1-way merges, each of which merges one set of places, the resultant model can be obtained as shown in Fig. 3.10(d).

3.2.3 Analysis of properties of the example system

By using Theorem 1, the net has the following P-invariants (not necessarily minimum) $\{p'_{11}, p_{13}, p_{14}, p_{16}, p_{24}\}$, $\{p''_{11}, p_{13}, p_{14}, p_{16}, p_{34}\}$, $\{p'_{12}, p_{13}\}$, $\{p'_{22}, p_{24}, p_{26}, p_{29}\}$, $\{p'_{32}, p_{34}, p_{36}, p_{39}\}$ and $\{p'_{23}, p_{24}, p_{29}, p_{34}, p_{39}\}$. If each of $p'_{11}, p'_{12}, p'_{22}, p'_{23}$ and p'_{32} is marked with one token initially, then we have the following equations

$$m(p'_{11}) + m(p_{13}) + m(p_{14}) + m(p'_{16}) + m(p_{24}) = 1 \quad (3.1)$$
$$m(p'_{11}) + m(p_{13}) + m(p_{14}) + m(p'_{16}) + m(p_{34}) = 1 \quad (3.2)$$
$$m(p'_{12}) + m(p_{13}) = 1 \quad (3.3)$$
$$m(p'_{22}) + m(p_{24}) + m(p_{26}) + m(p_{29}) = 1 \quad (3.4)$$
$$m(p'_{32}) + m(p_{34}) + m(p_{36}) + m(p_{39}) = 1 \quad (3.5)$$
$$m(p'_{23}) + m(p_{24}) + m(p_{29}) + m(p_{34}) + m(p_{39}) = 1 \quad (3.6)$$

We shall prove that the net is live, bounded and safe for this initial marking. The boundedness is easy to verify because each place is covered by some P-invariant, but checking liveness takes more effort, because we have to verify that each transition is live. Fortunately, with the help of the P-invariants this job is simplified. First consider the liveness of t_{11}.

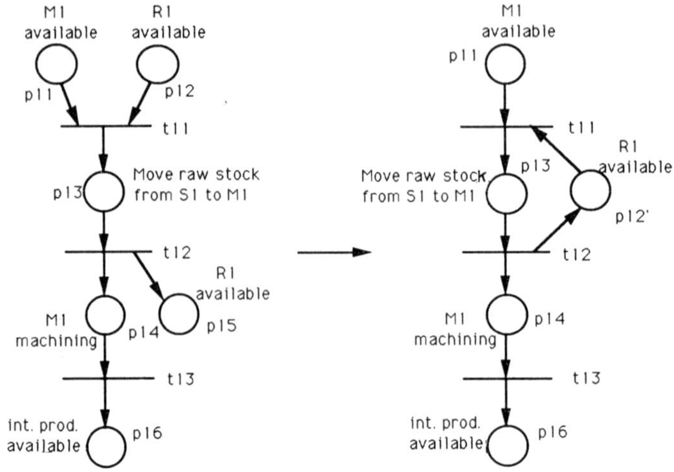

Figure 3.10 (a) Operations on M_1 (an initial merge).

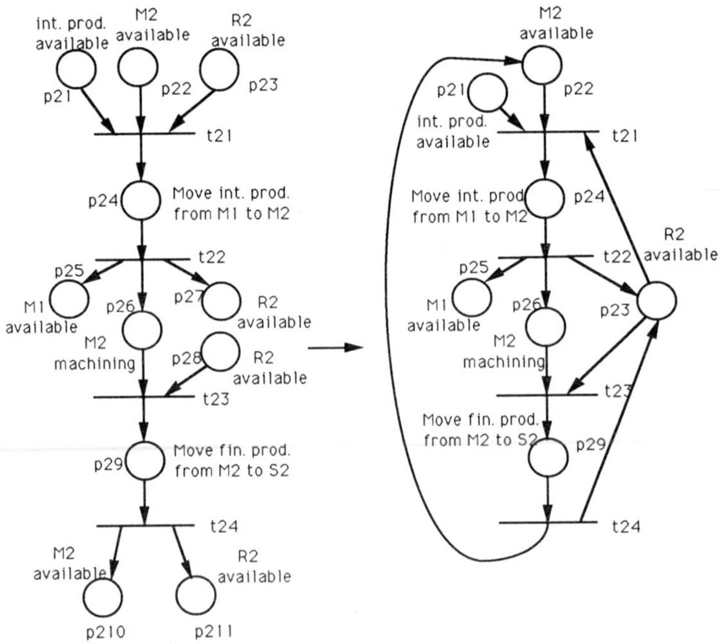

Figure 3.10 (b) Operations on M_2 (initial merges).

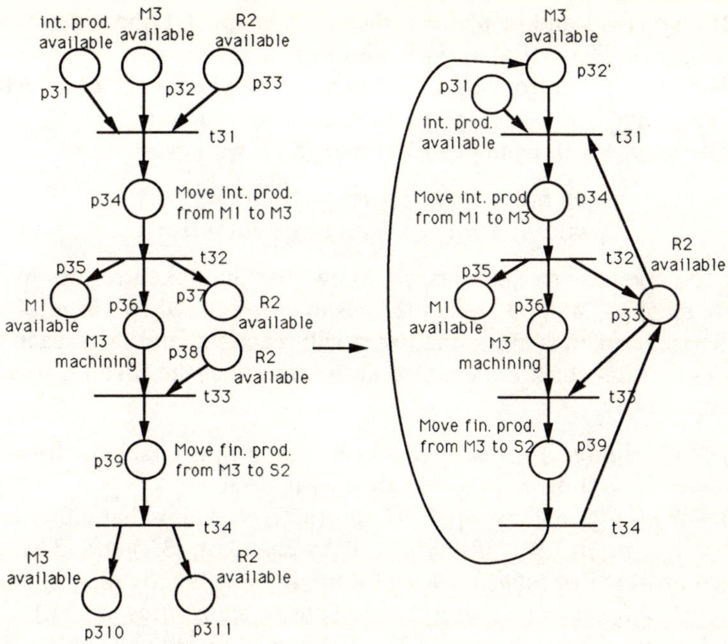

Figure 3.10 (c) Operations on M_3 (initial merges).

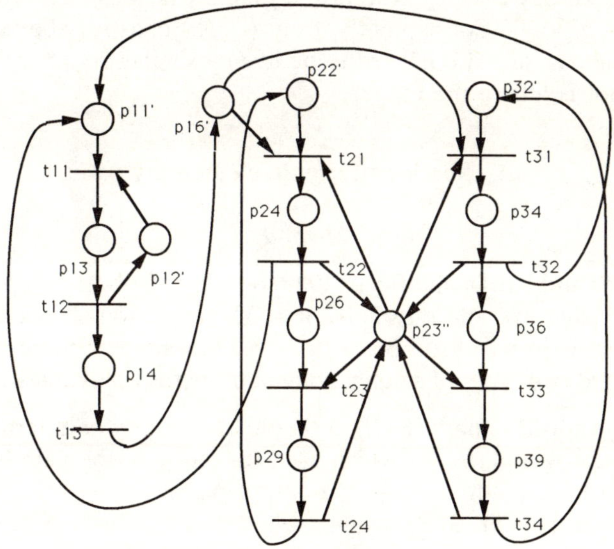

Figure 3.10 (d) Final model after 1-way merges.

(i) If $m(p'_{11}) = 1$ and $m(p'_{12}) = 1$, then t_{11} is enabled. Otherwise, if $m(p'_{11}) = 0$ or $m(p'_{12}) = 0$, then t_{11} is disabled.
(ii) For $m(p'_{12}) = 0$, from eq. (3.3), $m(p_{13}) = 1$, which enables t_{12}. After t_{12} fires, $m(p'_{12}) = 1$. If $m(p'_{11}) = 1$, then t_{11} is enabled.
(iii) For $m(p'_{11}) = 0$, from eqs (3.1) and (3.2), we have

$$m(p_{13}) + m(p_{14}) + m(p'_{16}) + m(p_{24}) = 1 \tag{3.7}$$
$$m(p_{13}) + m(p_{14}) + m(p'_{16}) + m(p_{34}) = 1 \tag{3.8}$$

From the above two equations, we know that one token resides in either p_{13}, p_{14}, p'_{16} or p_{24}, and one token resides in either p_{13}, p_{14}, p'_{16} or p_{34}. For the case where a token is in p_{13}, the token will reach p_{14}, and then reach p'_{16} by firing t_{12} and t_{13}, which are enabled at the time that the token is in p_{13} and p_{14}, respectively.

(iii.1) Now, if $m(p'_{22}) = 1$ and $m(p'_{23}) = 1$, then t_{21} is enabled. However, if $m(p'_{22}) = 0$ or $m(p'_{23}) = 0$, then t_{21} is disabled.
(iii.2) For $m(p'_{22}) = 0$, by eqs (3.4) and (3.7), we know that either $m(p_{26}) = 1$ or $m(p_{29}) = 1$. If $m(p_{26}) = 1$, by eqs (3.6), (3.7) and (3.8), either $m(p'_{23}) = 1$ or $m(p_{39}) = 1$. But if $m(p_{39}) = 1$ then by firing t_{34}, $m(p''_{23}) = 1$. After t_{23} fires, $m(p_{29}) = 1$. Thus, once t_{24} fires, $m(p'_{22}) = 1$.
(iii.3) For $m(p'_{23}) = 0$, by eqs (3.6), (3.7) and (3.8), we have either $m(p_{29}) = 1$ or $m(p_{39}) = 1$, which enables t_{24} and t_{34}, respectively. So, eventually $m(p'_{23}) = 1$.

Therefore, for any case, p_{24} will have a token, which enables t_{22}, and after t_{22} fires, $m(p'_{11}) = 1$. Consequently, from (i), (ii) and (iii) above, t_{11} is live (t_{12}, t_{13}, t_{21} and t_{22} are also proved to be live by similar argument). The rest of the proof is left to the reader.

3.2.4 Sharing of simple elementary paths

Method

Beck (1985) and Krogh and Beck (1986) developed a bottom-up technique for synthesizing live and safe Petri nets. Their method shares simple elementary paths in which no place or transition appears more than once. They defined two types of simple elementary paths for synthesis purposes:

1. **Solitary transition path** (STP): a simple elementary path terminated on both ends by a place, for which each transition in the path has exactly one input place and exactly one output place.
2. **Solitary place path** (SPP): a simple elementary path terminated on both ends by a transition, for which each place in the path is an input place for exactly one transition and an output place for exactly one transition.

The synthesis procedure starts with a collection of simple elementary circuits (SECs), Petri nets constructed from finite-length paths with coincident initial and terminal places (e.g. Figs 3.11(a), (b) and (c)). These SECs correspond to basic activity cycles in the system with the assumption that the system is designed to perform a repetitive operation. Next, we choose one SEC as the initial Petri net, and combine one of the remaining SECs with the initial Petri net along an STP or an SPP. We repeat the combining step until all SECs are included.

Figure 3.11 illustrates the synthesis of a Petri net by joining three SECs. First, SECs in Figs 3.11(a) and (c) are combined along a common STP $\{p_1, t_1, p_2\}$ and then the resultant net is combined with Fig. 3.11(b) along a common SPP $\{t_2, p_3, t_3\}$. The final combined net is shown in Fig. 3.11(d).

The Petri net obtained in this way will be live and safe with respect to any initial marking for which there is exactly one token in each of the *P*-invariants of the system. This conclusion is supported by Krogh and Beck's theorems. For example, if we mark each of p_1 and p_5 above with exactly one token, then the combined net in Fig. 3.11(d) is live and safe. Beck (1985) and Krogh and Beck (1986) also show that after each synthesis step, the *P*-invariants of the combined Petri net can be easily calculated.

The result they provide, i.e. that liveness and safeness are preserved, is very strong, but the limitation of their method is such that only safe places are allowed. This leads to some difficulties in modeling some resources such as buffers. In this case, instead of using a place to denote a buffer, we have to use a set of places to describe all its possible states which might be large if the buffer is bounded by a large number. Their method also places more restriction on modeling shared resources than the former two methods.

3.2.5 Model synthesis examples

Example 2

In this example, we apply Krogh and Beck's method to model the following simple automated manufacturing system (see Fig. 3.12) which produces two products. The two product flows within the system are independent except that R_3 is shared between them. Again, the same assumptions as those in Example 1 apply to the raw stock storages and finished product storages. In the description below, the names in the parentheses are for product 2. The procedure to make product 1 (product 2) is depicted as follows: robot R_1 (robot R_2) takes raw stock 1 (stock 2) from storage S_1 (storage S_2) and loads machine M_1 (machine M_2). Then M_1 (M_2) starts machining. After M_1 (M_2) is finished, robot R_3 unloads M_1 (M_2) and moves product 1 (product 2) to storage S_3 (storage S_4).

In the following figures, basic activity cycles are modeled as SECs. Two or more simple elementary paths to be merged are depicted using same place names and same transition names along the paths. Figure 3.13(a) describes a cycle where raw stock 1 is machined on M_1. Figure 3.13(b)

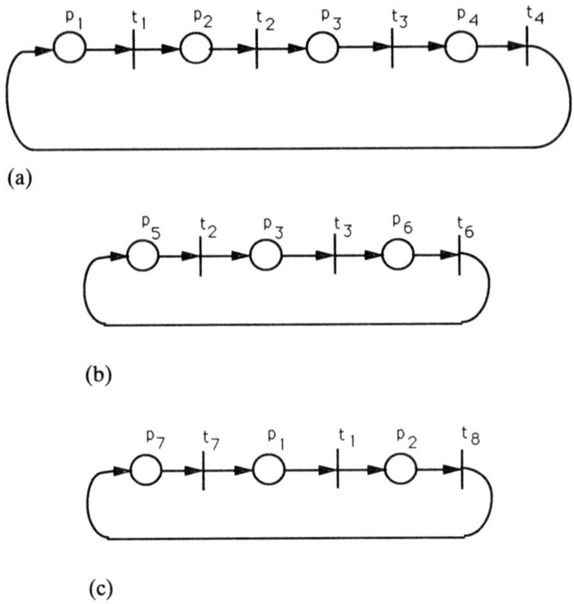

Figure 3.11 (a)–(c) Three simple elementary circuits.

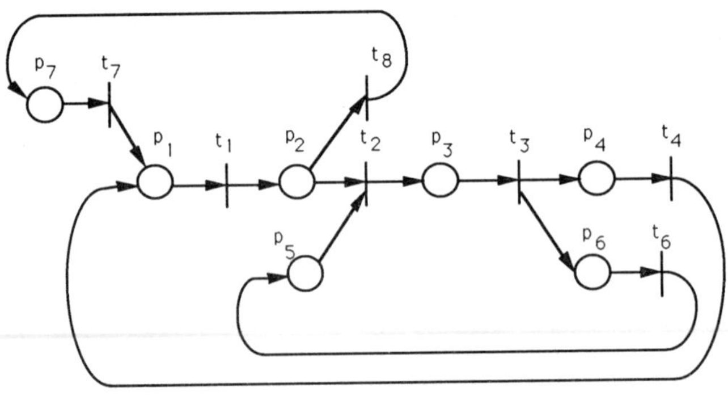

(d)

Figure 3.11 (d) Three SECs combined along an STP and an SPP.

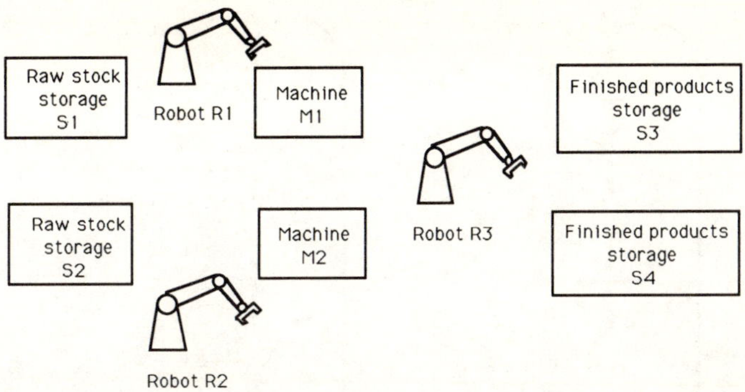

Figure 3.12 A simple automated manufacturing system.

denotes a similar cycle for M_2. Figures 3.13(c)–(f) describe the moving cycles for robots R_1, R_2 and R_3.

Figure 3.13(g) shows the final model after merging simple elementary paths (STP or SPP). Table 3.1 lists these merges. By using their theorems, the model in Fig. 3.13(g) is live and safe if we mark it in such a way that each of its P-invariants contains exactly one token. For example, if each of the places $p_{11}, p_{21}, p_{31}, p_{41}, p_{51}$ is marked with exactly one token, then the model is live and safe.

Krogh and Beck's result is extended by Koh and DiCesare (1990), who invented a similar synthesis method which can be applied to model nets with bounded places and generalized Petri nets, i.e. Petri nets with multiple arcs.

3.2.6 Summary of bottom-up synthesis techniques

In terms of interactions in a concurrent environment, bottom-up methods provide some freedom in specifying the system. At the beginning of the synthesis, we can treat the system as a composite of independent subsystems and neglect any interaction. This has the advantage of ease in describing the system, since the subsystems usually have real-life correspondences such as robots, machines and transport devices. The tradeoff is that we may lose some control in the behavior of the composed net so that some important properties (liveness, boundedness, etc.) may not be obtained. Nevertheless, using bottom-up synthesis methods can still be advantageous, because the analysis of the final net is simplified. At each synthesis step, some important properties of the resultant net can be verified easily from the information known from the individual subnets.

From the bottom-up synthesis methods addressed above, we see that the invariant method is often used to analyze the properties of the combined net after each synthesis step. However, the disadvantage of using invariants is that they do not convey complete information about the net,

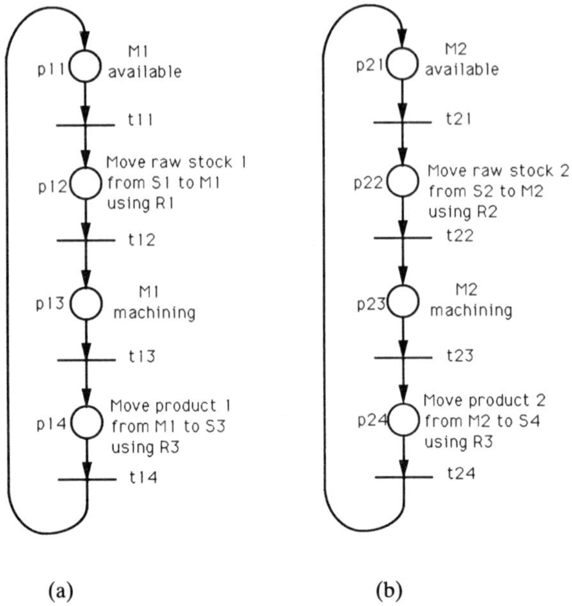

Figure 3.13 (a) Cycle $S_1 - M_1 - S_3$; (b) cycle $S_2 - M_2 - S_4$.

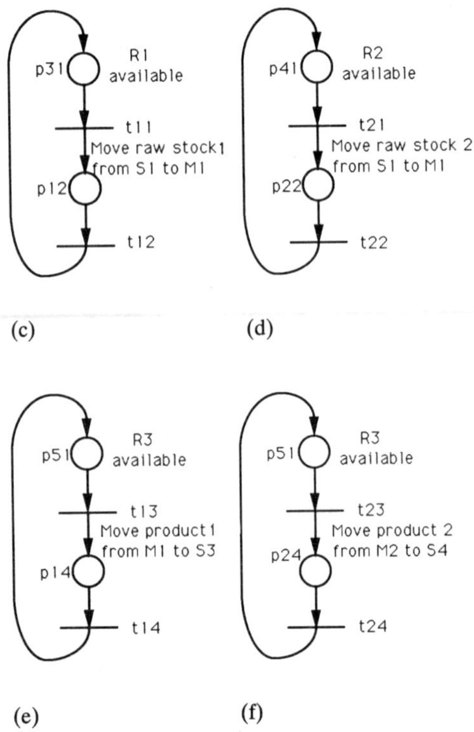

Figure 3.13 (c)–(f) Moving cycles of Robot R_1, R_2 and R_3.

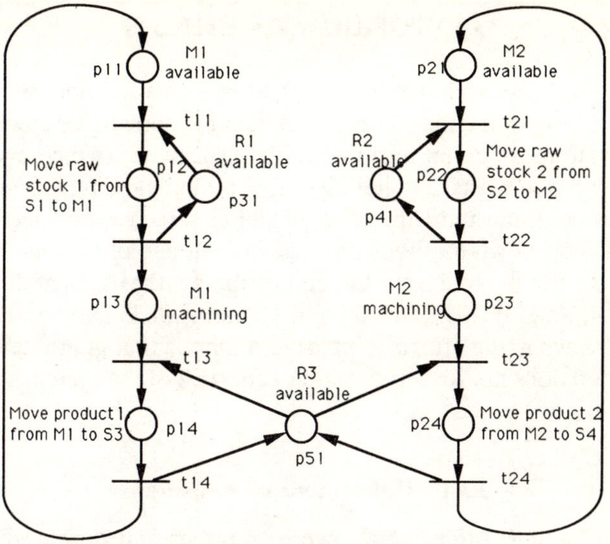

(g)

Figure 3.13 (g) Final model of Example 2.

Table 3.1 Merges for the nets in Figure 3.13

SEC's merged	*Common path*	*STP or SPP*
(a), (c)	$[t_{11}, p_{12}, t_{12}]$	SPP
(a), (c), (e)	$[t_{13}, p_{14}, t_{14}]$	SPP
(a), (c), (e), (f)	$[p_{51}]$	STP
(a), (c), (e), (f), (b)	$[t_{23}, p_{24}, t_{24}]$	SPP
(a), (c), (e), (f), (b), (d)	$[t_{21}, p_{22}, t_{22}]$	SPP

which makes investigating some properties (e.g. liveness and reversibility) difficult. Besides, these methods usually do not guarantee that the resultant net will preserve important properties. The exception is that of Beck (1985) and Krogh and Beck (1986), in which liveness and safeness are preserved. Nevertheless, these bottom-up methods have provided successful solutions in some applications (Beck, 1985; Krogh and Beck, 1986; Narahari and Viswanadham, 1985). Top-down synthesis as described in the next section provides a complementary methodology.

3.3 TOP-DOWN METHODS

Top-down synthesis usually begins with an aggregate model of the system and neglects low-level detail. Then, refinement is done in a stepwise manner to incorporate more detail into the model. There are two commonly used schemes for refinements: expanding places and expanding transitions. The refinements continue until the level of detail satisfies the specification of the system. Top-down methods have the advantage of viewing the system globally from the beginning to the end of the synthesis. In addition, many researchers (Valette, 1979; Suzuki and Murata, 1983; Zhou and DiCesare, 1990; 1991) have made efforts to provide methods that guarantee that each synthesis step does not lose important properties of the system so that final analysis will not be necessary.

3.3.1 Refinement of transitions

Valette (1979) first studied top-down techniques and proposed a method for stepwise refinement of transitions for analyzing Petri nets. The argument is that if it is assumed that the firing of a transition is not instantaneous (relaxing the original definition of Petri nets), and is made up of two steps, then it is possible to associate a transition with a complex operation that can later be depicted in detail by means of another Petri net. In order to ensure that important properties of a net are preserved after each refinement step, Valette defined the conditions with which a transition can be replaced by a Petri net.

A **well-formed block** can be substituted for a transition which is not 2-enabled (Valette, 1979) in a net without losing the properties of boundedness, safeness and liveness. A transition t is said to be 2-enabled if and only if there exists a marking m reachable from the initial marking such that $\forall p \in P, m(p) \geq 2$ fi$I(p, t)$. A block is defined as a net with only one input transition, called the **initial transition**, and only one output transition, called the **final transition**. The **associated Petri net** of a block is defined as the net which results from adding a place, called an idle place, to the block such that the output (input) transition of the idle place is the initial (final) transition (see Fig. 3.14). A well-formed block is a block for which the following three conditions are satisfied:

1. its associated Petri net is live;
2. the initial marking of the associated Petri net is the only marking such that the idle place is not empty;
3. the only transition enabled by this initial marking is the initial transition.

For example, Fig. 3.15(a) shows a Petri net where t_3, which is not 2-enabled, is to be replaced by the well-formed block in Fig. 3.15(b). The resultant Petri net is shown in Fig. 3.15(c).

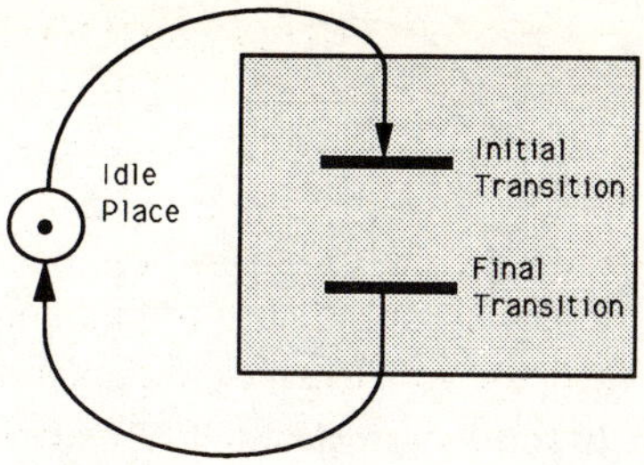

Figure 3.14 A well-formed block and its associated Petri net (Valette, 1979).

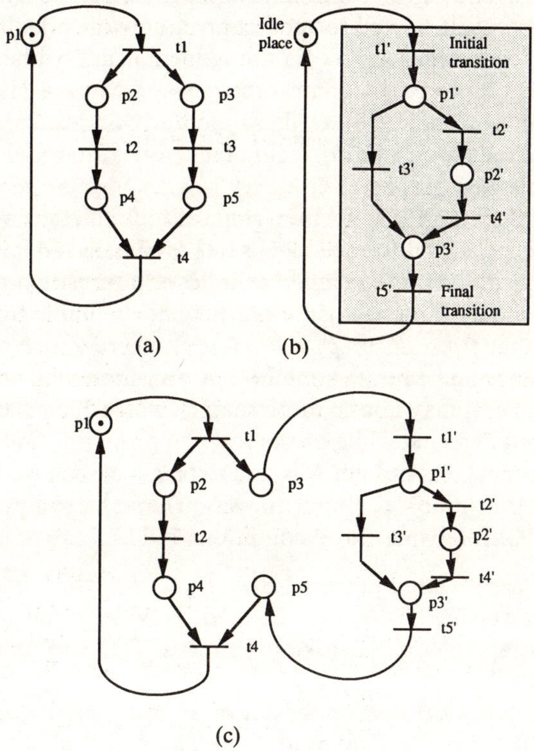

Figure 3.15 Example of refinement: (a) Petri net; (b) well-formed block and its idle place; (c) combined Petri net (Valette, 1979).

Valette's method can be considered a generalization of the results of Bruno and Altman (1971) who developed a theory of **asynchronous control networks** for modeling the control structure of digital systems. These are similar in modeling power to Petri nets. They defined a number of basic control modules, such as WYE, JUNCTION, SEQUENCE, ITERATION and SELECT. These modules can be treated as specializations of well-formed blocks if the modules are represented as Petri nets. Bruno

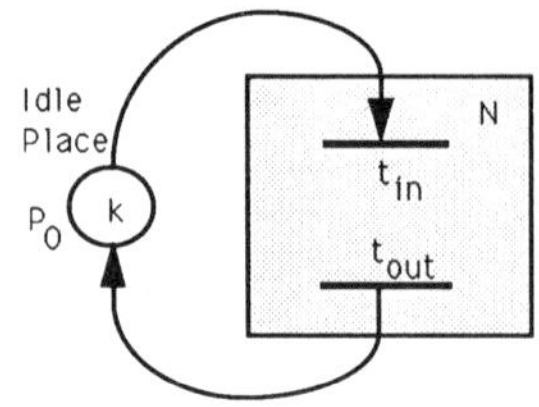

Figure 3.16 A Petri net N and its associated $B(N, t_{in}, t_{out}, k)$ (Suzuki and Murata, 1983).

and Altman investigated the conditions under which combinations of basic control modules result in well-formed control networks, i.e. networks not exhibiting deadlocks. Similar work for the reduction and synthesis of marked graphs has been explored by Johnsonbaugh and Murata (1981), Murata and Koh (1980), Murata (1980). In these methods, Series, Parallel, Unique circuit and *Y–V* reduction and expansion techniques are provided to obtain live and safe marked graphs.

Suzuki and Murata (1983) further generalized Valette's work and extended the concept of well-formed blocks to ***k*-well-behaved** Petri nets. They also extended 2-*enabled* to *k-enabled* as follows: a transition t is said to be k-enabled if and only if there exists a marking m reachable from the initial marking such that $\forall p \in P,\ m(p) \geq k \cdot I(p, t)$ where $k \in N$.

By using Suzuki and Murata's method, a transition which is not K + 1-enabled ($k \geq 1$; i.e. it may not be more than k-enabled) can be replaced by a ***k*-well-behaved** Petri net while preserving the properties of boundedness, safeness and liveness. A Petri net N is said to be k-well-behaved with respect to two distinct transitions, t_{in} (input transition) and t_{out} (output transition), if and only if the following three conditions hold:

1. t_{in} is live in its associated Petri net $B(N, t_{in}, t_{out}, k)$, which is similar to the concept of associated Petri nets defined by Valette but the idle place may contain up to k initial tokens (refer to Fig. 3.16 and Suzuki and Murata (1983)).
2. Let $L(B(N, t_{in}, t_{out}, k))$ denote the set of all firing sequences in $B(N, t_{in},$ tout, $k)$ starting from the initial marking and $\#(\sigma, t)$ denote the number of firing of t in the firing sequence σ. Then, for each $\sigma_1 \in L(B(N, t_{in}, t_{out}, k))$ such that $\#(\sigma_1, t_{in}) > \#(\sigma_1, t_{out})$, there exists $\sigma_2 \in (T - \{t_{in}\})^+$ such that $\sigma_1\sigma_2 \in L(B(N, t_{in}, t_{out}, k)$ and $\#(\sigma_1, t_{in}) = \#(\sigma_1\sigma_2, t_{out})$.
3. $\#(\sigma, t_{in}) > \#(\sigma, t_{out})$ for any $\sigma \in L(B(N, t_{in}, t_{out}, k))$.

Condition (1) means that t_{in} can never 'get blocked'. Conditions (2) and (3) say that t_{in} can 'get ahead' of t_{out} in firing, and t_{out} can always 'catch up'.

Consider the example below. It is easy to verify that transition t_0 in Fig. 3.17(a) is 1-, 2- and 3-enabled but not k-enabled for $k \geq 4$. The Petri net (excluding the idle place) in Fig. 3.17(b) is 1-, 2- and 3-well-behaved *PN* with respect to t_{in} and t_{out}. Hence, transition t_0 can be replaced by the Petri

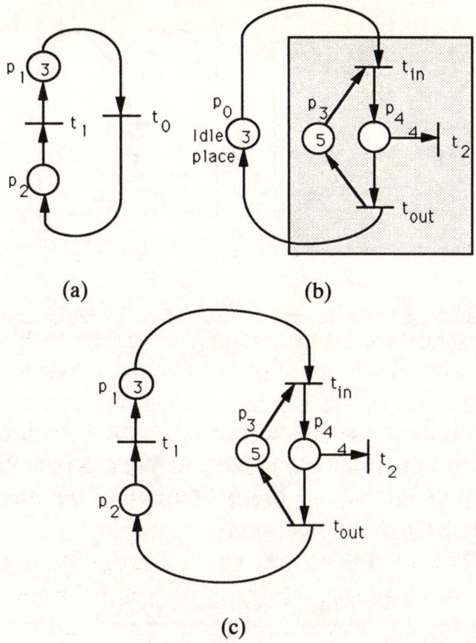

Figure 3.17 Example of refinement: (a) Petri net; (b) 3-well-behaved *PN* and its idle place; (c) combined Petri net (Suzuki and Murata, 1983).

net in Fig. 3.17(b), which results in Fig. 3.17(c). Note that the numbers in the places stand for token numbers, and the numbers beside the arcs denote numbers of arcs (multiple arcs).

Although the results of Valette, Suzuki and Murata are very useful for constructing a net with important properties, applications of their top-down methods are limited to the systems that can be described as independent modules, i.e. there are only shared resources inside blocks. The reason for this is that a block has only two entries.

3.3.2 Refinement of places

Suzuki and Murata (1983) also present a refinement method for places, by which the place to be refined, p_0, is first replaced by two places, p_{01} and p_{02}, and by a transition t_0 such that the following conditions are satisfied:

1. the input (output) transitions of p_{01} (p_{02}) are the input (output) transitions of p_0;
2. p_{01} (p_{02}) is the only input (output) place of t_0;
3. the initial marking of p_{01} is equal to the initial marking of p_0;
4. the initial marking of p_{02} is zero (see Fig. 3.18).

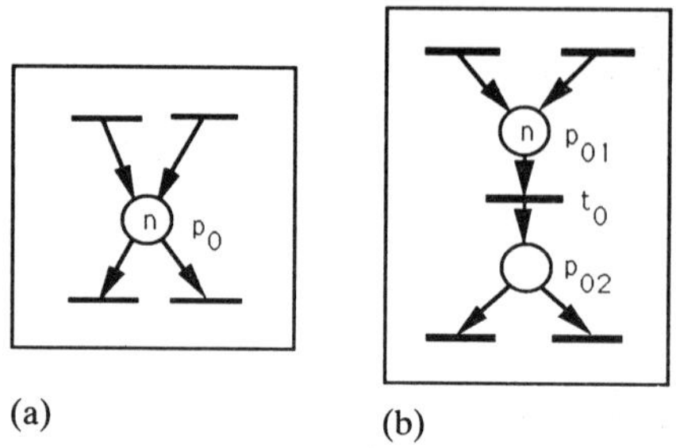

Figure 3.18 First step of the refinement of a place (Suzuki and Murata, 1983): (a) a Petri net; (b) p_0 is replaced by p_{01} and p_{02}.

After these conditions are satisfied, the refinement technique for transitions can be applied to t_0. Suzuki and Murata have proven that the method for refining places also preserves the properties of boundedness, safeness and liveness after each refinement step.

3.3.3 Model synthesis example

Example 3

We use a simple automated machining and assembly system to demonstrate the application of these methods. This system only makes one type of product which needs two machining operations and one assembly operation. As shown in Fig. 3.19, the system has one assembly station A_1, one robot R_1, and two conveyors for transferring raw stock type 1 and type 2 to two machines, machine M_1 and machine M_2, respectively. Similarly to Example 1, we have the following two assumptions for the system:

1. the supply of raw stock 1 and 2 in storage S_1 and storage S_2 is never exhausted;
2. the finished product will be taken away so that there is no overflow problem with storage S_3.

The production procedure is depicted as follows:

1. the conveyor C_1 (C_2) transfers raw stock type 1 (type 2) to M_1 (M_2);
2. M_1 (M_2) starts machining;
3. after M_1 and M_2 finish their operations, the robot R_1 takes part 1 from M_1 to A_1, and then takes part 2 from M_2 to A_1;
4. R_1 begins the assembly on A_1 using part 1 and part 2;
5. after R_1 finishes the assembly on A_1, it moves the product to storage S_3.

Because we refine transitions, the activities in the system are defined initially as transitions. At the first step of the modeling, an abstract model of the system is specified as described in Fig. 3.20(a) which states that the pro-

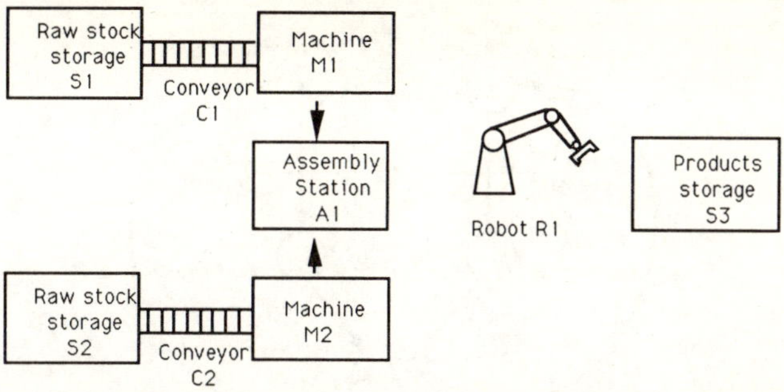

Figure 3.19 A simple automated machining and assembly system.

duction procedure needs an M_1 machine operation and an M_2 machine operation, followed by unloading of parts and an A_1 assembly operation. It is easy to verify that this model is live and bounded (safe).

Then we substitute, respectively, the well-formed blocks that denote M_1 operation, M_2 operation, unloading of parts, and A_1 operation in Figs 3.20(b), (c), (d) and (e) for the transitions t_{11}, t_{12}, t_{13} and t_{14} in Fig. 3.20(a). For this simple system, one level of refinement is sufficient.

The final model, which is live and bounded (safe), as shown in Fig. 3.20(f), is obtained after these refinements. The shaded areas in the figure are the well-formed blocks of Figs 3.20(b), (c), (d) and (e).

It is not easy to apply Valette, Suzuki and Murata's top-down methods to model the system in Example 1. The problem is that we have difficulties in specifying the interactions with shared resources by using refinement. Nevertheless, in situations in which shared resources exist within well-formed blocks, their methods are still very useful.

3.3.4 Summary and limitations of top-down methods

The motivation for top-down synthesis methods is that by using some principles (modularization, information hiding, etc.), we restrict the synthesis methods so that certain 'good' behaviors can be guaranteed in the final synthesized model. These good behaviors may be qualitative properties such as liveness or boundedness, or just characteristics such as structured design and modularity. This kind of methodology has been used widely in sequential environments such as structured programming techniques for simplifying costly debugging, maintenance and verification. However, if we want to preserve properties, the same principles are not easily applied to the concurrent environments with highly shared resources, because of the complexity of interactions within these environments. The reason is simply that interactions among the subsystems are coupled throughout all steps of synthesis, which makes it more difficult to specify the system using the top-down approach.

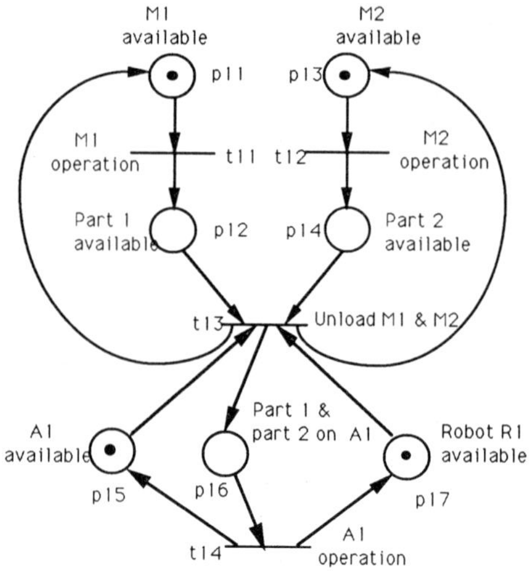

Figure 3.20 (a) First level of the system in Example 3.

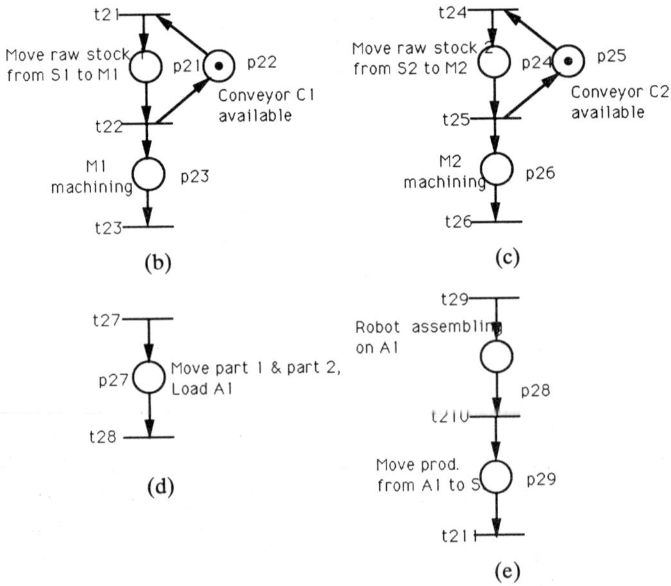

Figure 3.20 (b) Well-formed block for t_{11}; (c) well-formed block for t_{12}; (d) well-formed block for t_{13}; (e) well-formed block for t_{14}.

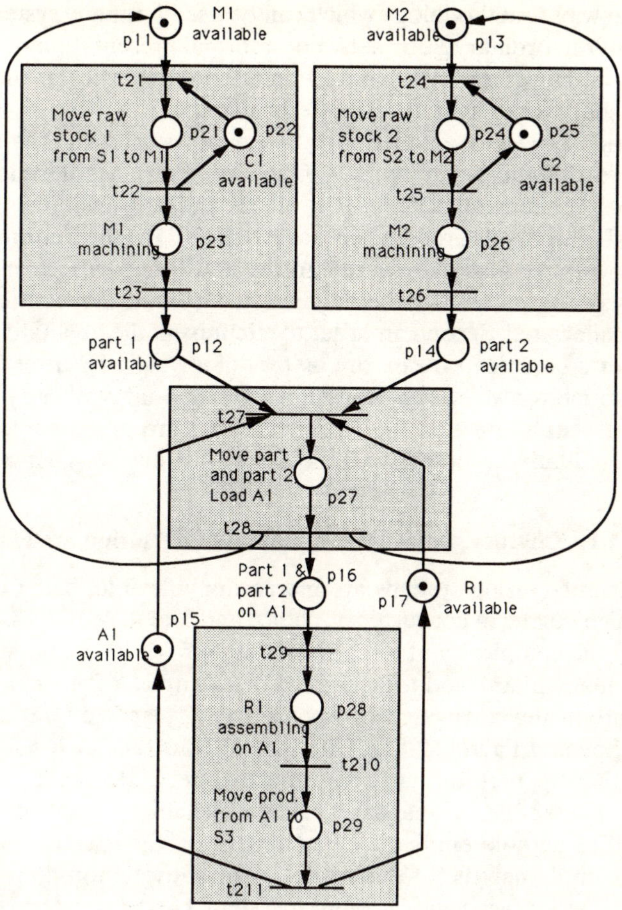

Figure 3.20 (f) Final model after refinements of transitions.

On the other hand, if the system can be designed in such a way that shared resources are separated and embodied within modules, then the top-down methods that we have mentioned still provide some solutions. Actually, their concept of well-formed blocks (or well-behaved *PN*) decouples further interactions between subsystems inside the block and subsystems outside the block. Future extensions of their methods might consider multiple input and output entries for more design flexibility.

3.4 HYBRID SYNTHESIS: COMBINING TOP-DOWN AND BOTTOM-UP

An outgrowth of the research in bottom-up and top-down methods for synthesis is work in the area of hybrid synthesis. This section presents in an informal way a methodology which can be used for many systems whose models result in ordinary Petri nets. For a formal presentation of this work see Zhou and DiCesare (1991) and Zhou (1990). In particular this method begins to address the shared resource problem.

This design process is divided into two major stages: (a) the top-down stage where designers start with a first-level Petri net description and use stepwise refinement of operation places and/or transitions to include more detail until the desired level is achieved; and (b) the bottom-up stage where the resource places are added to the net. If needed, (a) and (b) can be exercised alternately. In this manner the complexity of the detailed problem is alleviated. Second, in order to circumvent the qualitative analysis problem for a complex system, the methodology includes a set of mutual exclusion structures which can be used in the proposed synthesis procedure. Both the net structure and the initial marking are designed so that the desirable qualitative properties of the final Petri net are guaranteed.

3.4.1 Characteristics of complex manufacturing systems

Modern manufacturing systems are increasingly flexible, being characterized by a high degree of concurrency, choice and shared resources. Problems arise when the complexity of a real-world system leads to a large Petri net which has many places and transitions. For example, a Petri net modeling a factory producing corrugated fiberboard has 92 places, 59 transitions and 174 arcs (Lee and Favrel, 1985). One way to construct such a large net is using bottom-up methods and merging subnets. However, it may be practically impossible to analyze it using reachability graph or invariant methods. The authors emphatically comment on the extreme complexity of the structural analysis for this model. Alternatively, top-down methods are powerful when faced with a complex system. However, when confronted with shared resources introduced at the detail level of synthesis, the analysis problem again becomes practically insurmountable.

3.4.2 A hybrid synthesis approach for manufacturing systems

As discussed earlier a manufacturing system can be specified as a set of activities (operations) and resources and process plans that describe and relate the two. In the hybrid methodology there is a need to divide the resources into two types: resources whose number is fixed at design time, such as robots, machines and conveyors; and resources whose number can vary at design time – for example, pallets, fixtures and parts or jobs to be processed. The second class of resources plays a special role in the resource-sharing environment. Their number needs to be determined such that the system will be neither deadlocked nor starved. This is done by finding the appropriate number of initial tokens in these resource places in the net.

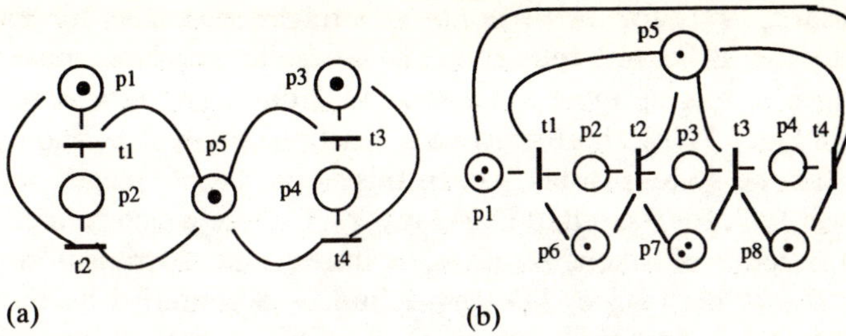

Figure 3.21 Two mutual exclusion examples: (a) a parallel mutual exclusion, $m_0 = (1, 0, 1, 0, 1)^T$; (b) a sequential mutual exclusion, $m_0 = (m_0(p_1), 0, 0, 0, 1, 1, 2, 1)^T$.

In this methodology a token in a resource place means that the corresponding resource is available. When a buffer with capacity greater than one is involved, bounded places can be used to model it. The places are grouped into three classes according to their function in the net and the nature of their initial marking. Activity (operation) places are defined as *A* places and have no tokens in the initial marking. It is a realistic assumption that no activity is occurring at system start-up. The two classes of resources are defined as *B* places for fixed resources and *C* places for variable resources. The initial marking for *B* places is fixed according to system design and the initial marking for *C* places is to be determined in order to guarantee boundedness, liveness and reversibility. The sets of *A*, *B* and *C* places are referred to as P_A, P_B and P_C, respectively.

Two mutual exclusion structures are used in this approach. These are defined as parallel mutual exclusions and sequential mutual exclusions and examples of these structures are given in Fig. 3.20. A parallel mutual exclusion (PME) models a resource shared by different processes in parallel. Each parallel process is bounded by one or more start transitions and stop transitions and in general there could be several activities between start and stop. Figure 3.21(a) is an example where p_5 represents the shared resource and transition pairs (t_1, t_2) and (t_3, t_4) define two parallel processes. The case where there exists one resource and one parallel process is a trivial but useful PME and models a non-shared resource. The theory for parallel mutual exclusions (Zhou and DiCesare, 1991; Zhou, 1990) guarantees the same liveness, boundedness and reversibility for the Petri net with the shared resource as existed for the net without the shared resource place added. In the example given in Fig. 3.21(a) the net with p_5 has the same liveness, boundedness and reversibiity properties as the net without p_5.

A sequential mutual exclusion (SME) models a resource shared by sequentially related groups of processes. Figure 3.21(b) is an example of an SME where p_5 represents the shared resource and transition pairs (t_1, t_2) and (t_3, t_4) define two sequentially related processes. The theory (Zhou and DiCesare, 1991) for SMEs is not as straightforward as for PME. In addition to the structural requirement, the initial marking must be restricted in order to guarantee liveness, boundedness and reversibility. For example, in Fig. 3.21(b), there must be at least one token in p_1 and no more than two for the system to be live. In this net p_1 is a C place.

The lower limit for the initial marking for a C place is determined by the minimum number of tokens required to insure that all transitions in the SME will fire at least once. The upper limit is determined by the **token capacity**. The token capacity concept is essentially identical to the **deviation bound** (Silva, 1987).

The token capacity between two transitions t and t' in a marked Petri net is the maximum number of times transition t can fire from the initial marking without firing t'. The token capacity of two consecutive transition pairs in an SME is the maximum number of times the second transition of the first pair can fire before the first transition of the second pair must fire. In the example in Fig. 3.21(b) the token capacity between t_2 and t_3 is two. Sequential mutual exclusion theory (Zhou and DiCesare, 1991; Zhou, 1990) guarantees that for an SME the net with the shared resource will have the same properties of liveness, boundedness and reversibility as the net without the shared resource provided the initial marking of C places is within the bounds described in the preceding paragraph.

3.4.3 Synthesis procedure

Step 1 Start with a bounded, live and reversible Petri net as a first-level model of a system that is working when all major resources are available, and determine P_A, P_B and P_C where the memberships of P_A and P_B are subject to change while P_C is fixed and its members have a variable number of initial tokens.

Step 2 Using top-down methods, decompose the activity sequence into the desired detail. Note that P_A expanded at each refinement step.

Step 3 Add the non-shared (dedicated) resource places at each activity where required. The arcs are linked to the resource places so that the conditions for a simple PME are satisfied. At each addition only P_B is augmented.

Step 4 Add the buffer-B places (free buffer slots) complementing the existing buffer-A places (occupied buffer slots), form Petri net buffer modules if any buffers exist. Each added buffer place will have initial tokens whose number equals the capacity of the buffer.

Step 5 Add the shared resource places which form PMEs. The resource places are B-places.

Step 6 Add the shared resource places which form an SME and calculate the token capacities between their PMEs and the minimum marking of the P_C places to guarantee liveness.

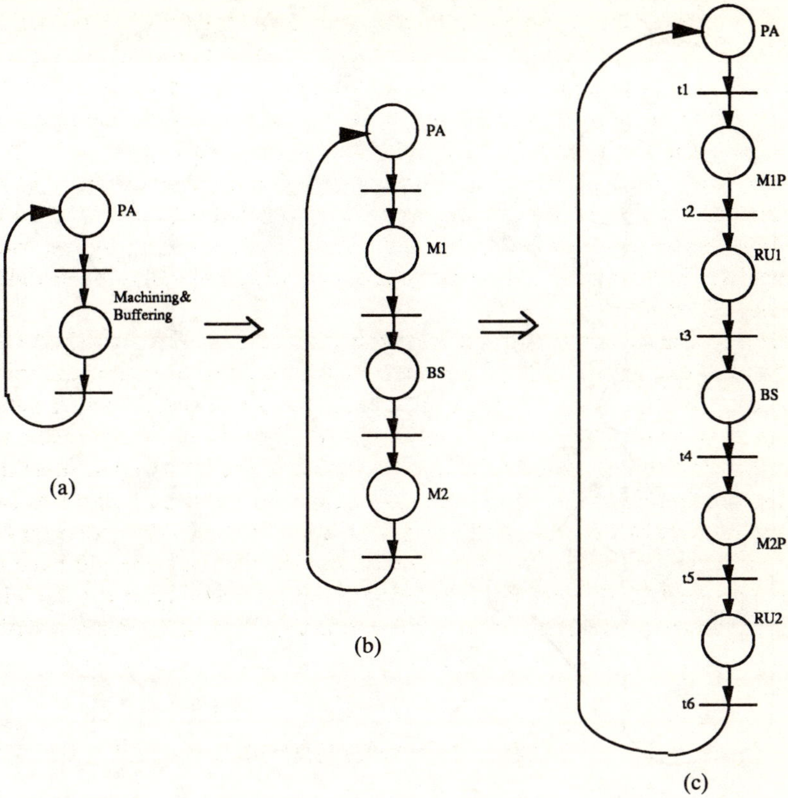

Figure 3.22 Illustration of top-down part of synthesis: (a) first-level model; (b) and (c) successive stepwise refinement.

Following these steps, a bounded, live and reversible Petri net model is synthesized.

The refinement of activity places and the addition of resource places can be alternately applied. This means that no strict order between them needs to be followed.

3.4.4 Model synthesis example

Take the synthesis of the system shown in Fig. 3.1 as an example to illustrate this methodology.

Step 1 Figure 3.22(a) shows the first-level Petri net model of the system where all activities are represented by one place, an *A* place, labeled Machining and Buffering, and the availability of pallets is represented by a *C* resource place. With appropriate initial marking this net is live, bounded and reversible.

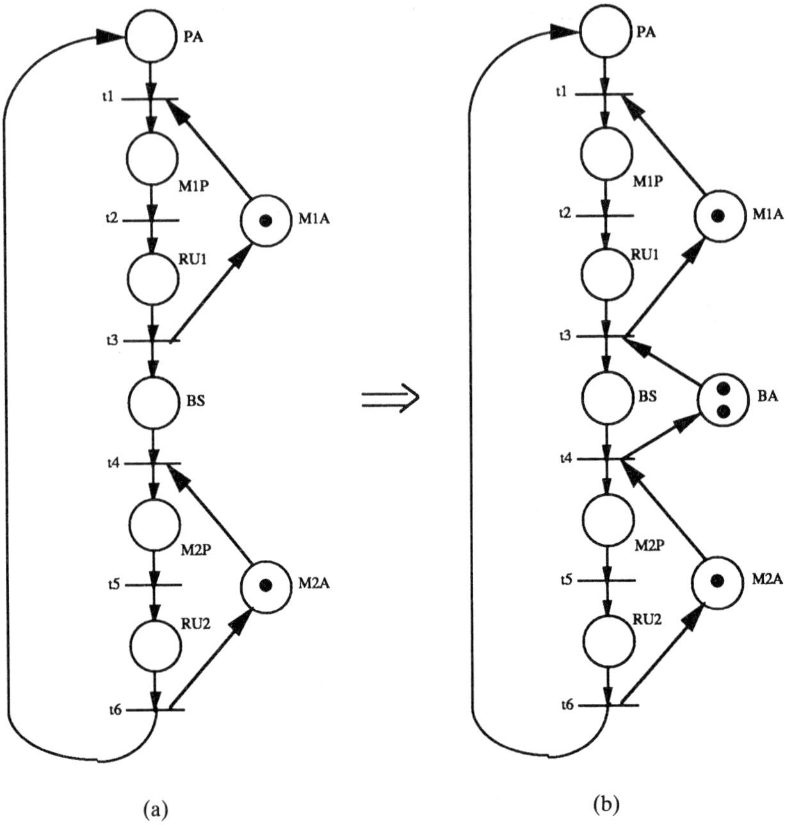

Figure 3.23 Addition of non-shared resources.

Step 2 Figure 3.22(b) shows the first stage of decomposition where the single activity, representing machining and buffering, is refined into three activities, processing on machine 1 represented by place M_1, buffering represented by place BS, and processing on machine 2 represented by place M_2 each with appropriate start and stop transitions.

Figure 3.22(c) shows the next level of activity refinement where machining activities are refined to include robotic unloading as represented by places RU_1 and RU_2. This is the desired level of detail for modeling this system.

Step 3 Figure 3.23(a) shows the addition of the non-shared resources. Machine-available places M_1A and M_2A are added. Note that each machine is released after the machine is unloaded by the robot. Each of these resource places will be initially marked by 1 token to indicate that the resource is available.

Step 4 The buffer space available place, *BA* is added as shown in Fig. 3.23(b). The initial marking of this place is 2 tokens to indicate the buffer capacity of 2.

Step 5 There are no shared resource places which form PMEs in this system. So none is added.

Step 6 The shared resource place, *RA*, which forms an SME is added as shown in Fig. 3.4(b). The SME is defined by the shared resource place *RA* and the transitions pairs (t_2, t_3) and (t_5, t_6). The token capacity between t_3 and t_5 is 3. This is the upper limit for the initial marking for the *C* place *PA*. The minimum marking for *PA* is 1. Within these initial marking bounds for *PA*, the net will be live.

The net synthesized by the above procedure is guaranteed to be live, bounded and reversible. As mentioned earlier, the reader should refer to Zhou and DiCesare (1991) and Zhou (1990) for a formal presentation of this hybrid synthesis methodology.

3.4.5 A complex shop-floor example

The following example is introduced to show that properly applied hybrid synthesis methods can be used to model moderately complex manufacturing systems while preserving desirable properties. For more detail on the synthesis of this model see Zhou *et al.* (1990). The flexible manufacturing system (FMS) developed at Rensselaer Polytechnic Institute and shown in Fig. 3.24 is in fact an FMS physical simulator which is a 1/6 scale model of an automated factory shop floor and can manufacture a family of products. It originally was developed to illustrate the benefits of constructing a physical simulation of a manufacturing system before committing to a full-scale production facility. It is also used for research on the design and implementation of control for automated manufacturing systems using PNs.

The system takes raw input stock of two types, machines them into desired shapes, and then assembles these two finished parts into a product. One part starts as a rectangular solid block and is milled and drilled on the CNC milling machine and ends as a geometrically shaped part with a cylindrical hole in its top. The other part starts as a cylindrical solid peg and is turned by the CNC lathe into a complex cylindrical/conical-shaped peg whose base fits into the hole of the finished block part. The robotic insertion of the peg into the hole of the block is the assembly process. The family of products comprises various sizes and shapes of blocks and pegs assembled in this manner.

3.4.6 Layout of the FMS

The major components of the system as illustrated in Fig. 3.24 are:

1. The CNC mill and drill machine.
2. The CNC lathe.

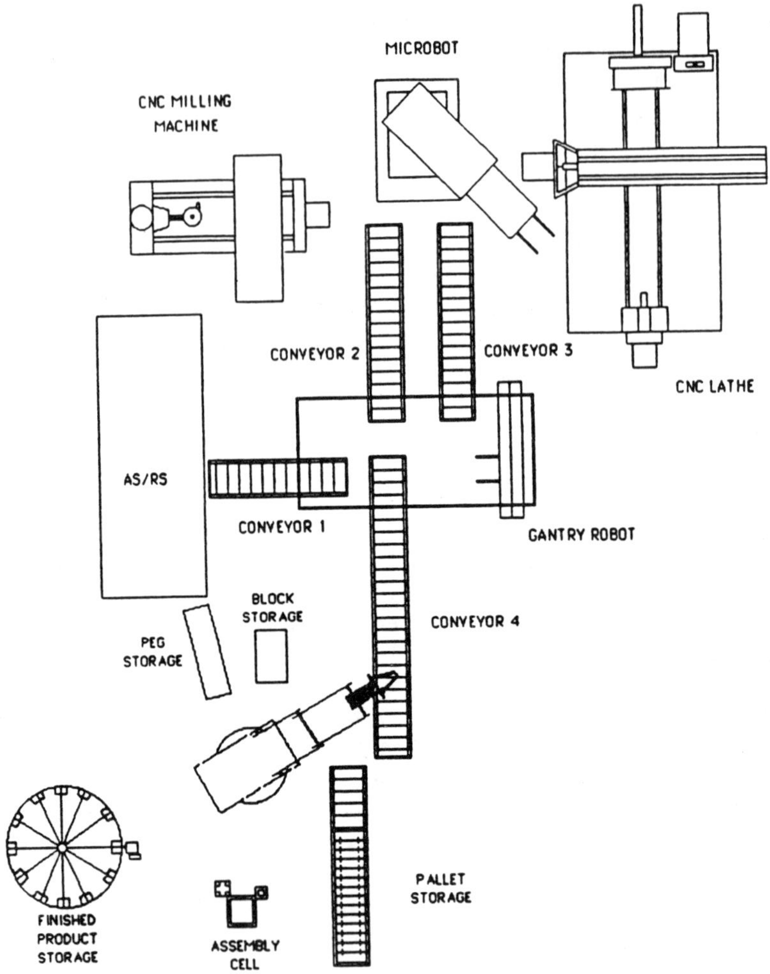

Figure 3.24 The layout of the FMS simulator.

3. The Microbot (M) robot is a shared resource used to load and unload the materials between the CNC lathe and Conveyor (C) 3, and between the CNC milling machine and C_2.
4. The Automated Material Transfer System (AMTS) is a set of shared resources which include four two-way conveyors (one pallet at a time) with presence sensors at each end and a Gantry (G) robot. The AMTS is in charge of the transfer of the materials between the different stations. The pallets used are universal, i.e. they can hold all materials, parts and products.

   G robot: transfers pallets, with either raw stock or a finished part, between the four conveyors.

C_1: transfers pallets with either raw stock or a finished part between the ASRS and the G robot.

C_2: transfers pallets with either raw stock or a finished part between the CNC milling machine and the G robot.

C_3: transfers pallets with either raw stock or a finished part between the CNC lathe and the G robot.

C_4: transfers pallets with either raw stock from the raw material storage area to the G robot or finished parts from the G robot to the assembly area.

5. The Automated Storage and Retrieval System (ASRS) is a buffer where pallets with raw materials and intermediate parts can be stored. It is linked to the other components of the system via C_1 of the AMTS. The ASRS structure is composed of 5×4 storage locations including 19 usable pallet-storage bins. An elevator carries loaded pallets to and from the ASRS.
6. The Scorbot (S) robot is a shared resource and has several functions: move pallets between the empty pallet storage and C_4 in the AMTS; move pegs from the peg storage and blocks from the block storage onto pallets on C_4; move blocks and pegs from pallets on C_4 to the assembly cell and assemble them; and store the final product in the carousel (finished product storage).

The final Petri net for this system is shown in Fig. 3.25 and Table 3.2.

The place representing the number of pallets is p_1. This net is guaranteed to be live, bounded and reversible with $2 \leq m_0(p_1) \leq 20$, which means we have between 2 and 20 pallets in the system at any time.

3.4.7 Summary and limitations of the hybrid method

This section briefly introducing a hybrid method illustrates the opportunity and possibility for development of synthesis methods that combine top-down and bottom-up methods while preserving the desired properties throughout the design process. On the other hand this method has limitations. Both parallel and sequential mutual exclusion deal with a single resource shared by different processes. When there are many identical resources in a single place, the problem may become a resource allocation problem as discussed in the literature. However, when there are many distinct resources, whose numbers are more than one and that are shared by different processes, the synthesis of a Petri net with desirable properties becomes a more challenging problem. The theory cannot be easily extended to such a general case.

The state explosion problem still exists when a system is specified by a Petri net. Although top-down modular design alleviates the complexity to some extent, the bottom-up design stage is still needed in order to deal with the interactions among processes in a system. The placement of arcs

Table 3.2 Place labels for the example system

p_1: Pallet available
p_2: *S* robot available
p_3: C_4 available
p_4: Move a pallet from the storage to C_4
p_5: Empty pallet on C_4
p_6: Raw pegs available
p_7: Raw blocks available
p_8: Move a raw peg from storage to pallet on C_4
p_9: Get a peg second
p_{10}: Get a block first
p_{11}: Move a raw block from the storage to the pallet on C_4
p_{12}: Move a peg-carrying pallet down C_4
p_{13}: Move a block-carrying pallet down C_4
p_{14}: A peg-carrying pallet on C_4 available
p_{15}: A block-carrying pallet on C_4 available
p_{16}: *G* robot available
p_{17}: C_3 available
p_{18}: C_1 available
p_{19}: C_2 available
p_{20}: Move a raw-peg-carrying pallet from C_4 to C_3
p_{21}: Move a raw-peg-carrying pallet from C_4 to C_1
p_{22}: Move a raw-block-carrying pallet from C_4 to C_1
p_{23}: Move a raw-block-carrying pallet from C_4 to C_2
p_{24}: A raw-peg-carrying pallet on C_3
p_{25}: Move a raw-peg-carrying pallet up C_1 and elevator into the ASRS
p_{26}: Move a raw-block-carrying pallet up C_1 and elevator into the ASRS
p_{27}: A raw-block-carrying pallet on C_2
p_{28}: The ASRS has empty slots
p_{29}: Store a raw-peg-carrying pallet in ASRS
p_{30}: Store a raw-block-carrying pallet in ASRS
p_{31}: *M* robot available (note p_{31} appears twice on Fig. 3.25)
p_{32}: Move a raw-peg-carrying pallet up C_3 to Lathe
p_{33}: Raw-peg-carrying pallets available
p_{34}: Raw-block-carrying pallets available
p_{35}: Move a raw-peg-carrying pallet up C_2 to Mill
p_{36}: Load Lathe with raw peg
p_{37}: Lathe available
p_{38}: Retrieve a raw peg from ASRS
p_{39}: The ASRS elevator available
p_{40}: Retrieve a raw peg from ASRS
p_{41}: Mill available
p_{42}: Load the Mill with a raw block

Table 3.2 (cont.)

p_{43}: Machining a raw peg
p_{44}: Move a raw-peg-carrying pallet down C_1
p_{45}: Move a raw-block-carrying pallet down C_1
p_{46}: Machine raw block
p_{47}: Unload finished peg to pallet on C_3
p_{48}: Move raw-peg-carrying pallet from C_1 to C_3
p_{49}: Move raw-block-carrying pallet from C_1 to C_2
p_{50}: Unload finished block to pallet on C_2
p_{51}: Move finished-peg-carrying pallet down C_3
p_{52}: Move finished-peg-carrying pallet down C_2
p_{53}: Move finished-peg-carrying pallet from C_3 to C_4
p_{54}: Move raw-peg-carrying pallet from C_3 to C_1
p_{55}: Move raw-block-carrying pallet from C_2 to C_1
p_{56}: Move finished-block-carrying pallet from C_2 to C_4
p_{57}: Finished-peg-carrying pallet on C_4 available
p_{58}: Move finished-peg-carrying pallet up C_1 and the elevator into the ASRS
p_{59}: Move finished-block-carrying pallet up C_1 and the elevator into the ASRS
p_{60}: Assembly station empty
p_{61}: A finished-block-carrying pallet on C_4 available
p_{62}: Move a finished-peg-carrying pallet up C_4
p_{63}: Store finished-peg-carrying pallet in ASRS
p_{64}: Store finished-block-carrying pallet in ASRS
p_{65}: Move finished-block-carrying pallet up C_4
p_{66}: Move finished peg into the Assembly Station
p_{67}: Finished-peg-carrying pallets in ASRS available
p_{68}: Finished-block-carrying pallets in ASRS available
p_{69}: Place finished block in Assembly Station
p_{70}: Assemble peg-block product
p_{71}: Retrieve finished peg from ASRS
p_{72}: Retrieve finished block from ASRS
p_{73}: Put pallet from finished-block-carrying pallet in pallet storage
p_{74}: Move assembled product to output storage carousel
p_{75}: Move finished-peg-carrying pallet down C_1
p_{76}: Move finished-block-carrying pallet down C_1
p_{77}: Finished block in the Assembly Station
p_{78}: Put pallet from a finished-peg-carrying pallet in the pallet storage
p_{79}: Move finished-peg-carrying pallet from C_1 to C_3
p_{80}: Move finished-block-carrying pallet from C_1 to C_2

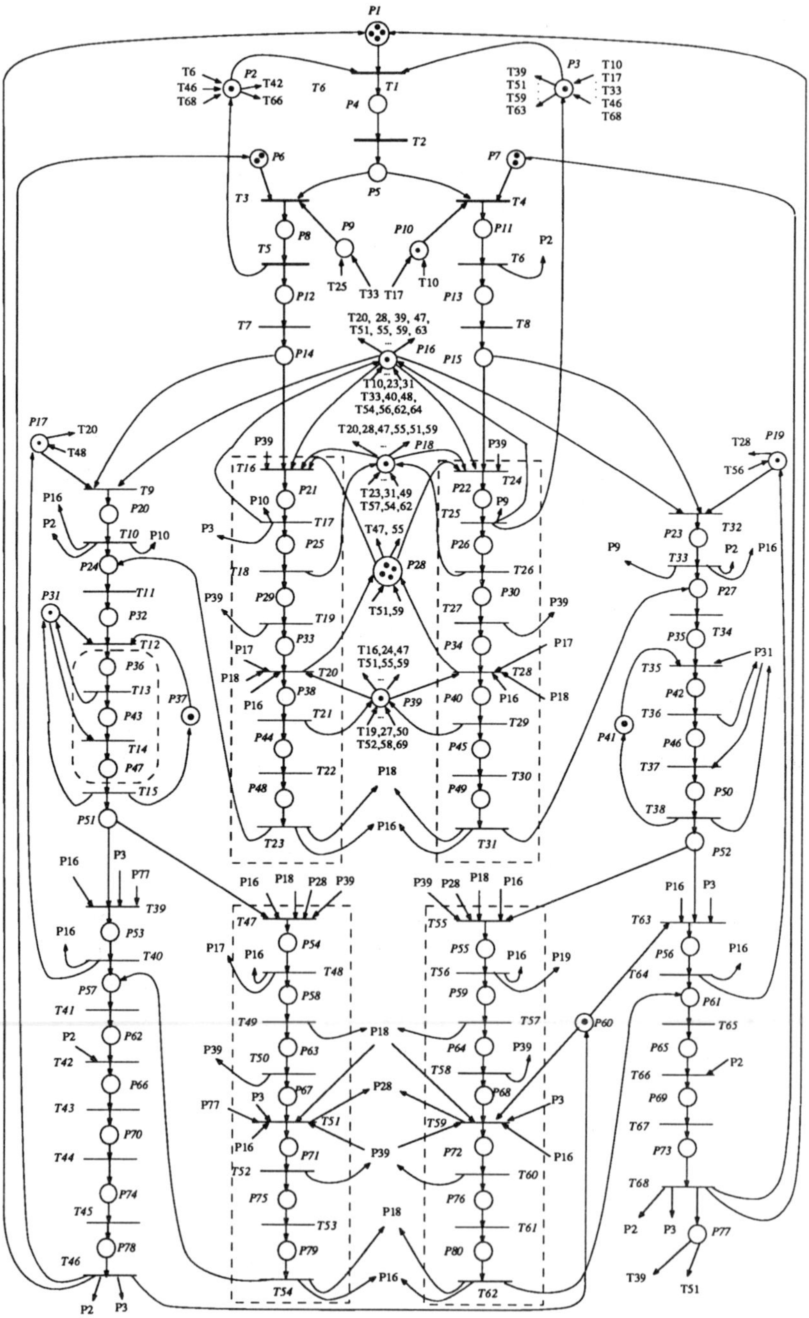

Figure 3.25 Petri net for the FMS simulator.

to form a PME or an SME may be a very difficult task in a complex system. In the bottom-up stage, a large amount of validation work may be needed when this approach is used to model a system with complicated interactions.

The synthesis approach developed here is limited to ordinary Petri nets even though it may be extended. Furthermore, this method is far from a complete solution for every kind of Petri net or system. When multiple or multi-layer distinct shared resources are considered, the synthesis becomes more difficult.

3.5 SUMMARY

This chapter has reviewed synthesis techniques for Petri nets and provided some manufacturing examples. The synthesis methods are divided into three categories: bottom-up, top-down and hybrid. Top-down techniques have the advantage of viewing the system globally, which may generate more structured designs. For systems without highly shared resources, methods exist to preserve important properties of the net after the synthesis steps by means of restricting the synthesis rules such that concurrent interactions among the subnets are properly controlled. However, it is not easy to find rules and methodologies that are appropriate for some applications. In other words, developing techniques that satisfy the requirement of both the guaranteeing net's properties and the flexibility of specifying systems (e.g. systems with highly shared resources) is very difficult. On the other hand, bottom-up techniques have the advantage of convenience for specifying systems (especially on describing the concurrent interactions such as resource sharing), but in most cases these methods do not guarantee that the synthesized net preserves important properties.

A current and future direction is the study of hybrid synthesis, which combines the advantages of bottom-up and top-down methods. The shared resource problem is a focus in this research and a preliminary methodology has been presented to show the value of this approach.

BIBLIOGRAPHY

Agerwala, T. A complete model for representing the coordination of asynchronous processes. *Hopkins Computer Research Report*, no. 32, Computer Science Program, Johns Hopkins University, Baltimore, MD, July 1974.

Agerwala, T. Putting Petri nets to work. *IEEE Computer*, December 1979, pp. 85–94.

Agerwala, T. and Choed-Amphai, Y. A synthesis rule for concurrent systems. *Proceedings of the 15th Design Automation Conference*, Las Vegas, pp. 305–311, June 1978.

Al-Jaar, R. Y. and Desrochers, A. Petri nets in automation and manufacturing. To appear in *Advances in Automation and Robotics*, G. N. Saridis (ed.), vol. 2, JAI Press, 1991.

Andre, C. Boeri, F. and Marin, J. Synthèse et realisation des systèmes logiques à evolution simultanées. *PAIRO*, vol. 10, pp. 67–86, April 1976.

Ayache, J. M., Courtiat, J. P. and Diaz, M. REBUS, a fault-tolerant distribution system for industrial real-time control. *IEEE Transactions on Computers*, vol. C-31, no. 7, pp. 637–674, 1982.

Baer, J. L. and Ellis, C. S. Model, design and evaluation of a compiler for a parallel processing environment. *IEEE Transactions on Software Engineering*, vol. SE-3, no. 6, pp. 394–405, 1977.

Beck, C. L. Modeling and simulation of flexible control structures for automated manufacturing systems. M.S. thesis and Robotics Institute Technical Report, Carnegie-Mellon University, Pittsburgh, Pennsylvania, 1985.

Berthelot, G. Checking properties of nets using transformations. *Advances in Petri Nets 1985* (*Lecture Notes in Computer Science 222*), Springer-Verlag, pp. 19–40, 1985.

Berthelot, G. Transformations and decompositions of nets. *Advances in Petri Nets 1986* (*Lecture Notes in Computer Science 254*), Springer-Verlag, pp. 359–376, 1986.

Berthelot, G. and Terrat, R. Petri nets theory for the correctness of protocols. *IEEE Transactions on Communication*, vol. COM-30, no. 12, pp. 2497–2509, 1982.

Bruno, J. and Altman, S. M. A theory of asynchronous control networks. *IEEE Transactions on Computers*, vol. C-20, no. 6, June 1971.

Commoner, F., Holt, A. W., Even, S. and Pnueli, A. Marked directed graphs. *Journal of Computer and Systems Science*, vol. 15, pp. 511–523, 1971.

Crockett, D., Desrochers, A., DiCesare, F. and Ward, T. Implementation of a Petri net controller for a machining workstation. *Proceedings IEEE Conference Robotics Automation*, NC, April 1987.

Diaz, M. Modeling and analysis of communication and cooperation protocols using Petri net based models. *Comput. Net.*, vol. 6, 1982.

Dong, S. T. The modelling, analysis and synthesis of communication protocols. Ph.D. thesis, University of California, Berkeley, 1983.

Genrich, H. J. and Lautenbach, K. System modelling with high-level Petri nets. *Theoret. Comput. Sci.*, vol. 13, pp. 109–136, 1981.

Jensen, K. Colored Petri nets and the invariant method. *Theoret. Comput. Sci.*, vol. 14, pp. 317–336, 1981.

Jensen, K. Coloured Petri nets. *Advances in Petri Nets 1986*, vol. 254, Springer-Verlag, pp. 248–299, 1986.

Johnsonbaugh, R. and Murata, T. Additional methods for reduction and expansion of marked graphs. *IEEE Transactions on Circuits and Systems*, vol. CAS-28, no. 10, October 1981.

Kasturia, E., DiCesare, F. and Desrochers, A. Real time control of multilevel manufacturing systems using colored Petri nets. *Proceedings IEEE Conference on Robotics and Automation*, PA, April 1988.

Kluge, W. E. and Lautenbach, K. The orderly resolution of memory access conflicts among competing channel processes. *IEEE Transactions on Computers*, vol. C-31, no. 3, pp. 194–207, 1982.

Koh, I. and DiCesare, F. Transformation methods for generalized Petri nets and their applications in flexible manufacturing systems. *Proceedings 2nd International Conference CIM*, Rensselaer Polytechnic Institute, Troy, NY, May 1990.

Kowalk, W. and Valk, R. On reduction of parallel programs. *Lecture Notes in Computer Science*, no. 71, Springer-Verlag, pp. 356–369, 1979.

Krogh, B. H. and Beck, C. L. Synthesis of place/transition nets for simulation and control of manufacturing systems. *Proceedings IFIP Symposium Large Scale Systems*, Zurich, August 1986.

Kwong, Y. S. On reduction of asynchronous systems. *Theoret. Comput. Sci.*, vol. 5, pp. 25–50, 1977.

Lee, K. H. and Favrel, J. Hierarchical reduction and decomposition of graphs for system analysis. *Proceedings IEEE Conference on Systems, Man, and Cybernetics*, October 1984.

Lee, K. H. and Favrel, J. Hierarchical reduction method for analysis and decomposition of Petri nets. *IEEE Transactions Systems, Man, and Cybernetics*, vol. SMC-15, no. 2, pp. 272–280, March/April 1985.

Lee, K. H. Favrel, J. and Baptiste, P. Generalized Petri net reduction method. *IEEE Transactions Systems, Man, and Cybernetics*, vol. SMC-17, no. 2, pp. 297–303, March/April 1987.

Lipton, R. J. Reduction: a method of proving properties of parallel programs. *J. Ass. Comput. Mach.*, no. 3, pp. 561–567, 1981.

Martinez, J., Alla, H. and Silva, M. Petri nets for the specification of FMSs. *Modelling and Design of Flexible Manufacturing Systems.* Elsevier, Amsterdam, 1986.

Martinez, J., Muro, P. and Silva, M. Modeling, validation and software implementation of production systems using high level Petri nets. *Proceedings International Conference IEEE Robotics and Automation*, Raleigh, NC, 1987.

Martinez, J., Muro, P. R., Silva, M., Smith, S. F. and Villarroel, J. L. Merging artificial intelligence techniques and Petri nets for real time scheduling and control of production systems. *Technical Report GISI-1/88*, University of Zaragoza, Spain, January 1988.

Mayr, E. W. and Meyer, A. R. The complexity of the finite containment problem for Petri nets. *J. Ass. Comput. Mach.*, no. 3, pp. 561–567, 1981.

Merlin, P. M. Specification and validation of protocols. *IEEE Transactions on Communication*, vol. COM-27, no. 11, pp. 1671–1680, 1979.

Miller, R. E. A comparison of some theoretical models of parallel computations. *IEEE Transactions on Computers*, vol. C-22, no. 8, pp. 710–717, August 1973.

Molloy, M. K. Performance analysis using stochastic Petri nets. *IEEE Transactions on Computers*, vol. C-31, no. 9, pp. 913–917, September 1982.

Murata, T. and Koh, J. Y. Reduction and expansion of live and safe marked graphs. *IEEE Transactions on Circuits and Systems*, vol. CAS-27, no. 1, January 1980.

Murata, T. Synthesis of decision-free concurrent systems for prescribed resources and performance. *IEEE Transactions on Software Engineering*, vol. SE-6, no. 6, November 1980.

Murata, T. Petri nets: properties, analysis and applications. *Proceedings IEEE*, vol. 77, no. 4, April 1989.

Narahari, Y. and Viswanadham, N. A Petri net approach to the modelling and analysis of flexible manufacturing systems. *Annals of Operations Research*, vol. 3, pp. 449–472, 1985.

Peterson, J. L. Petri nets. *Computing Surveys*, vol. 9, no. 3, pp. 223–252, September 1977.

Peterson, J. L. *Petri Net Theory and the Modeling of Systems*, Prentice-Hall, Englewood Cliffs, NJ, 1981.

Ramamoorthy, C. V. and Ho, G. S. Performance evaluation of asynchronous concurrent systems using Petri nets. *IEEE Transactions on Software Engineering*, vol. SE-6, no. 5, pp. 440–449, 1980.

Reisig, W. Petri nets with individual tokens. *Theoret. Comput. Sci.*, vol. 41, pp. 185–213, 1985.

Silva, M. Sur le concept de macro place et son utilisation pour l'analyse des reseaux de Petri. *PAIRO Automatique*, vol. 15, no. 4, pp. 335–345, 1981.

Silva, M. Toward a synchronic theory for P/T nets. In *Concurrency and Nets*, (K. Voss, H. J. Genrich G. Rozenberg, eds). Spinger-Verlag, New York, pp. 435–460, 1987.

Suzuki, I. and Murata, T. A method for stepwise refinement and abstraction of Petri nets, *Journal of Computer Systems Science*, vol. 27, pp. 51–76, 1983.

Valette, R. Analysis of Petri nets by stepwise refinement, *Journal of Computer Systems Science*, vol. 18, pp. 35–46, 1979.

Valette, R., Courvoisier, M., Demmou, H., Bigou, J. M. and Desclaux, C. Putting Petri nets to work for controlling flexible manufacturing systems. *Proceedings International Symposium Circuits and Systems*, Kyoto, Japan, 1985.

Velilla, S. and Silva, M. The SPY: a mechanism for safe implementation of highly concurrent systems. *15th IFAC/IFIP Workshop Real-Time Programme*, Valencia, Spain, May 1988.

Villarroel, J. L. Martinez, J. and Silva, M. GRAMAN: a graphic system for manufacturing system design. *IMACS International Symposium Systems Modeling and Simulation*, Cetraro, Italy, September 1988.

Voss, K. Using predicate/transition-nets to model and analyze distributed database systems. *IEEE Transactions Software Englewood*, vol. SE-6, no. 6, pp. 539–544, 1980.

Zhou, M. C. A theory for the synthesis and augmentation of Petri nets in automation. Ph.D. dissertation, Department of Electrical, Computer, and Systems Engineering, Rensselaer Polytechnic Institute, Troy, NY, May 1990.

Zhou, M. C. and DiCesare, F. Hybrid synthesis of Petri nets for manufacturing systems: a case study. *The Third International Symposium on Robotics and Manufacturing*, Vancouver, July 1990.

Zhou, M. C. and DiCesare, F. Parallel and sequential mutual exclusions for Petri net modeling of manufacturing systems with shared resources. To appear in *IEEE Transactions Robotics and Automation*, vol. 7, no. 4, 1991.

Zhou, M. C. DiCesare, F. and Desrochers, A. A. A top-down modular approach to synthesis of Petri net models for manufacturing systems. *Proceedings IEEE. International Conference and Automation*, pp. 534–539, Scottsdale, AZ, May 1989.

Zhou, M. C. DiCesare, F. and Rudolph, D. Control of a flexible manufacturing system using Petri nets. *Proceedings IFAC Congress Conference*, vol. 8, pp. 38–43, Tallinn, Estonia, July 1990.

4

Performance evaluation of manufacturing systems

J. M. Proth

4.1 INTRODUCTION

Evaluating a manufacturing system consists of providing some information about its behavior, the most important information being the productivity of the system. According to the approach adopted in Chapter 2, we consider the cyclic production case and the general case separately. Within these two parts of the study, we will consider the deterministic case and the stochastic case. As the reader will see, the most important results are in the deterministic cyclic production case. We propose also an efficient approach for evaluating the productivity in the stochastic cyclic production case. The general case is more difficult to treat. Nevertheless, we propose a simple way to evaluate the system, assuming that the control is chosen and that the modules composing the system have been chosen in an adequate manner.

Because the manufacturing systems working on a cyclic basis can often be modeled using event graphs, we will highlight the most important properties of the strongly connected event graphs in the next section. Section 4.3 will present the main results available for the most common manufacturing systems when they operate on a cyclic basis. Section 4.4 is devoted to the general case.

4.2 EVENT GRAPHS: DEFINITION AND PROPERTIES

4.2.1 Fundamentals

An event graph is a Petri net such that each place has exactly one input and one output transition. Figure 4.1 is an example of an event graph. An event graph is timed if a firing time is assigned to each transition. It is assumed here that a firing is initiated as soon as the transition is enabled, provided that the transition is not already in a firing state.

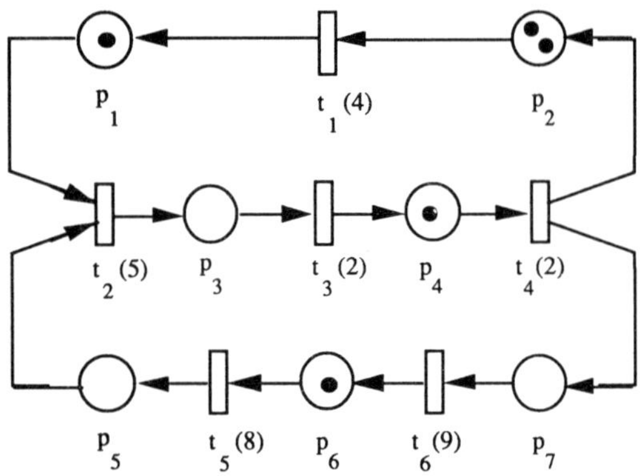

Figure 4.1 A strongly connected timed event graph.

An event graph (like a Petri net in general) is said to be strongly connected if there is a directed path joining any node A to any node B of the graph. The event graph presented in Fig. 4.1 is strongly connected.

We also define an elementary circuit in a strongly connected event graph as a directed path that goes from one node, i.e. a place or a transition, back to the same node, while any other node is not repeated. For instance, Fig. 4.1 exposes two elementary circuits, namely:

$$\gamma_1 = \langle p_1, t_2, p_3, t_3, p_4, t_4, p_2, t_1, p_1 \rangle$$

and:

$$\gamma_2 = \langle t_2, p_3, t_3, p_4, t_4, p_7, t_6, p_6, t_5, p_5, t_2 \rangle$$

The following result is due to Commoner *et al.* (1971).

Result 1

The total number of tokens in any elementary circuit is invariant by transition firing. (We admit that when a transition fires, it is as though there is one more token in each of its input places.)

Let us again consider the example provided in Fig. 4.1, and assume that a transition fires as soon as it is enabled. Assume also that the system starts at time 0. At time 0, there are 4 tokens in γ_1 and 2 tokens in γ_2. After 6 units of time, the system is as shown in Fig. 4.2.

A token in a transition means that the transition is firing. The numbers enclosed in the boxes indicate the remaining firing time. In this case, we have to count:

- two tokens in p_1;
- two tokens in p_2, because p_2 contains one token and it is the input place of a transition which is firing;

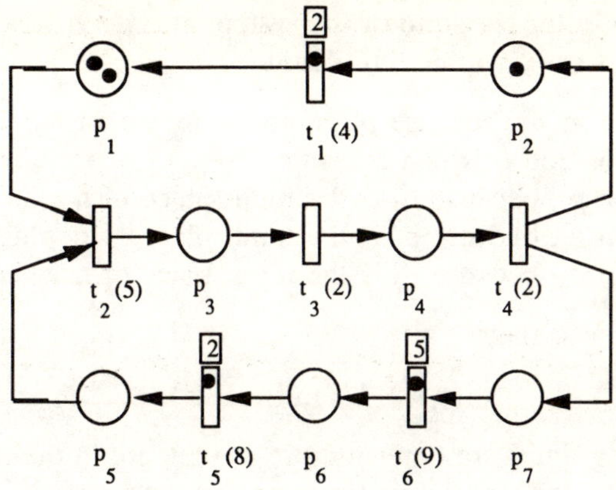

Figure 4.2 State of the system at time 6.

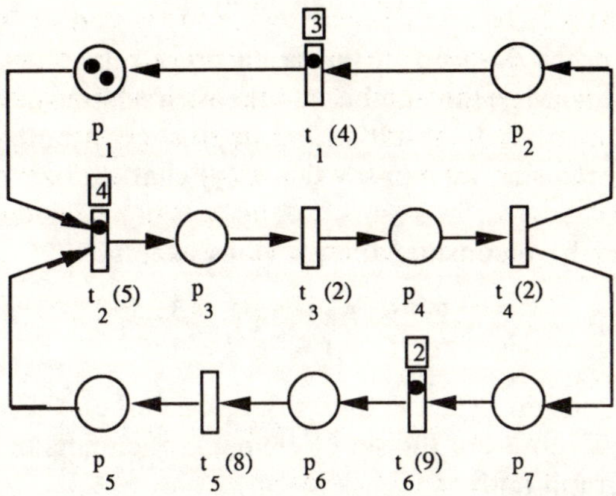

Figure 4.3 State of the system at time 9.

- one token in p_7, because p_7 is the input place of t_6 which is firing;
- one token in p_6, because p_6 is the input place of t_5 which is firing.

The total number of tokens in γ_1 is then:

$n_1(\gamma_1)$ = (number of tokens in p_1) + (number of tokens in p_2) + (number of tokens in p_3) + (number of tokens in p_4) $= 2 + 2 + 0 + 0 = 4$

Similarly, the number of tokens in γ_2 is:

$n_2(\gamma_2)$ = (number of tokens in p_3) + (number of tokens in p_4) + (number of tokens in p_5) + (number of tokens in p_6) + (number of tokens in p_7) $= 0 + 0 + 0 + 1 + 1 = 2$

Let us now consider the state of the system at time 9. It is represented in Fig. 4.3. In this case, we have to count:

- three tokens in p_1, because p_1 contains one token and it is the input place of transition t_2 which is firing;
- one token in p_2, because p_2 is the input place of t_1 which is firing;
- one token in p_5, because p_5 is the input place of t_2 which is firing;
- one token in p_7, because p_7 is the input place of t_6 which is firing.

Again, we can see that:

$$n_1(\gamma_1) = 4 \quad \text{and} \quad n_2(\gamma_2) = 2$$

When the firing times are deterministic as they are in the above example, we can compute, for any elementary circuit γ, the following ratio called the cycle time of the circuit:

$$C(\gamma) = \mu(\gamma)/M(\gamma) \tag{4.1}$$

where $\mu(\gamma)$ denotes the sum of the firing times related to the transitions belonging to γ, and $M(\gamma)$ the number of tokens circulating in γ. Because $M(\gamma)$ is a constant according to result 1, and $\mu(\gamma)$ is a constant when the firing times are deterministic, we can say that $C(\gamma)$ characterizes the elementary circuit γ.

For instance, if we consider the previous example:

$$C(\gamma_1) = 13/4 = 3.25$$
$$C(\gamma_2) = 26/2 = 13$$

Of course, the notion of cycle time disappears when the firing times are stochastic. Let Γ represent the set of elementary circuits of a strongly connected event graph and:

$$C^* = \max_{\gamma \in \Gamma} C(\gamma) \tag{4.2}$$

Any $\gamma \in \Gamma$ such that $C(\gamma) = C^*$ is a **critical circuit**. In the previous example, γ_2 is the critical circuit. These circuits are the ones that actually bind the speed of the system. This aspect is presented in result 2, due to Chretienne (1983).

Result 2

Under an operational mode where transitions fire as soon as they are enabled, the firing rate of each transition in steady state is given by:

$$\rho = 1/C^*$$

Let us come back to the previous example. In that case:

$$C^* = \max\,(3.25, 13) = 13$$

and:

$$\rho = 1/13 \simeq 0.077$$

This result means that, if we observe what happens at any point of the graph, we can see that 0.077 tokens pass through that point during each unit period of time (on the average, and assuming that the conditions of result 2 are verified).

As a consequence, if we want to increase the speed (i.e. the throughput) of the system, we have to add one (or several) token(s) to the critical circuits. Adding tokens in other circuits would be worthless. For instance, adding one token in place p_6 in Fig. 4.1 would change the cycle time of γ_2 into:

$$C(\gamma_2) = 26/3 = 8.666$$

and thus:

$$C^* = \max (3.25, 8.666) = 8.666$$

It leads to:

$$\rho = 1/C^* \simeq 0.1154$$

It means that, on the average, 0.1154 tokens pass through any point of the circuit in steady state, which is an improvement compared to the previous situation. The last important result is as follows.

Result 3

A strongly connected event graph is guaranteed to be deadlock-free if and only if every elementary circuit contains at least one token. For instance, assume that we put only one token in p_1 and one token in p_2 in the strongly connected event graph represented in Fig. 4.1. In that case, we see that t_2 will never fire and that the system will block as soon as the firing of t_1 ends. The above results will be used in section 4.3.

4.2.2 Optimization of deterministic strongly connected event graphs

An initial marking of the event graph being given, it is easy to evaluate its performance using result 2, assuming that all the elementary circuits are known. Some simple algorithms have been described in the literature (see Laftit *et al.* (1992) for instance) that can be applied to obtain all the elementary circuits of a strongly connected event graph. Note that the performance evaluation we propose in this section is based on the assumption that transitions fire as soon as they are enabled, which is a convenient and realistic assumption for manufacturing systems.

However, the analysis of a deterministic manufacturing system operating on a cyclic basis needs more than the ability of evaluating the performance of its event graph model. To optimize such a system, we must be able to minimize a linear function of the place markings under the constraint of reaching a given performance.

Thus, the problem at hand can be set as follows:

$$\text{minimize} \sum_{i=1}^{n} u_i x_i \tag{4.3}$$

s.t.

$$\rho = 1/C^* \geq \alpha$$

x_i $(i = 1, 2, \ldots, n)$ is a non-negative integer, where:

x_i is the initial number of tokens in place i $(i = 1, 2, \ldots, n)$
α is the required performance
ρ is the firing rate as expressed in result 2.

Note: Because a transition can fire only if it is not already in a firing execution, we must choose α less than or equal to $\text{Min}_{j=1,\ldots,9}(1/\theta_j)$, where θ_j is the firing time of transition t_j.

The linear function to be minimized should not depend on the state of the system. In other words, this function should be invariant by any transition firing, otherwise its minimization would be meaningless. Let us consider the incidence matrix A given by:

$$a_{i,j} = \begin{cases} 1 & \text{if } t_j \in {}^o p_i \\ -1 & \text{if } t_j \in p_i^o \\ 0 & \text{otherwise} \end{cases} \tag{4.4}$$

for $i = 1, 2, \ldots, n$ and $j = 1, 2, \ldots, q$, where:

n is the number of places
q is the number of transitions
${}^o p_i$ is the set of input transitions of p_i (it is singleton in the case of an event graph)
p_i^o is the set of output transitions of p_i (it is singleton in the case of an event graph)

Let M_0 be the initial marking, i.e. $M_0 = \langle x_1, x_2, \ldots, x_n \rangle$, and let M be the marking reached, starting from M_0, when applying a feasible sequence σ of transition firings. The following relation holds:

$$M_t = M_0^t + AV_\sigma^t \tag{4.5}$$

where $V_\sigma = \langle v_1, v_2, \ldots, v_q \rangle$, v_i $(i = 1, 2, \ldots, q)$ is the number of occurrences of transition t_i in the sequence σ. The vector V_σ is known as the characteristic vector of σ. Let $U = \langle u_1, u_2, \ldots, u_n \rangle$ be a vector such that:

$$UA = 0 \tag{4.6}$$

and $u_i \geq 0$ for $i = 1, 2, \ldots, n$. U is a p-invariant (also called p-semiflow) and:

$$UM_o^t = UM^t \tag{4.7}$$

for any marking M reachable from M_0 through a sequence of transition firing σ (we say that $M \in R\,(M_0)$). The proof of relation (4.7) is straightforward by multiplying both sides of relation (4.5) by U:

$$UM^t = UM_0^t + (UA)V_\sigma^t$$

and by considering the definition of U (see eq. (4.6)) which implies that $(UA)V_6^t = 0$

Example

Let us consider again the event graph presented in Fig. 4.1. For this Petri net, according to eq. (4.4), the incidence matrix is:

$$A = \begin{array}{c|cccccc|} & t_1 & t_2 & t_3 & t_4 & t_5 & t_6 \\ p_1 & 1 & -1 & 0 & 0 & 0 & 0 \\ p_2 & -1 & 0 & 0 & 1 & 0 & 0 \\ p_3 & 0 & 1 & -1 & 0 & 0 & 0 \\ p_4 & 0 & 0 & 1 & -1 & 0 & 0 \\ p_5 & 0 & -1 & 0 & 0 & 1 & 0 \\ p_6 & 0 & 0 & 0 & 0 & -1 & 1 \\ p_7 & 0 & 0 & 0 & 1 & 0 & -1 \end{array}$$

The relation $UA = 0$ can be rewritten as:

$$\begin{aligned} u_1 - u_2 \qquad\qquad\qquad\qquad\qquad &= 0 \\ -u_1 \qquad + u_3 \qquad - u_5 \qquad\qquad &= 0 \\ -u_3 + u_4 \qquad\qquad\qquad &= 0 \\ u_2 \qquad - u_4 \qquad\qquad + u_7 &= 0 \\ u_5 - u_6 \qquad &= 0 \\ u_6 - u_7 &= 0 \end{aligned}$$

From these equations, it turns out that:

$$u_1 = u_2 \tag{4.8}$$
$$u_3 = u_4 \tag{4.9}$$
$$u_5 = u_6 = u_7 \tag{4.10}$$

and the system of equations is reduced to:

$$-u_1 + u_3 - u_5 = 0 \tag{4.11}$$

Any set of positive values which verify eqs (4.8) to (4.11) is a p-invariant. For instance:

$$\begin{aligned} U_1 &= \langle 0, 0, 1, 1, 1, 1, 1 \rangle \\ U_2 &= \langle 1, 1, 2, 2, 1, 1, 1 \rangle \end{aligned} \tag{4.12}$$

are two p-invariants among an infinity of others. Thus, the linear function to minimize in problem (4.3) would be, for instance:

$$U_1 X^t = x_3 + x_4 + x_5 + x_6 + x_7$$

or:

$$U_2 X^t = x_1 + x_2 + 2x_3 + 2x_4 + x_5 + x_6 + x_7$$

or any linear function UX^t, where U is a p-invariant.

Now, let us go back to the problem given by eq. (4.3). Assume that the set Γ of elementary circuits is known ($\Gamma = \langle \gamma_1, \gamma_2, \ldots, \gamma_r \rangle$), and that the linear function to minimize is based on a p-invariant. The problem can be rewritten as:

$$\text{minimize} \sum_{i=1}^{n} u_i x_i \tag{4.13}$$

s.t.

$$\sum_{i=1}^{n} b_{ki} x_i \geq \alpha \cdot \mu(\gamma_k)$$

x_i is a non-negative integer for $i = 1, 2, \ldots, n$, where:

$$b_{ki} = \begin{cases} 1 & \text{if place } p_i \text{ belongs to } \gamma_k \\ 0 & \text{otherwise} \end{cases}$$

Example

Consider the event graph presented in Fig. 4.1 and the objective function derived from U_2 (see relation (4.13)). Assume that $\alpha = 0.1$.

The problem to solve is as follows:

$$\text{minimize } (x_1 + x_2 + 2x_3 + 2x_4 + x_5 + x_6 + x_7)$$

s.t.

$$\begin{aligned} x_1 + x_2 + x_3 + x_4 &\geq 13 \times 0.1 = 1.3 \\ x_3 + x_4 + x_5 + x_6 + x_7 &\geq 26 \times 0.1 = 2.6 \end{aligned} \tag{4.14}$$

$x_1, x_2, \ldots, x_7$ are non-negative integers. An optimal solution to this problem is:

$$x_1 = 2, \quad x_2 = x_3 = x_4 = 0, \quad x_5 = 3, \quad x_6 = x_7 = 0$$

The major drawback of the previous approach is the fact that the number of elementary circuits is unpredictable, and usually grows very fast with the size of the strongly connected event graph. To overcome this difficulty two new approaches have recently been developed (see Laftit *et al.*, 1992). They are presented hereafter.

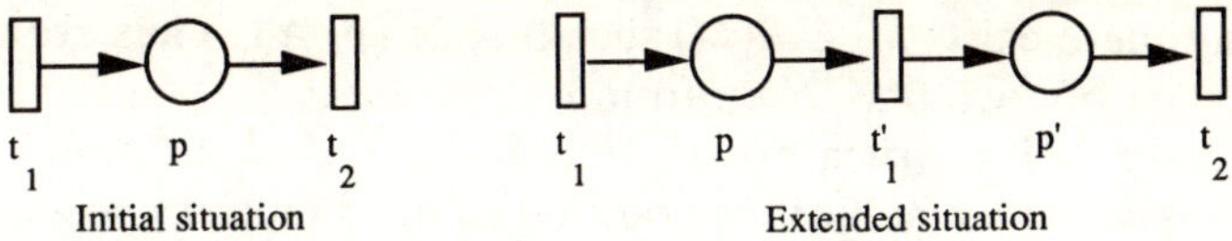

Figure 4.4 Extension of the event graph.

(a) The adjustment heuristic algorithm (AHA)

This algorithm is a heuristic, which means that it does not guarantee the optimal solution. Nevertheless, a large number of problems have been solved using AHA, and it appears that the optimal solution has been reached in 94% of the cases. In the remaining cases, the difference between the value of the objective function obtained using AHA and the optimal one was less than 5% of the optimum.

To introduce the AHA, it has first been proved that there exists an optimal solution to the problem at hand such that each place contains at most two tokens. But, it is always possible to extend a strongly connected event graph in the way shown in Fig. 4.4, where the time assigned to t_1' is zero. In that case, it is possible to put one token in place p and one token in place p' instead of putting two tokens in p. Thus, the problem at hand can always be reduced to a problem whose optimal solution is such that any place contains at most one token. This property will also be used in the next algorithm. We then define:

$$\omega(M, p, 1/\alpha) = \underset{\gamma \in \Gamma_p}{\mathrm{Min}}[M(\gamma) - \alpha \cdot \mu(\gamma)]$$

where Γ_p is the set of elementary circuits containing p. The following property has been proved:

Property 4.1. Let $\mathfrak{M}(\alpha)$ be the set of markings M such that $\omega(M, p, 1/\alpha) \in [0, 1)$ for $p \in P$ (P is the set of places). Then:

1. Any $M \in \mathfrak{M}(\alpha)$ is a feasible solution of the problem.
2. Let $M \in \mathfrak{M}(\alpha)$ and M_1 be such that $M_1 \leq M$ and $M_1 \neq M$ (i.e. $M_1(p) \leq M(p)$, $\forall p \in P$ and there exists at least one $p^* \in P$ such that $M_1(p^*) < M(p)$). Then, M_1 is not feasible and thus cannot be part of an optimal solution to the problem at hand.
3. $\mathfrak{M}(\alpha)$ contains at least one optimal solution to the problem at hand.

The AHA is based on the previous property. It starts with a feasible marking and removes iteratively tokens from places until a solution belonging to $M(\alpha)$ is reached. The initial feasible marking can be one token in each place (the feasibility of such a marking has been proved), or any other feasible solution. At each iteration, the heuristic removes one token from a place p such that $\omega(M, p, 1/\alpha) \geq 1$ **and** $M(p) \geq 1$. If $M(p) = 0$, we know that there exists $M^1 \in R(M)$ such that $M^1(p) \geq 1$. Thus, we compute M_1 and then remove one token from p.

If several $p \in P$ are such that $\omega(M, p, 1/\alpha) \geq 1$, then the basic idea is to remove tokens which reduce the criterion value as much as possible, while preserving further removal of tokens and, as a consequence, the chance of reaching a marking very close to the optimal one is increased. To reach this goal, we introduced a secondary criterion denoted by $g_{\cup}(M)$ and defined as follows:

$$g_{\cup}(M) = \sum_{p \in P} \omega\,(M, p, 1/\alpha)$$

Let M_p^1 be the marking obtained by removing one token from p and:

$$\Delta_p = [g_{\cup}(M) - g_{\cup}(M_p^1)]/u_p \quad \text{assuming that } \omega(M, p, 1/\alpha) \geq 1,$$

Where u_p is the gain in the criterion value. We choose to remove one token from the place p^* such that:

$$\Delta_p^* = \min_{q \in F} \Delta_q \tag{4.15}$$

where $F = \{p/p \in P \text{ and } \omega(M, p, 1/\alpha) \geq 1\}$.

The amount of computations is proportional to the number of places. Finally, the algorithm can be formalized as follows:

Step 1 Generate a feasible marking M (for instance, put one token in each of the places).

Step 2 If $\omega(M, p, 1/\alpha) < 1$ for any $p \in P$, then stop the computation and keep M as the solution to the problem.

Step 3 If $\omega(M, p, 1/\alpha) \geq 1$ for at least one $p \in P$, compute $p^* \in P$ such that $\Delta_p^* = \min_{q \in F} \Delta_q$ (see relation (4.15)).

Step 4 If $M(p^*) \geq 1$ set:

$$\begin{cases} M(p^*) = M(p), \ \forall p \in P, \ p \neq p^* \\ M(p^*) = M(p^*) - 1 \end{cases}$$

If $M(p^*) = 0$, compute a marking M^1 reachable from M and such that $M^1(p^*) \geq 1$, and set:

$$\begin{cases} M(p) = M^1(p), \ \forall p \in P, \ p \neq p^* \\ M(p^*) = M^1(p^*) - 1 \end{cases}$$

Step 5 Go to step 2.

Example

Let us come back to the event graph presented in Fig. 4.1 and the objective function derived from U_2. Assume that $\alpha = 0.1$ This problem is set in eq. (4.14). We first put one token in each of the seven places, i.e.:

$$x_1 = x_2 = x_3 = x_4 = x_5 = x_6 = x_7 = 1\text{: this is the marking } M$$

We then compute:

$$\begin{aligned}
\omega(M, p_1, 10) &= 4 - 0.1 \times 13 = 2.7\\
\omega(M, p_2, 10) &= 4 - 0.1 \times 13 = 2.7\\
\omega(M, p_3, 10) &= \min[4 - 0.1 \text{ x } 13,\ 5 - 0.1 \times 26] = \min[2.7, 2.4] = 2.4\\
\omega(M, p_4, 10) &= 2.4\\
\omega(M, p_5, 10) &= 5 - 0.1 \times 26 = 2.4\\
\omega(M, p_6, 10) &= 2.4\\
\omega(M, p_7, 10) &= 2.4
\end{aligned}$$

Then we can choose to remove one token from any place. To make this decision, we first consider the marking $M^1_{p_1}$ obtained by removing one token from p_1:

$$\begin{aligned}
&\omega(M^1_{p_1}, p_1, 10) = 3 - 0.1 \times 13 = 1.7 = \omega(M^1_{p_{11}}, p_2, 10)\\
&\omega(M^1_{p_1}, p_3, 10) = \omega(M^1_{p_1}, p_4, 10) = \min[3 - 0.1 \times 13, 5 - 0.1 \times 26] = 1.7\\
&\omega(M^1_{p_1}, p_5, 10) = \omega(M^1_{p_1}, p_6, 10) = \omega(M^1_{p_1}, p_7, 10) = 2.4
\end{aligned}$$

and:

$$\Delta_{p_2} = [17.4 - 14]/1 = 3.4$$

In the same manner, we compute:

$$\begin{aligned}
\Delta_{p_3} &= [17.4 - 10.4]/2 = 3.5\\
\Delta_{p_4} &= [17.4 - 10.4]/2 = 3.5\\
\Delta_{p_5} &= [17.4 - 12.4]/1 = \Delta_{p_6} = \Delta_{p_7} = 5
\end{aligned}$$

Finally, we choose to remove one token from p_1 because:

$$\Delta_{p_1} = \min_{p \in P} \Delta_p$$

Let M be the new marking

$$M = \langle 0, 1, 1, 1, 1, 1, 1, 1 \rangle$$

Let us now restart from this new marking. In that case:

$$\begin{aligned}
&\omega(M, p_1, 10) = \omega(M, p_2, 10) = 3 - 0.1 \times 13 = 1.7\\
&\omega(M, p_3, 10) = \omega(M, p_4, 10) = \min[3 - 0.1 \times 13,\ 5 - 0.1 \times 26] = 1.7\\
&\omega(M, p_5, 10) = \omega(M, p_6, 10) = \omega(M, p_7, 10) = 5 - 0.1 \times 26 = 2.4
\end{aligned}$$

Again, the criterion is greater than one for all the places. Thus, we can choose to remove one token from any place. To make this decision, we compute Δ_{p_i} for $i = 1, 2, \ldots, 7$.

Let $M^1_{p_1}$ be this marking obtained by firing t_1 and removing from M the token contained in place p_1. In that case:

$$\begin{aligned}
&\omega(M^1_{p_1}, p_1, 10) = \omega(M^1_{p_1}, p_2, 10) = 2 - 0.1 \times 13 = 0.7\\
&\omega(M^1_{p_1}, p_3, 10) = \omega(M^1_{p_1}, p_4, 10) = \min[2 - 0.1 \times 13,\ 5 - 0.1 \times 26] = 0.7\\
&\omega(M^1_{p_1}, p_5, 10) = \omega(M^1_{p_1}, p_6, 10) = \omega(M^1_{p_1}, p_7, 10) = 5 - 0.1 \times 26 = 2.4
\end{aligned}$$

And:

$$\Delta_{p_1} = [(4 \times 1.7 + 3 \times 2.4) - (4 \times 0.7 + 3 \times 2.4)] = 4$$

Now let $M^1_{p_2}$ be the marking obtained by removing from M the token contained in p_2.

Similarly we obtain, starting from $D^1_{p_2}$:

$$\Delta^1_{p_2} = 4$$

Let $M^1_{p_3}$ be the marking obtained by removing from M the token contained in p_3:

$$\omega(M^1_{p_3}, p_1, 10) = \omega(M^1_{p_3}, p_2, 10) = 2 - 0.1 \times 13 = 0.7$$
$$\omega(M^1_{p_3}, p_3, 10) = \omega(M^1_{p_3}, p_4, 10) = \min[2 - 0.1 \times 13, 4 - 0.1 \times 26] = 0.7$$
$$\omega(M^1_{p_3}, p_5, 10) = \omega(M^1_{p_3}, p_6, 10) = \omega(M^1_{p_3}, p_7, 10) = 1.4$$

Thus:

$$\Delta_{p_3} = [(4 \times 1.7 + 3 \times 2.4) - (4 \times 0.7 + 3 \times 1.4)]/2 = 3.5$$

Starting from $M^1_{p_4}$, we obtain similarly:

$$\Delta_{p_4} = 3.5$$

Let $M^1_{p_5}$ be the marking obtained by removing from M the token contained in p_5:

$$\omega(M^1_{p_5}, p_1, 10) = \omega(M^1_{p_5} p_2, 10) = 3 - 0.1 \times 13 = 1.7$$
$$\omega(M^1_{p_5}, p_3, 10) = \omega(M^1_{p_5}, p_4, 10) = \min[3 - 0.1 \times 13, 4 - 0.1 \times 26] = 1.4$$
$$\omega(M^1_{p_5}, p_5, 10) = \omega(M^1_{p_5}, p_6, 10) = \omega(M^1_{p_5}, p_7, 10) = 4 - 0.1 \times 26 = 1.4$$

Thus:

$$\Delta_{p_5} = [(4 \times 1.7 + 3 \times 2.4) - (2 \times 1.7 + 5 \times 1.4)] = 14 - 10.4 = 3.6$$

Similarly we obtain:

$$\Delta_{p_6} = \Delta_{p_7} = 3.6$$

Finally:

$$\Delta_{p_3} = \Delta_{p_4} = \min_{i \in \{1,2\ldots,7\}} \Delta_{p_i}$$

and we can remove one token from one of the places p_3 or p_4. Thus, the new marking M will be:

$$M = \langle 0, 1, 1, 0, 1, 1, 1 \rangle$$

We then restart from this new marking and compute:

$$\omega(M, p_1, 10) = \omega(M, p_2, 10) = 2 - 0.1 \times 13 = 0.7$$
$$\omega(M, p_3, 10) = \omega(M, p_4, 10) = \min[2 - 0.1 \times 13, 4 - 0.1 \times 26] = 0.7$$
$$\omega(M, p_5, 10) = \omega(M, p_7, 10) = \omega(M, p_7, 10) = 4 - 0.1 \times 26 = 1.4$$

Here, we have to remove one token from either p_5 or p_6 or p_7. To make this decision, we compute Δ_{p_5}, Δ_{p_6} and Δ_{p_7}, because the value of the criterion ω (. . .), is less than one for all other places.

Let $M^1_{p_5}$ be the marking obtained by removing from M the token contained in p_5. Then:

$$\omega(M^1_{p_5}, p_1, 10) = \omega(M^1_{p_5}, p_2, 10) = 2 - 0.1 \times 13 = 0.7$$
$$\omega(M^1_{p_5}, p_3, 10) = \omega(M^1_{p_5}, p_4, 10) = \min[2 - 0.1 \times 13,\ 4 - 0.1 \times 26] = 0.7$$
$$\omega(M^1_{p_5}, p_5, 10) = \omega(M^1_{p_5}, p_6, 10) = \omega(M^1_{p_5}, p_7, 10) = 3 - 0.1 \times 26 = 0.4$$

Thus:

$$\Delta_{p_5} = [(4 \times 0.7 + 3 \times 1.4) - (4 \times 0.7 + 3 \times 0.4)] = 3$$

Similarly we obtain:

$$\Delta_{p_6} = \Delta_{p_7} = 3$$

Thus, we can choose to remove one token from any of the three places, and the new marking will be:

$$M = \langle 0, 1, 1, 0, 0, 1, 1 \rangle$$

At this step, we obtain:

$$\omega(M_1, p_1, 10) = \omega(M_1, p_2, 10) = 2 - 0.1 \times 13 = 0.7$$
$$\omega(M_1, p_3, 10) = \omega(M_1, p_4, 10) = \min[2 - 0.1 \times 13,\ 3 - 0.1 \times 26] = 0.4$$
$$\omega(M_1, p_5, 10) = \omega(M_1, p_6, 10) = \omega(M_1, p_7, 10) = 3 - 0.1 \times 26 = 0.4$$

All the previous values being less than one, we keep the last marking which leads to the following value of the criterion:

$$1 \times 0 + 1 \times 1 + 2 \times 1 + 2 \times 0 + 1 \times 0 + 1 \times 1 + 1 \times 1 = 5$$

(b) The convex optimization algorithm

It has been proved that there always exists a periodic solution (denoted by POM for Periodic Operational Mode) which reaches the same performance as the previous operational mode (denoted by EOM for Earliest Operational Mode), i.e. the mode when a transition fires as soon as it is enabled, **with the same initial marking.**

The convex optimization algorithm consists of solving the POM problem and keeping the initial marking of this solution as the initial marking of the EOM problem. This algorithm leads to the optimal solution of the problem at hand.

We assume that there exists an optimal solution to the EOM problem, and thus to the POM problem, which is such that each place contains at most one token. In other words, we assume that the transformation represented in Fig. 4.4 has been made on the initial event graph model.

Using the same notations as those of problem (4.13), the problem to solve, denoted by $\mathfrak{P}_0$, is as follows:

$$\text{minimize} \sum_{i=l}^{n} u_i (x_i + y_i) \tag{4.16}$$

s.t.

$$S_{°p}(1) + \tau_{°p} - S_{p°}(1) + (x_p + y_p)/\alpha \leq 2/\alpha, \quad \forall p \in P \tag{4.17}$$
$$x_p, y_p \in \{0, 1\}, \ \forall p \in P \tag{4.18}$$

where:

P is the set of places
x_i (resp. y_i) is the number of tokens in place p_i (resp. p_i'), referring to Fig. 4.4
^{o}p (resp. p^{o}) is the input (resp. output) transition of $p \in P$
S_t is the instant of the first initiation of transition t
τ_t is the firing time of transition t

We further define the problem $\mathfrak{P}_k$ as follows:

$$\text{minimize} \sum_{i=l}^{n} u_i(x_i^k + y_i^k) \tag{4.19}$$

s.t. constraint (4.17) and

$$x_p, y_p \in [0, 1] \tag{4.20}$$

Problem $\mathfrak{P}_0$ is a mixed linear programming problem, where the variables are the instants of the first initiation of transitions of the model and the number of tokens in the places, i.e. the $S_{°p}(1)$, $S_{p°}(1)$, x_p and y_p for $p \in P$. The convex optimization algorithm is based on property 4.2.

Property 4.2. Let $\langle x^*_{p_i,k}, y^*_{p_i',k}\rangle$, $i = 1, 2, \ldots, n$ be the optimal solution of problems (4.19), (4.17) and (4.20).

We define:

$$P_1 = \{p_i/p_i \in P \quad \text{and} \quad x^*_{p_i,k} = 1\}$$
$$P_2 = \{p_i'/p_i' \in P \quad \text{and} \quad y^*_{p_i',k} = 1\}$$

If:

$$\sum_{p_i \in P/P_1} u_{p_i}(x^*_{p_i,k})^k + \sum_{p_i \in P/P_2} u_{p_i'}(y^*_{p_i',k})^k < 1 \tag{4.21}$$

Then:

$$\lfloor x^*_{p_i,k} \rfloor \quad \text{and} \quad \lfloor y^*_{p_i',k} \rfloor \quad i = 1, 2, \ \ldots, n$$

is the optimal solution of the problem at hand (i.e. the EOM problem). [•] denotes the greatest integer less than or equal to •.

The basic idea behind the convex optimization algorithm is to solve a problem $\mathfrak{P}_k$ for k 'large enough' and to check if relation (4.21) holds. If yes, the optimal solution is $\lfloor x^*_{p_i,k} \rfloor$ and $\lfloor y^*_{p_i,k} \rfloor$, $i = 1, 2, \ldots, n$; otherwise, we have to increase the value of k and to recompute the solution of $\mathfrak{P}_k$ until relation (4.21) holds.

The algorithm used for solving $\mathfrak{P}_k$ can be described as a branch-and-bound procedure requiring the solutions to linear programming subproblems during the branch-and-bound search. Detailed information about this algorithm can be found in van Thoai and Tuy (1980) and Benson (1985).

An advantage of this algorithm is that, at each step of the computation, we have only to consider the coefficient of u_i related to the variables taking the value 1 and the variable having the greatest value among those, whose value is smaller than 1 and greater than 0. As a consequence, it is possible to solve problems $\mathfrak{P}_k$ with very large values of k, and thus to reach the optimal solution to problem $\mathfrak{P}_0$ at the first step of the computation.

4.3 MANUFACTURING SYSTEMS WORKING ON A CYCLIC BASIS

In this section, we consider deterministic and stochastic manufacturing systems working on a cyclic basis. In subsections 4.3.1 and 4.3.2, we apply the AHA and the convex optimization algorithm to a deterministic job-shop system and to a Kanban system. The same algorithms may apply to the assembly systems and, more generally, to any manufacturing system working on a cyclic basis and which can thus be modeled using a strongly connected event graph. The previous algorithms no longer apply in the stochastic case, simply because the cycle time no more characterizes the behavior of an elementary circuit. In section 4.3.3, we propose an approach to find the best compromise between the mean cycle time of a strongly connected stochastic event graph and the value of the objective function. The last subsection is devoted to the conclusions.

4.3.1 Optimization of deterministic job-shop operations

(a) The problem: definition and modeling

The problem consists of maximizing the productivity of a job-shop with a minimal work-in-process (WIP) level, assuming a cyclic manufacturing of the parts. This problem has been studied by Hillion and Proth (1989). It has been shown that the solution to this problem is the solution to an integer linear programming problem derived from the elementary circuits of the event graph model. For more information, we refer the reader to Hillion and Proth (1989). As we outlined before, this approach may result in a computational burden.

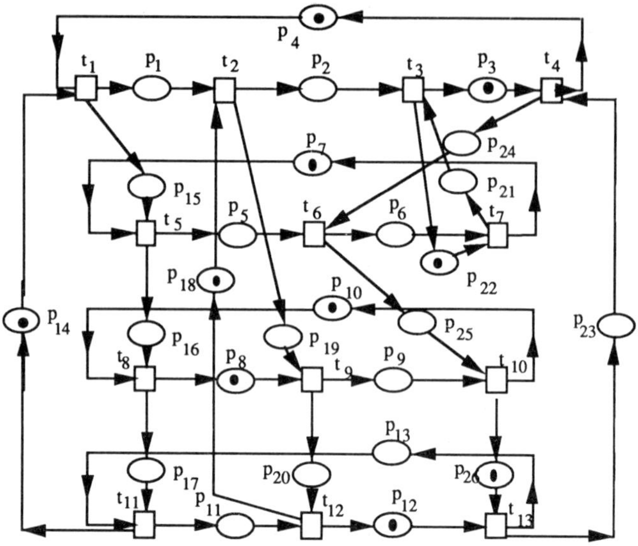

Figure 4.5 Event graph model of the job-shop.

We use the same model in this section. It is illustrated by the following simple example. We consider a job-shop composed of four machines M_1, M_2, M_3 and M_4, which can manufacture three types of parts denoted by R_1, R_2 and R_3. The production mix is 25%, 25% and 50% for R_1, R_2 and R_3, respectively. The production processes of the part-types are:

$$R_1\colon M_1(1), M_2(1), M_3(3), M_4(3)$$
$$R_2\colon M_1(1), M_4(1), M_3(2)$$
$$R_3\colon M_1(1), M_2(2), M_4(1)$$

If $s(M_i)$ $(i = 1, 2, 3, 4)$ denotes the fixed sequencing of the part types on machine M_i, we choose:

$$\sigma(M_1) = (R_1, R_2, R_3, R_3);\ \sigma(M_2) = (R_1, R_3, R_3);\ \sigma(M_3) = (R_1, R_2);$$
$$\sigma(M_4) = (R_1, R_2, R_3, R_3)$$

Furthermore, $a = 6$. According to Hillion and Proth (1989), the model related to this problem is given in Fig. 4.5. This model is a strongly connected event graph. In such an event graph, an elementary *p*-invariant is a vector having as many components as places in the event graph and where 1-values are assigned to the elements related to the places belonging to one of the elementary circuits and 0-values are assigned to the other elements. Thus, there are as many elementary *p*-invariants as elementary circuits.

For the problem at hand, there should be exactly one token in each command circuit in order to avoid simultaneous use of the same machine

Table 4.1 Marking at the different iterations for the job-shop problem

Step	*p**	*Places* *0000000001111–1111112222222* *1234567890123–4567890123456*	*Total number of tokens*
0		1111111111111–1111111111111	26
1	21	1111111111111–1111111011111	25
2	19	1111111111111–1111101011111	24
3	25	1111111111111–1111101011101	23
4	14	1111111111111–0111101011101	22
5	18	1111111111111–0111001011101	21
6	23	1111111111111–0111001010101	20
7	16	1111111111111–0101001010101	19
8	15	1111111111111–0001001010101	18
9	24	1111111111111–0001001010001	17
10	6	1111101111111–0001001010001	16
11	1	0111101111111–0001001010001	15
12	11	0111101111011–0001001010001	14
13	8	0111101011011–0001001010001	13
14	5	0111001011011–0001001010001	12
15	1	0020001011011–0100100101000	11
16	11	0011001101010–1000100010001	10

for various parts (a command circuit is an elementary circuit passing through the transitions related to a given machine). In this example, command circuits contain places p_{14} to p_{26}. So, the elementary p-invariants containing these places are highly penalized to avoid more than one token in these circuits. The objective function used in this case is then:

$$f_{\cup}(M_0) = \sum_{i=1}^{13} M_0(p_i) + 1000 \sum_{i=14}^{26} M_0(P_i)$$

Table 4.2 An optimal marking

Places	00000000011111111112222222 12345678901234567890123456
Optimal marking	00111100100101000100100100

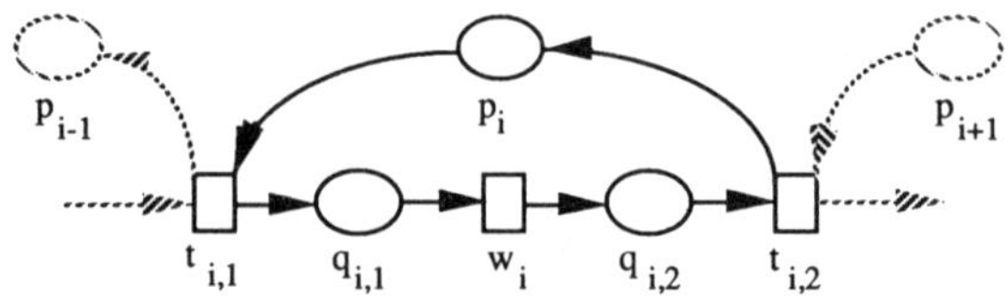

Figure 4.6 A stage of a Kanban system.

(b) Applying the adjustment heuristic algorithm

The tokens in the places of the command circuits are assigned as shown in Fig. 4.5. Furthermore, $\alpha = 6$ (largest cycle time among the cycle times of the command circuits). We start with one token in each place, i.e. $M_0(p_i) = 1$ for $i = 1, 2, \ldots, 26$. The results obtained using the adjustment heuristic algorithm are summarized in Table 4.1. The total number of tokens in the final marking is 10 while the total WIP (i.e. the number of tokens in places p_1 to p_{13}) is 6.

We observe that some discontinuities appear in Table 4.1. For instance, the state at step 15 cannot be obtained by removing one token from place 1, because place 1 does not contain any token. Thus, we compute a marking reachable from the one of step 14 and which contains one token, and then we remove this token. The same remark can be made for the next example.

(c) Applying the convex optimization algorithm

The solution is obtained by solving $\mathfrak{P}_6$. An optimal marking is given in Table 4.2. We can see that the total number of tokens is 10 while the total WIP is 6, as when using the adjustment heuristic algorithm. Nevertheless, the marking is not the same.

4.3.2 Optimization of deterministic Kanban system operations

(a) The problem: definition and modeling

As shown by Di Mascolo *et al.* (1989), an event graph can be used to model a Kanban system. In Fig. 4.6, we show how to model the *i*th stage of such a system. Transition w_i represents the operation performed at stage i, p_i

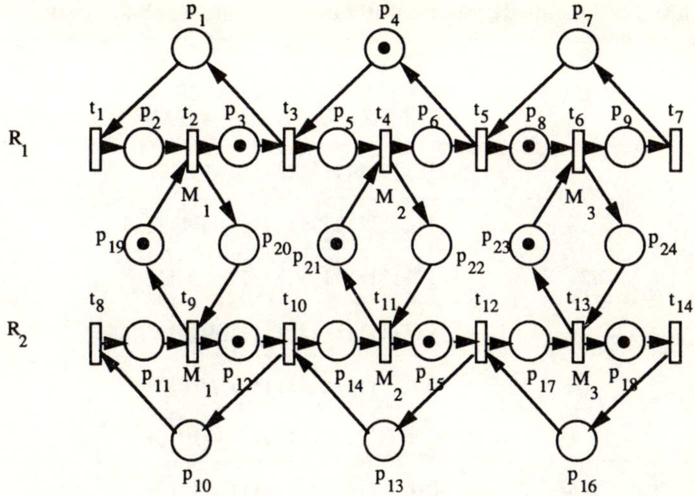

Figure 4.7 Model of a Kanban system with 3 machines and 2 part types.

contains as many tokens as free Kanbans, $q_{i,1}$ contains as many tokens as parts waiting to be manufactured, and $q_{i,2}$ contains as many tokens as parts ready for the next operation. These parts will be transferred to the next operation if some Kanbans are available (i.e. if there are some tokens in p_{i+1}). We consider the multi-part-type case. To illustrate the approach, we study a production line composed of three machines M_1, M_2 and M_3 which can manufacture two part-types denoted by R_1 and R_2. The manufacturing times of one part of R_1-type on M_1, M_2 and M_3 are 1, 2 and 1 respectively. The manufacturing times of one part of R_2-type on M_1, M_2 and M_3 are 1, 1 and 2 respectively. The parts enter the production line according to sequence $R_1, R_2, R_1, R_2, \ldots$. The event graph model of this system is given by Fig. 4.7. In this model, command circuits are $(p_{19}, t_2, p_{20}, t_9)$, $(p_{21}, t_4, p_{22}, t_{11})$ and $(p_{23}, t_6, p_{24}, t_{13})$. We aim at producing a pair of parts of types R_1 and R_2 every 3 units of time with a minimal number of Kanbans in the system. Furthermore, $a = 3$.

As in the case of job-shop, we consider a p-invariant U which assigns high penalization coefficients to places belonging to command circuits and 1 to the other places. The linear invariant function used in this example is as follows:

$$f_U(M_0) = \sum_{i=1}^{18} M_0(p_i) + 1000 \sum_{i=19}^{24} M_0(p_i)$$

(b) *Applying the adjustment heuristic algorithm*

The initial marking M_0 consists of putting one token in each place p_i, $i = 1, 2, \ldots, 24$. The marking at the different stages of the computation are

Table 4.3 Marking at the different iterations for the Kanban problem

Step	*p**	*Places* *000000000111111111–122222* *123456789012345678–901234*	*Total number of tokens*
0		111111111111111111–111111	24
1	19	111111111111111111–011111	23
2	21	111111111111111111–010111	22
3	23	111111111111111111–010101	21
4	1	011111111111111111–010101	20
5	1	010021111111111111–010101	19
6	5	010011111111111111–010101	18
7	5	010001111111111111–010101	17
8	8	010001101011111111–010101	16
9	8	010101000011111111–010101	15
10	12	010100001110111111–010101	14
11	15	010100001110110111–010101	13
12	16	010100001110110011–010101	12
13	12	010100100010110011–100101	11
14	15	100010100100010011–101001	10
15	16	001100010001001001–101010	9

given in Table 4.3. The marking obtained is optimal since there is just one token in each Kanban circuit and in each command circuit.

(c) Applying the convex optimization algorithm

The same optimal result as that provided by the adjustment heuristic algorithm is obtained by solving $\mathfrak{P}_6$.

4.3.3 Study of stochastic systems

There is a correlation between the initial distribution of the tokens in the system and its mean cycle time. In this subsection, we propose a way to find a good, if not optimal, compromise between the distribution of the tokens and the mean cycle time. Note that a given distribution of the tokens is related to a value of the objective function. Thus, the problem can be formulated as follows: find the best compromise between the objective function value (which should be as small as possible) and the mean cycle time (which should also be as small as possible).

To simplify the explanation, we restrict ourselves to the case of a job-shop. The goal is to find the best compromise between the total number of tokens, knowing that there is exactly one token in each command circuit, and the mean cycle time.

(a) Bounds of the mean cycle time

In this section, we assume that the number of tokens as well as their distribution in the strongly connected event graph are known at the initial time. Let M_0 be this marking. We provide a lower bound and an upper bound of the mean cycle time of the system.

We consider an operation mode of the event graph model, called **earliest operation mode** (EOM), for which transitions fire as soon as they are enabled, provided that they are idle. As shown by Chretienne (1983), this operation mode leads to the minimal mean cycle time. This mode is used in the following.

(i) The lower bound

Note that the cycle time which is the solution to the deterministic problem obtained by replacing the random variables generating the firing times by their mean values is a lower bound of the mean cycle time. Proposition 1 provides a better lower bound of the value of the mean cycle time denoted by π.

Proposition 4.1. The following inequality holds:

$$\pi \geq \max_{t \in T} E[\beta_t] = \bar{\pi} \tag{4.22}$$

where:

T is the set of transitions of the strongly connected event graph, and

$$\beta_t = \max_{\gamma \in \Gamma_t} C(\gamma),$$

where Γ_t is the set of elementary circuits which contain t and $C(\gamma) = \mu(\gamma)/n(\gamma)$, $\mu(\gamma)$ is the convolution of the random variables which generate the firing times of the transitions belonging to γ, and $n(\gamma)$ is the number of tokens in the elementary circuit γ.

(ii) The upper bound

M_0 being the initial marking, we derive a marking M_1 from M_0 by leaving empty in M_1 the places which are empty in M_0 and by reducing to one the number of tokens in the places which contain more than one token in M_0. Thus, $M_1(p) \leq M_0(p)$ for any $p \in P$, set of places of the strongly connected event graph. A fast operation mode running with the initial marking M_1 leads to a greater mean cycle time than the one obtained when starting from M_0.

Then, starting from M_1, we apply to the event graph the fast operation mode, but we block the tokens as soon as they reach a place already marked in M_1. We refer to this operation mode as the **constrained operation mode**. We know (see Chretienne (1983) and Hillion and Proth (1989)) that the new marking obtained is again M_1. Let us denote by $\bar{\pi}$ the mean cycle time obtained using the constrained operation mode when M_1 is the initial marking. We know (see Chretienne, 1983) that $\bar{\pi}$ is greater than the mean cycle time obtained using the fast operation mode starting from M_1 which, in turn, is greater than the mean cycle time obtained with the fast operation mode when the initial marking is M_0. Thus, $\bar{\pi}$ is an upper bound of the solution to our problem (i.e. the mean cycle time obtained starting from M_0 when using the fast operation mode). This upper bound is defined in the following proposition.

Proposition 4.2. The upper bound is given by:

$$\bar{\pi} = E[\max_{z \in Z} \mu(z)] \tag{4.23}$$

where:

Z is the set of oriented paths verifying the following properties:

- the origin and the extremity of any path is a marked place,
- there is no marked place between the origin and the extremity of the path;

$\mu(z)$ is the sum of the random variables which generate the firing times of the transitions of z.

(b) Reachability of a given cycle time

In this section, we consider a given value C_0 as the greatest mean value of the random variables which generate the firing times of the transitions of the strongly connected event graph Γ under consideration.

We denote by X_t the random variable related to transition t and by x_t^k the kth value taken by X_t (i.e. the kth firing time of transition t). We assume that:

(a) The mean value $m_t = E[X_t]$ and the standard deviation s_t exist for any $t \in T$;
(b) For any $t \in T$, the values $x_t^1, x_t^2, \ldots$ are taken independently from each other. (4.24)

Thus:

$$C_0 = \max_{t \in T} m_t$$

Let M_0 be the initial marking such that:

$$M_0(p) = N \quad \forall_p \in P$$

We define a so-called ***N*-periodic operation mode**, denoted by ***N*-POM**, as follows:

(i) Each transition $t \in T$ is fired N times under a fast operation mode. In other words, the N tokens of each place are used for living their output transition as soon as possible. Let $\theta_1^N(t)$ be the sum of the N first firing times of transition t for $t \in T$. Let $\theta_1^N = \max_{t \in T} \theta_1^N(t)$. Transitions which complete N firings before time θ_1^N (assuming that firings start at time 0) are frozen until time θ_1^N (i.e. further firings are not allowed until time θ_1^N). As a consequence, the marking is the same at time θ_1^N as at time 0.
(ii) We restart the same process from time θ_1^N on. Let θ_2^N be the time needed to reach again the marking M_0.
(iii) The process restarts from time $\theta_1^N + \theta_2^N$ and the third step takes a time θ_3^N, and so on. During each of the periods $[\theta_s^N, \theta_{s+1}^N]$, $s = 0, 1, 2, \ldots$, each transition $t \in T$ is fired exactly N times and $M_s(p) = N$ for any $p \in P$, M_s being the marking at time θ_s^N and $\theta_0^N = 0$.

Since some transitions are temporarily frozen when using the N-POM, the relating mean cycle time is greater than or equal to the one obtained when using the EOM, assuming that the initial marking is the same in both cases. As a consequence, if the strongly connected event graph reaches a mean cycle time less than C^* ($C^* > C_0$) when the N-POM is applied, it also reaches a mean cycle time less than C^* when the EOM is applied.

In the following, we focus our attention on the N-POM and derive some properties related to the EOM from this study. We define the positive value θ_k^N for $k = 1, 2, 3, \ldots$ as:

$$\theta_k^N = \max_{t \in T} \sum_{i=1}^{N} x_t^{(k-1)N+i} \tag{4.25}$$

We denote by θ^N the random variable which takes successively the values $\theta_1^N, \theta_2^N, \theta_3^N, \ldots$ The following properties of θ^N are easily derived from the ones of X_t, $t \in T$ (see eq. (4.3));

(a) The mean value $E[\theta^N]$ and the standard deviation $\sigma(\theta^N)$ of the random variable θ^N exist;
(b) θ^N takes the values $\theta_1^N, \theta_2^N, \theta_3^N, \ldots$ independently from each other;
(c) $\lim_{N \to +\infty} \left(\sum_{k=1}^{L} \theta_k^N / L \right) = E[\theta^N]$ with probability 1. (4.26)

A consequence of these properties is that the mean cycle time of the system exists when N-POM is applied. This cycle time is denoted by C_N and:

$$C_N = E[\theta^N]/N \tag{4.27}$$

The Proposition 4.3 is derived from the previous properties. It sets that C_N can be as close as we want to C_0 if N is large enough, but finite.

Result 4

If properties (4.24) hold, then:

$$C_0 \le C_N \le C_0 + [2/N^{1/3}] \sum_{t \in T} \sigma_t \tag{4.28}$$

Remark

The standard deviations σ_t being finite for any $t \in T$, the right-hand side of (4.28) tends to 0 as N tends to infinity. Thus, whatever $C^* > C_0$, it is possible to find N such that the mean cycle time of the strongly connected event graph is less than C^*. The value N can be obtained by solving the equation $(C_0 - C^*) + [2/y^{1/3}]\ \Sigma_{t \in T}\ \sigma_t = 0$. Which leads to $y^* = [2\ \Sigma_{t \in T}\ \sigma_t/(C^* - C_0)]^3$ and $N = \lceil y^* \rceil$ where $\lceil \bullet \rceil$ denotes the smallest integer greater than or equal to y^*.

(c) *Operating optimization of an event graph*

In the previous section, we proved that it is always possible to reach a mean cycle time less than C^* with a finite number of tokens, providing that $C^* > C_0$. Thus, the value of any p-invariant is finite. A p-invariant being given, we aim at finding a marking M_0 which minimizes the value of the p-invariant and leads to an average cycle time less than C^*.

The heuristic algorithm presented hereafter leads to a near-optimal solution to the problem. Its first phase consists of computing the optimal solution to the deterministic problem obtained by assigning to each transition the mean value of the related random variable. We use the algorithm presented in section 4.2.2 to solve this problem. The second phase of the algorithm is a step-by-step process. At each step of the process, we select the elementary circuit having the greatest mean cycle time (i.e. the elementary circuit such that the sum of the mean values of the firing times is the greatest). $\mathcal{P}$ being the set of places of this elementary circuit having the smallest coefficients in the p-invariant, we select as many elementary circuits as possible having a great cycle time and whose intersect with $\mathcal{P}$ is not empty, and we add one token in a place belonging to this intersection. The process stops when the average cycle time of the strongly connected event graph is less than C^*. In the following, G is the set of elementary circuits.

Algorithm

First phase: computation of an initial solution

1. Compute the optimal solution W_0 to the problem when using the mean value of the related random variable as the firing time of each transition.

   The optimal solution is the one which leads to C_0 for the deterministic problem. We use the algorithms presented in section 4.2.1 to solve this problem. The optimal solution W_0 is specified as the number of tokens in each place at the initial state. The related cycle time π_0 is a lower bound of the cycle time related to the optimal solution to the problem at hand.

2. Using the initial random variables to generate the firing times, simulate the system in order to obtain the mean cycle time Q_0 related to W_0.

   Of course, $Q_0 \geq \pi_0$.

3. If $Q_0 < C^*$, stop the computation.

Second phase: increase adequately the set of tokens

4. Let $\gamma_0 \in \Gamma$ be the elementary circuit having the greatest mean cycle time and $\mathcal{P}$ the set of places belonging to γ_0 and which have the smallest coefficients in the p-invariant, $\gamma_1 \in \Gamma - \{\gamma_0\}$ the elementary circuit having the greatest mean cycle time and at least one place in common with $\mathcal{P}$, $\gamma_2 \in \Gamma - \{\gamma_0 \cup \gamma_1\}$ the elementary circuit having the greatest cycle time and at least one place in common with $\mathcal{P} \cap \gamma_1$, and so on until we reach γ_q such that $\mathcal{P} \cap \gamma_1 \cap \gamma_2 \cap \ldots \cap \gamma_{q-1} \cap \gamma_q = \varnothing$.

   The goal of this step is to define one or more places belonging to as many elementary circuits having a great cycle time as possible and having a small coefficient in the criterion.

5. Add one token to a place P belonging to $\mathcal{P} \cap \gamma_1 \cap \gamma_2 \cap \ldots \cap \gamma_{q-1}$. Let W_0 be the new solution.
6. Simulate the system in order to obtain the mean cycle time Q_0 related to W_0.
7. If $Q_0 \leq C^*$, W_0 is the near-optimal (or optimal) solution; otherwise, go to 4.

According to result 4, this algorithm converges. We illustrate this algorithm by the following small example. We consider the strongly connected event graph presented in Fig. 4.8. We choose the following p-invariant:

$$f(M_0) = 3x_1 + x_2 + x_3 + x_4 + x_5 + x_6 + x_7 + x_8 + 2x_9 + x_{10}$$

where $x_1, x_2, \ldots, x_{10}$ are the number of tokens in $p_1, p_2, \ldots, p_{10}$ respectively for the marking M_0. The random variables X_1, X_2, X_3, X_4, X_5, X_6 are

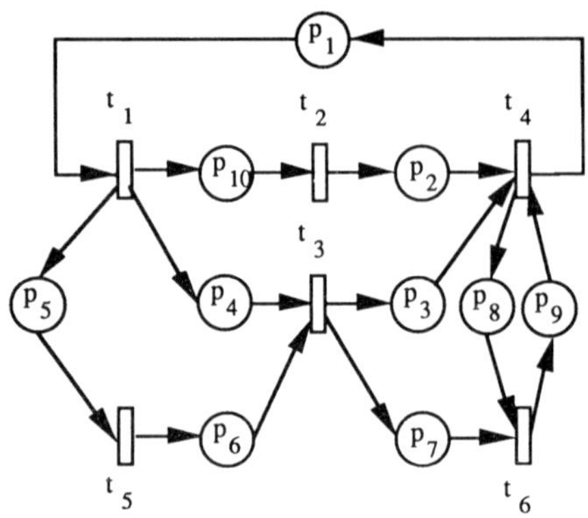

Figure 4.8 A strongly connected event graph.

Table 4.4 Steps of the second stage

Steps	*Put one more token in*	*Mean cycle time*	*Value of the p-invariant*
1	p_{10}	10.76	8
2	p_7	10.63	9
3	p_8	10.52	10
4	p_2	10.12	11
5	p_3	10.09	12

assigned to the transitions t_1, t_2, t_3, t_4, t_5 and t_6 respectively. Their distributions are the following:

$$X_1 : f_1(x) = \begin{cases} 1/10 & \text{if } x \in [0, 10] \\ 0 & \text{otherwise} \end{cases}$$

$$X_2 : f_2(x) = \begin{cases} (1/\beta^\alpha\, \Gamma(\alpha)) \cdot x^{\alpha-1} \exp(-x/\beta) & \text{if } x \geq 0 \\ 0 & \text{otherwise} \end{cases}$$

$a = 2$ and $b = 5$

$X_3 : f_3(x)$ is the same as f_2, but with $a = 3$ and $b = 1$

$X_4 = 5$ (constant)

$$X_5 : f_5(x) = \begin{cases} 10 \exp(-10x) & \text{if } x \geq 0 \\ 0 & \text{otherwise} \end{cases}$$

$$X_6 : P\{X_6 = 3\} = 1/2 \quad \text{and} \quad P(X_6 = 1) = 1/2$$

We can see that $C_0 = 10$ and we choose $C^* = 10.1$. The optimal solution of the deterministic problem consists of putting one token in each of the places p_2, p_3, p_4, p_5, p_9 and p_{10}. For this solution, the mean cycle time of the stochastic problem is 12.39 and the p-invariant is equal to 7. The following steps are given in Table 4.4.

Finally the optimal (or near-optimal) marking M_0 which leads to an average mean cycle time less than 10.1 and to a value of the p-invariant which is as small as possible is such that:

$$M_0(p_1) = 0;$$
$$M_0(p_2) = M_0(p_3) = 2;$$
$$M_0(p_4) = M_0(p_5) = 1;$$
$$M_0(p_6) = 0;$$
$$M_0(p_7) = M_0(p_8) = M_0(p_9) = 1;$$
$$M_0(p_{10}) = 2.$$

4.3.4 Conclusion

In the deterministic case, we pointed out three algorithms to reach a given performance while minimizing an invariant criterion (which represents, in our examples, the minimization of the WIP).

The first algorithm needs the computation of the elementary circuits of the event graph model, and thus results in a computational burden for real-life problems. The adjustment algorithm provides a limited set of feasible solutions containing at least one optimal solution. This heuristic algorithm leads to a near optimal, if not optimal, solution to the problem. The convex optimization algorithm needs more computation than the previous one, but leads to optimal solutions in a reasonable amount of time.

In the stochastic case, a good compromise between the objective function (which must be as small as possible) and the mean cycle time (which must also be as small as possible) is obtained using a step-by-step approach. A simulation step is included in this approach. We are still far from the analytical formulation of upper and lower bounds of the mean cycle time close enough to each other to provide a useful evaluation of the related system.

The work presented in this section is of great interest for evaluating manufacturing systems assuming a cyclic manufacturing of the parts, since it has been shown that not only job-shops and Kanban systems, but also assembly systems, can be modeled using event graphs.

4.4 THE GENERAL CASE

In Chapter 2 we showed that it is possible to take into account decisions made outside the system by means of control places, to model the arrival of raw material by means of input transitions and, finally, to model parts leaving the system using output transitions. These concepts have been used to model simple modules, and we observed that the control place markings had to comply with some constraints in order to guarantee some 'good' properties to the module models at hand. Let us first recall what these 'good' properties are from a manufacturing system point of view.

4.4.1 Qualitative properties of a Petri net model

Important properties of the whole Petri net model from a manufacturing point of view include:

- **liveness**, which ensures that each operation modeled in the manufacturing system can be executed, whatever the state of the system is;
- **boundedness**, which ensures that work-in-process (WIP) in the system is upper bounded (i.e. there is no excessive accumulation of WIP);
- **reachability**, which ensures that the system can always reach any of its states. This property is especially important for highly automated systems in which system restarts can only be made from a given state;
- **persistence**, which means that firing a transition does not preclude a firing which was previously possible. Such a Petri net is said to be **decision-free**. A manufacturing system modeled by a Petri net having this property is a system which can evolve without the need to make any decision during its operation.

These qualitative properties have to be preserved by marking the control places in an adequate manner. An adequate control place marking is called **feasible marking**. To be more specific in the definition of a feasible marking, we have to introduce some additional definitions about the Petri net model.

4.4.2 Some more definitions

A Petri net model enriched by control places, input transitions, output transitions and the related arcs is referred to hereafter as an Extended Petri Net (EPN). Before being enriched, a Petri net is called a Regular Petri Net (RPN).

In addition to the EPNs and the RPNs, we define the Reduced Petri Net related to a given control place marking M^C, denoted by RPN-M^C. An RPN-M^C is derived from an EPN, the marking of which is M^C, by:

- removing the input and the output transitions, as well as the arcs having these transitions at one end;
- removing the arcs, one end of which is the output transition of an **empty** control place and the other end of which is a place subject to conflicts, as well as the related transition, if it is isolated;
- removing the control places, as well as the arcs having these places at one end.

For instance, let us consider the model presented in Fig. 4.9 and let us denote by $(a_1, a_2, \ldots, a_6)$ a control place marking where $a_i = 1$ if C_i contains one token and $a_i = 0$ otherwise. Figures 4.10, 4.11, 4.12 and 4.13 present the RPN-M^C related to four possible markings.

The RPN-M^Cs corresponding to $M_5^C = (0, 0, 0, 0, 1, 0)$ and $M_6^C = (0, 0, 0, 0, 0, 1)$ are similar to the ones of Figs 4.12 and 4.13, respectively. If we

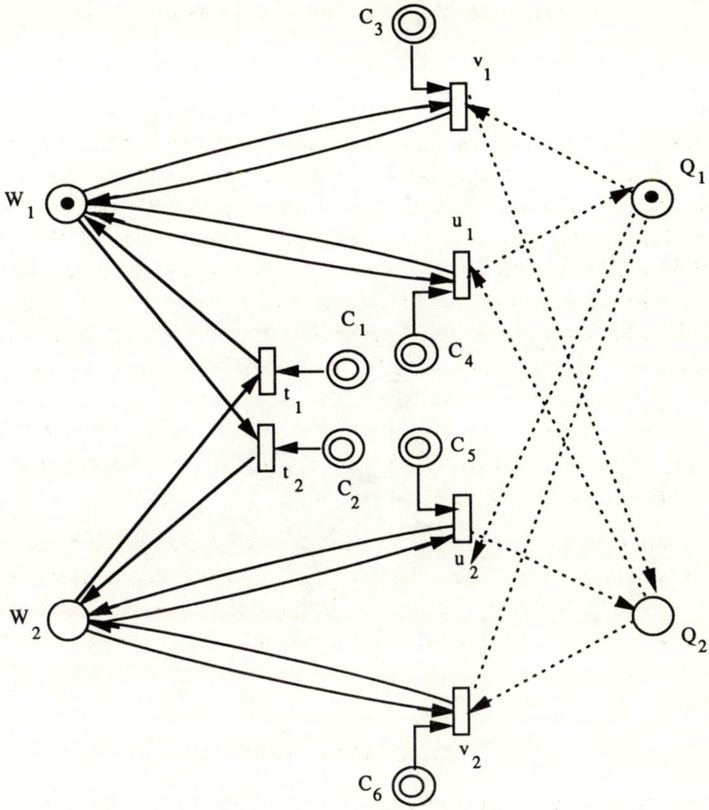

Figure 4.9 Crane model.

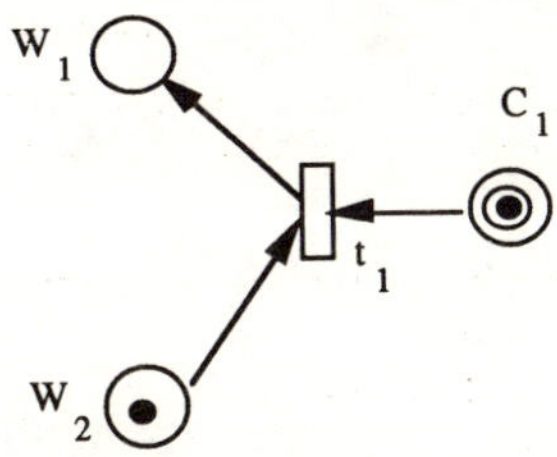

Figure 4.10 $M_1^C = (1, 0, 0, 0, 0, 0)$.

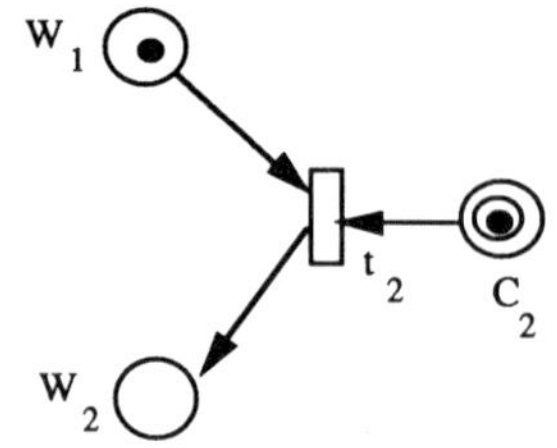

Figure 4.11 $M_2^C = (0, 1, 0, 0, 0, 0)$.

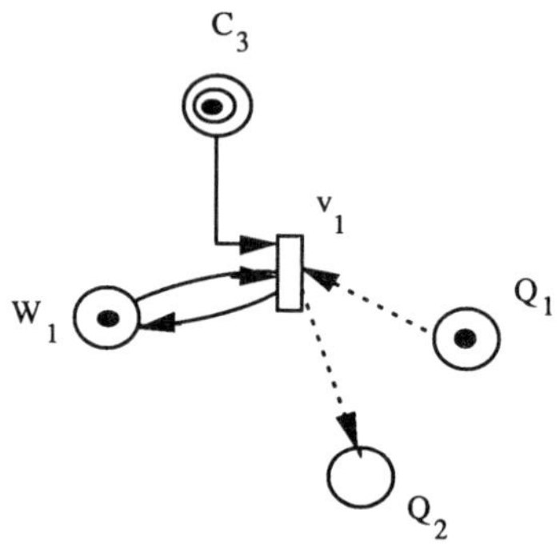

Figure 4.12 $M_3^C = (0, 0, 1, 0, 0, 0)$.

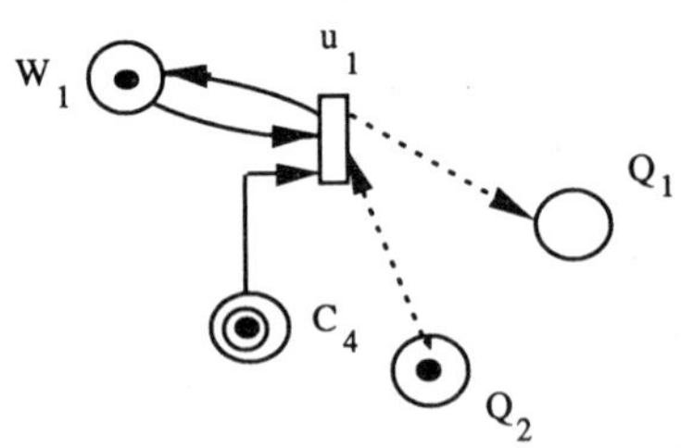

Figure 4.13 $M_4^C = (0, 0, 0, 1, 0, 0)$.

consider an RPN-M^C, we can see that its initial marking (i.e. the state place marking when marking M^C is applied) is always the same. For instance:

- the initial marking of RPN-M_1^C consists of one token in W_2;
- the initial marking of RPN-M_2^C consists of one token in W_1;
- the initial marking of RPN-M_3^C consists of one token in Q_1 and W_1;
- etc.

We can see that the final marking of an RPN-M^C (i.e. the state place marking when the RPN-M^C blocks after removing M^C) is also always the

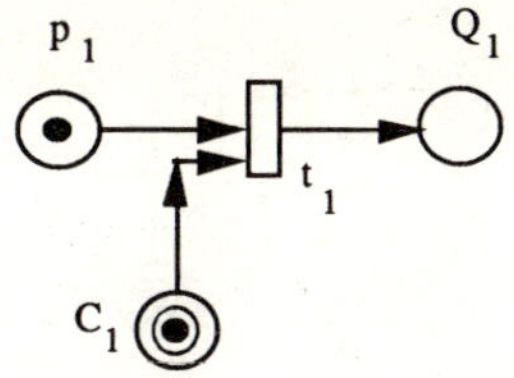

Figure 4.14 RPN-M^C of the first type.

same. Based on these definitions, we can define what should be a module model and the related feasible control place markings.

4.4.3 Module model and the related feasible control place markings

A module model is said to be **well defined** if the following characteristics hold:

(i) A set of control place markings is defined;
(ii) For each of the previous control place marking M^C:

- the initial marking of the RPN-M^C is known (and bounded),
- the final marking of the RPN-M^C is known,
- each transition of the RPN-M^C fires at least once when moving from the initial marking to the final marking,
- the RPN-M^C is decision-free for the initial marking,
- the RPN-M^C is bounded for the initial marking.

These control place markings are referred to as feasible control place markings.

Thus, if the EPN marking restriction to the control places is a feasible control place marking M^C and the EPN marking restriction to RPN-M^C is the initial marking of the RPN-M^C, then the marking will evolve toward the final marking of the RPN-M^C. Depending on the type of model, either the final marking will be reached after a finite amount of time, after which the system will block, or the system will stop at the final marking only after the tokens are removed from the control places.

Figure 4.14 shows an RPN-M^C of the first type. The initial marking consists of one token in place P_1. If M^C consists of putting one token in C_1, then t_1 fires and the final marking consists of one token in Q_1, and none in P_1. In this case, t_1 fires only once, after which the system blocks.

The model shown in Fig. 4.15 is of the second type: transition t_2 fires indefinitely as long as there is one token in C_2. But if this token is removed, the system blocks and the final marking, which is the same as the initial one, consists of one token in P_2. In both cases, we start from the initial marking and reach the final marking after the tokens are removed from the control places.

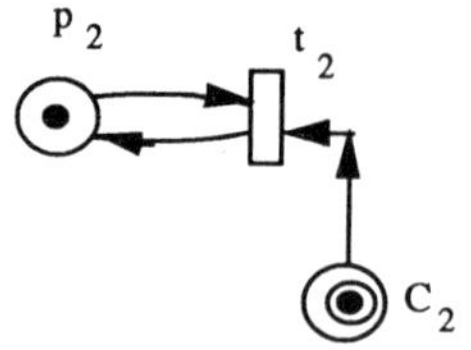

Figure 4.15 RPN-M^C of the second type.

4.4.4 The complete model

Assume that a complex manufacturing system has been split up into modules, and that each of these modules has been modeled. Let $\mathcal{H} = (H_1, H_2, \ldots, H_s)$ be this set of module models. We assume that the elements of $\mathcal{H}$ are well defined, according to the definition of subsection 4.4.3. The model of the whole system is obtained by merging output transitions of some module models to input transitions of other module models. Let $\mathcal{M}$ be this model. A feasible control place marking of the whole model is the union of feasible control place markings to the related module models.

The whole model $\mathcal{M}$ is said to be feasible for the problem at hand if there is at least one infinite sequence of feasible control place markings such that each output transition of a marked control place is enabled for any of these markings. In other words, the initial marking of the ith RPN-M^C is contained in the RPN marking at the end of the $(i-1)$th stage (i.e. the stage where the $(i-1)$th control place marking of the infinite sequence is applied).

Applying the infinite sequence of feasible control place markings to such a model guarantees that the RPN derived from the model is decision-free, bounded and live, assuming that the input places are marked at any time. Reachability in the RPN is not guaranteed in general, but it is easy to show that a home state usually exists.

Finally, a model can be seen that has a sequence of Petri net models having 'good' properties (from the point of view of manufacturing management), the links between two consecutive models being the RPN marking when one switches from the control place marking relative to the current model to the control place marking relative to the next one.

The goal of the O-DMS (Outside-Decision Making System) is to decide on the switching from a given feasible control place marking $M^C_{i_1}$ to the next one, say $M^C_{i_2}$. To make this decision, the O-DMS has to check that:

(i) the current RPN-$M^C_{i_1}$ is at its final marking;
(ii) the RPN-$M^C_{i_2}$ initial marking is included in the RPN marking at the end of the previous stage;
(iii) the marking of the input places is such that RPN-$M^C_{i_2}$ is not blocked.

The module models are connected to one another by merging the output transitions of some of them to the input places of others according to

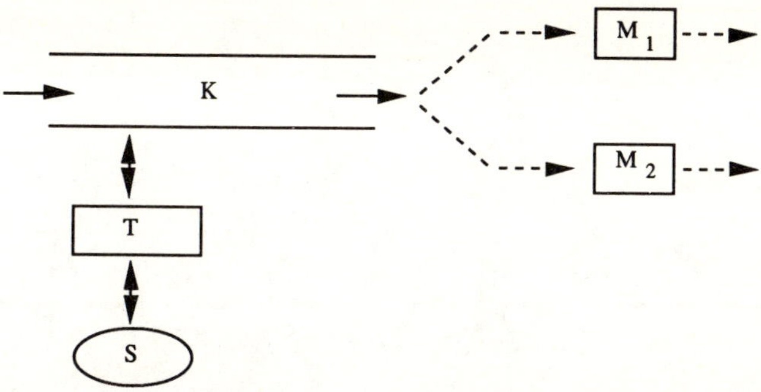

Figure 4.16 A comprehensive example.

system requirements. As a consequence, the input transitions of the module models (except those which concern the entrance of the manufacturing system) are replenished by the tokens provided by the output places of other module models. Furthermore, at each point in time, one can change or not change the feasible control place markings of the module models and, sometimes, several new control place markings are possible. The changing times and the next feasible control place markings are chosen at the O-DMS level according to the criterion to be optimized (or the compromise to be reached between several criteria). The O-DMS may be either a computer program, some skilled experts, or both. The decisions are made according to the state of the system. The algorithm applied to reach the decision depends on the objective function.

4.4.5 A comprehensive example

The system considered is depicted in Fig. 4.16. Parts arrive on a conveyor K, which can hold s parts. The output of K feeds two machines M_1 and M_2. If the conveyor is full and the machines are busy, then W_1 parts can be moved into a storage resource S using a crane T. If the conveyor is empty, W_2 parts can be carried back on the conveyor from S.

The set made of the conveyor K, the crane T and the storage resource S, can be modeled as shown in Fig. 4.17. The values in parentheses are firing times. In this example, we consider a three-position conveyor ($s = 3$). Places P_1, P_2 and P_3 represent the three positions. The firing times of transitions t_1, t_2, t_3 and t_4 are the times needed to move a part from outside the system to the first storage position, from the first storage position to the second, from the second storage position to the third, and from the third storage position to outside the system, respectively. The conveyor is in working order when there is one token in C_0. When the conveyor is empty, we can find one token in each of the places Q_1, Q_2 and Q_3.

If there is one token in C_1 and one token in P_1 and P_2, T_1 fires and (i)

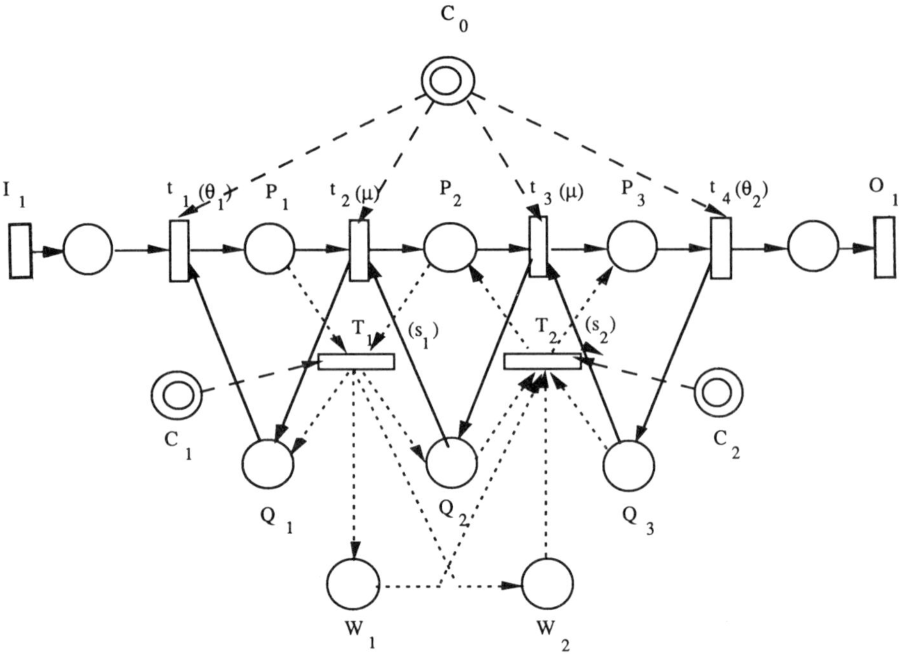

Figure 4.17 Conveyor, crane and storage facility: the KTS model.

one token appears in each of the places Q_1 and Q_2, and (ii) one more token appears in each of the places W_1 and W_2. Firing T_1 represents the transportation of two parts which are in the two first places of the conveyor into the storage facility.

Putting one token in the control place C_2 results in firing transition T_2 if there is one token in Q_2 and Q_3, and at least one token in W_1 and W_2. Firing T_2 represents the operation of carrying back two parts from the storage facility and putting them in the last two places of the conveyor. The constraint on the marking of the control places consists of putting at most one token in the set of control places C_0, C_1, C_2 at each point in time.

The model presented in Fig. 4.17 will be summarized in the remainder of this chapter as indicated by Fig. 4.18. Machines M_1 and M_2 are modelled as represented in Figs 4.19 and 4.20. It can be noticed that the reduced model is an event graph (i.e. the circuit (t, R)). The model for machine M_2 is identical (Fig. 4.20). The global model is presented on Fig. 4.21.

4.4.6 Conclusion

The method which has been proposed is a bottom-up approach which consists in: (i) splitting the whole system into simple, thus tractable, modules; (ii) modeling each module using extended Petri nets (in the sense

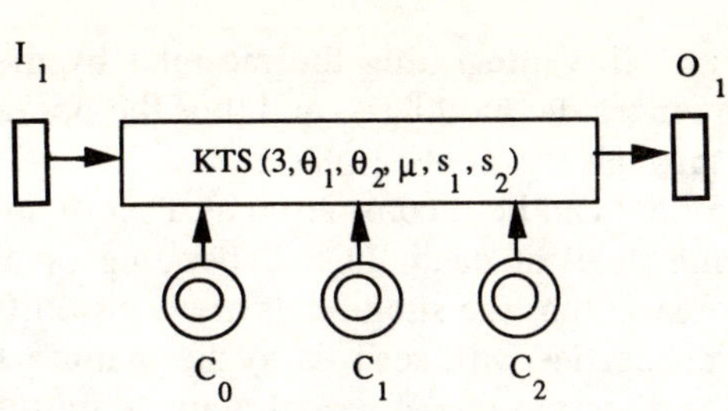

Figure 4.18 Representation of the KTS model.

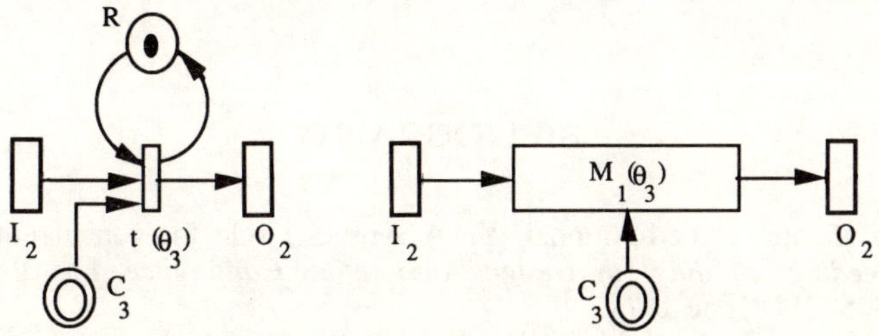

Figure 4.19 Model of M_1 and representation of the model.

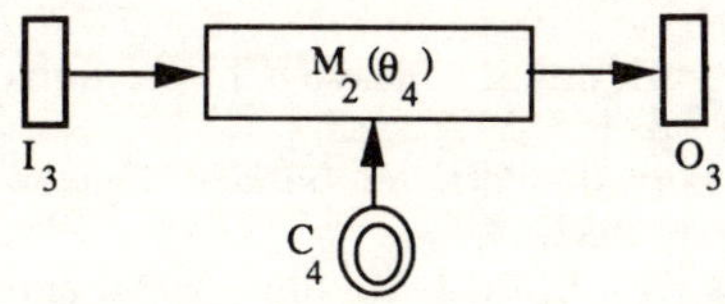

Figure 4.20 Representation of the model of M_2.

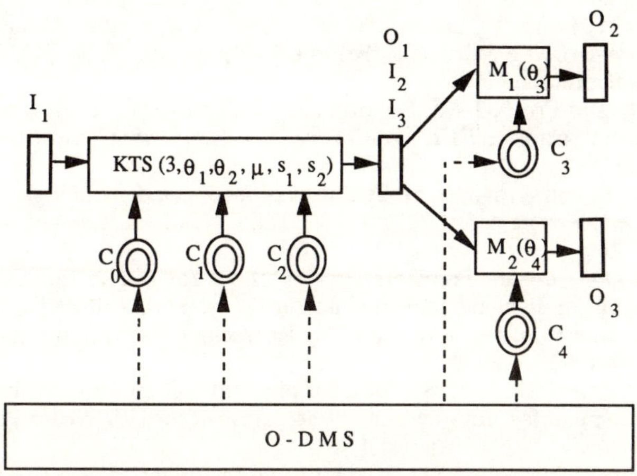

Figure 4.21 Global model.

defined above); and (iii) integrating the modules by means of a control system which influences the modules – and thus the operating mode of the whole system – through the control places.

Each module model can be reconsidered as a set of submodels, at most one of them being 'live' at each time, depending on the control place markings to make sure that the submodels are decision-free and that they have the 'good' properties with regards to the manufacturing system requirements. The integration of the model is made by the control system, which acts on the control place markings, under the constraints applied to these markings, in order to optimize the system – or to find the best compromise between the criteria being considered.

BIBLIOGRAPHY

Agerwala, T. and Choed-Amphai, Y., A synthesis rule for concurrent systems. *Proceedings of the 15th Design Automation Conference*. Las Vegas, NV, pp. 305–311, June 1978.

Alaiwan, H. and Toudic, J. M. Recherche des semi-flots, des verrous et des trappes dans less réseaux de Petri. *Techniques et Science Informatiques*, vol. 4, no. 1, pp. 103–112, 1985.

Baccelli, F. Ergodic theory of stochastic Petri networks, *INRIA Research Report No. 1037*, May 1989.

Benson, H. P. A finite algorithm for concave minimization over a polyhedron. *Nav. Res. Logis.* vol. 32, pp. 165–177, 1985.

Chretienne, P. Les réseaux de Petri temporisés. Université de Paris VI, Paris, France, Thèse d'Etat, 1983.

Claver, J. F., Harhalakis, G., Proth, J. M., Savi, V. M. and Xie, X. L. A step-wise specification of a manufacturing system using Petri nets. *Conference IEEE Systems, Man and Cybernetics*, Charlotte, Virginia, October 1991.

Commoner, F., Holt, A., Even, S. and Pnueli, A. Marked directed graphs. *Journal of Computer and System Science*, vol. 5, no. 5, 1971.

Di Mascolo, M., Frein, Y., Dallery, Y. and David, R. *A Unified Modeling of Kanban Systems Using Petri Nets*. Technical Report No. 89-06, LAG. Grenoble, France, September 1989.

Hillion, H. P. and Proth, J. M., Performance evaluation of job-shop systems using timed event-graphs. *IEEE Transactions on Automatic Control*, vol. 34, no. 1, pp. 3–9, 1989.

Holloway, L. E. and Krogh, B. H. Synthesis of feedback control logic for a class of controlled Petri nets. *IEEE Transactions on Automatic Control*, vol. 35, no. 5, May 1990.

Koh, I. and DiCesare, F. Transformation methods for generalized Petri nets and their application in flexible manufacturing systems. *Proceedings of the 2nd International Conference of the CIM*, Rensselaer Polytechnic Institute, Troy, NY, May 1990.

Krogh, B. H. and Beck, C. L. Synthesis of place/transition nets for simulation and control of manufacturing systems. *Proceedings of the IFIP Symposium of Large Scale Systems*, Zurich, August 1986.

Laftit, S., Proth, J. M. and Xie, X. Optimization of invariant criteria for event graphs. *IEEE Transactions on Automatic Control*, May 1992.

Mu Der Jeng and DiCesare, F. A review of synthesis techniques for Petri nets. *Proceedings International Conference on Computer Integrated Manufacturing*, Troy, NY, pp. 348–355, 1990.

Murata, T. Petri nets: properties, analysis and applications. *Proceedings of the IEEE*, vol. 77. no. 4, April 1989.

Narahari, Y. and Visvanadham, N. A Petri net approach to the modeling and analysis of flexible manufacturing systems. *Annals of O.R.*, vol. 3, pp. 449–472, 1985.

Peterson, J. L. *Petri Nets Theory and the Modeling of Systems*. Prentice-Hall, Englewood Cliffs, NJ, 1981.

Ramamoorthy, C. V. and Ho, G. S. Performance evaluation of asynchronous concurrent systems using Petri nets. *IEEE Transactions on Software Engineering*, vol. SE-6, no. 5, pp. 440–449, 1980.

Ramchandani, C. Analysis of asynchronous concurrent systems by timed Petri nets. *Laboratory of Computer Science, Massachusetts Institute of Technology*, Cambridge, MA. Tech. Rep. 120, 1974.

Sifakis, J. A unified approach for studying the properties of transition systems. *Theoretical Computer Science*, vol. 18, pp. 227–258, 1982.

Van Thoai, N. and Tuy, H. Convergent algorithms for minimizing a concave function. *Mathematics of Operations research*, vol. 5, no. 4, pp. 556–566, 1980.

Warland, J. *An Introduction to Queueing Networks*. Prentice-Hall, Englewood Cliffs, NJ, 1988.

5

Petri nets for manufacturing information systems

G. Harhalakis and F. B. Vernadat

5.1 INTRODUCTION

The information system is a central component of any manufacturing system in the sense that it provides support to its two other fundamental subsystems: the physical system (or operational system, i.e. where the physical actions take place using enterprise resources) and the decision system (i.e. where strategic and tactical decisions are made). The relationships among these three subsystems are illustrated by Fig. 5.1. While discussions of the previous chapters of the book are mainly concerned with modeling and analysis of the decision and physical systems of manufacturing, this chapter focuses more on the information system of a manufacturing enterprise.

Information system modeling and analysis have been the focus of extensive research and development both in the data processing community (with rapid and successful advances on data representation and data manipulation as used in database technology) as well as in software engineering (concerning structured analysis and structured design methodologies, formal specification methods and rapid prototyping of information systems). However, tremendous attention has been paid to data modeling rather than to true information system design and analysis. This is acknowledged by the recognition of well-established and widely used data models such as the hierarchical data model (Tsichritzis and Lochovsky, 1976), the relational data model (Codd, 1970) or the entity-relationship model (Chen, 1976; Teorey *et al.*, 1986). Furthermore, various semantic data models have been developed to capture more of the semantics of data, and thus to better address the information modeling issues (Hull and King, 1987). The problem is that all these models essentially concentrate on the representation of the static properties of data and information (i.e. structural properties and associated integrity rules) for the purpose of consistent data structuring, data querying and data storage. However, information analysis in manufacturing systems also requires the modeling of:

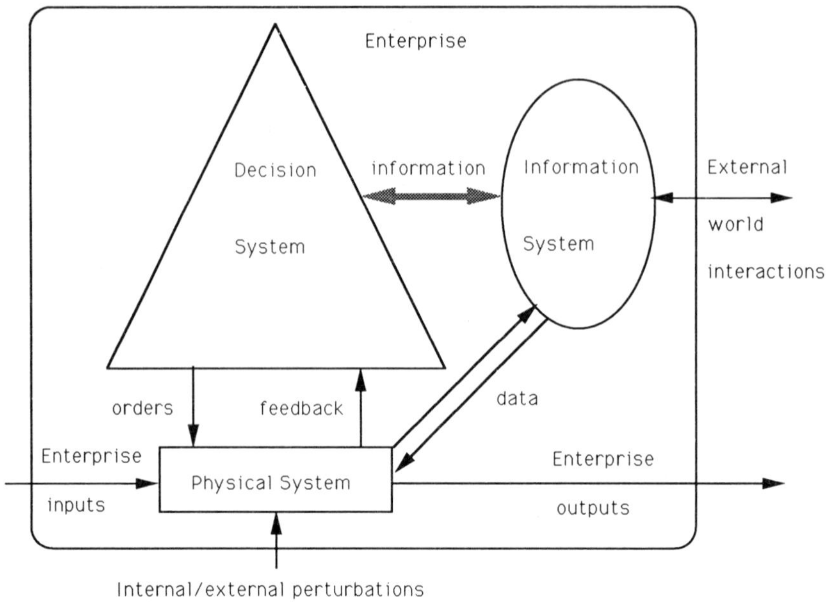

Figure 5.1 Subsystems of a manufacturing system.

(i) dynamic properties of data and information (e.g. events or time-dependent properties) in terms of processes and transactions for data accesses, updates and interoperability; and
(ii) company objectives and policies, i.e. enterprise knowledge in the form of management declarative rules and business rules.

Currently, research trends in database technology and software engineering are concentrating on the object-oriented approach (Atkinson *et al.*, 1990; Coad and Yourdon, 1990; Meyer, 1988; Wirfs-Brock and Johnson, 1990) to cover the static and dynamic aspects of information processing in an integrated way. This is certainly an adequate approach for data modeling and final implementation of the model. Nevertheless, it is our conviction that at the organization and even at the conceptual modeling levels Petri nets provide a more suitable framework to analyze dynamic properties of data and information and to implement company policies. Petri nets can be profitably coupled to existing data modeling approaches to extend their analytical power as shown in this chapter.

When manufacturing information systems are considered, two distinct problems can be identified. One is a *database design* problem, i.e. defining and implementing relevant data storage structures optimized for application systems or users' needs in terms of data access, data sharing and limited data redundancy. Another one is a *knowledge-base design* problem, i.e. defining and implementing relevant manufacturing policies and business rules governing enterprise operations in an orderly way which

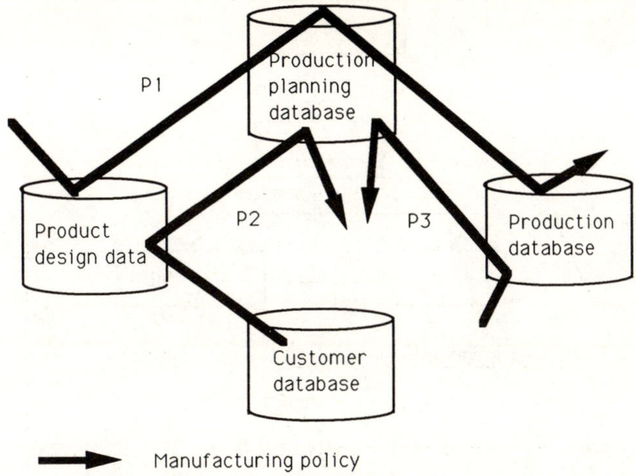

Figure 5.2 Manufacturing databases and company policies.

will guarantee overall consistency of the various databases and related data processing. This is illustrated by Fig. 5.2.

The chapter is thus organized as follows. After an introduction on the basics of manufacturing information systems and high-level Petri nets, two distinct, but complementary, methodologies are presented. The first one called *M** is concerned with the analysis and design of integrated manufacturing information systems, usually implemented by several databases. The other one concentrates on analysis, modeling and implementation of global company policies, assuming the existence of local databases, as a *knowledge-based* system managing the associated global integrity rules.

5.2 BASICS OF MANUFACTURING INFORMATION SYSTEMS

Basics and comprehensive discussions of data modeling, data manipulation, data querying and database technology are now widely available in many textbooks. For in-depth discussion, we recommend the authoritative books by Date (1986) and Ullman (1982).

5.2.1 Data and information

The information system of a manufacturing enterprise is made of all pieces of data and information used/stored/processed for the needs of users and applications of the enterprise. It includes any kind of data items, data stores, administrative or technical documents, forms, verbal orders, management directives, company policies, phone calls, signals, etc.

The information system is the memory of the manufacturing system, containing its valuable data, know-how and expertise. It is made of *data*

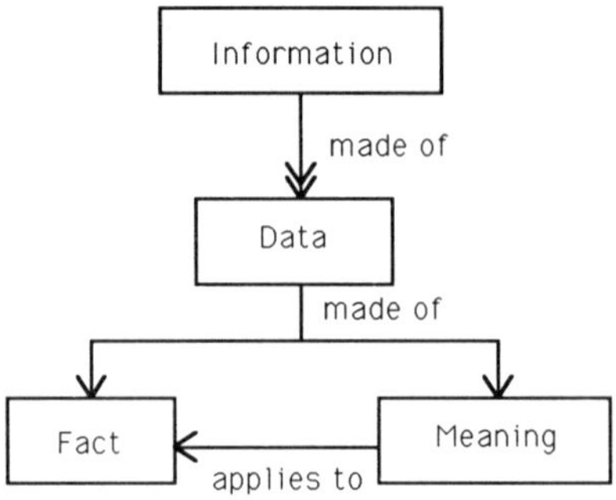

Figure 5.3 Data and information.

and *information* as well as *constraints* and *rules*. It is usually computerized in the form of one or more *databases*.

Database. A database is a large collection of interrelated operational data of some enterprise stored and structured to be shared and queried by the various users and applications of the enterprise (Date, 1986).

Data. A datum is a fact. It is a value. For instance, 1.62 is a datum.

Information. Information is made of at least two elements: a fact and a meaning. Information is the basis for communication. For instance, '1.62 m is my height' is information. Information is a recursive concept in the sense that information can be made of information (and, thus, of several facts and meanings). Figure 5.3 illustrates the relationships among data and information (Appleton, 1984).

Constraints. Constraints are used to restrict the domain of values of some pieces of information. For instance, the age of a person must be an integer value between 0 and 120. Some constraint may be defined as predicates; for instance UNIQUE (Social_Security_Number) means that any social security number must be unique.

Rules. Rules are used to express statements. They are usually made of two parts: a condition clause and an action clause. They can be represented using the following condition–action form:

ON (conditions) DO (actions)

where 'conditions' is a (possibly empty) list of conditions linked by logical operators (AND, OR, XOR, NOT) and 'actions' is a list of actions to be taken if the condition clause evaluates to true. If the condition clause is empty, the rule is said to be *imperative*. Otherwise, it is said to be *conditional*. Formally, a condition can be defined as a predicate.

What must be modeled. The basic aspects of an information system that should be modeled include:

- *Structural knowledge*, that is the entities of the manufacturing system, their characteristics and the relationships among them.

- *Declarative knowledge*, that is rules and constraints applicable to full entities or to some of their characteristics only and constraining the legal states of the manufacturing system. This class of knowledge can be further subdivided into: *integrity constraints* and *declarative rules*. Integrity constraints limit the range of possible values of characteristics of entities and relationships (a further subdecomposition of manufacturing integrity constraints can be found in Famili and Vernadat, 1986). Declarative rules are used to describe company policies and business rules.
- *Procedural knowledge*, that is the way in which basic operations should be organized into data processing units (or processes) and how they must be applied to structural knowledge to support activities performed in the physical and decision systems of the manufacturing system.
- *Information system dynamics*, that is the evolution of the information system from one state to another state when processes are executed according to the generation of occurrences of external events, as well as communications among processes and between processes and users.

A similar classification was first proposed by Albano *et al.* (1985).

Usually, structural knowledge and part of declarative knowledge are modeled during information system design and analysis using extended entity-relationship models such as in DATAID (Ceri, 1983), MERISE (Tardieu *et al.*, 1983), *M** (Di Leva *et al.*, 1987), TSER (Hsu and Rattner, 1990), the relational data model (Teorey *et al.*, 1986) or using structurally object-oriented models. Procedural knowledge is in many cases modeled by means of data-flow diagrams (DFD) like in SSAD (Gane and Sarson, 1979) or in the form of database transactions (Ngu, 1989). Two excellent books on these topics are by Martin and McClure (1985) and Pressman (1987). Information system dynamics is usually neglected or weakly addressed in many information system design methodologies with the exception of the object-oriented approach (Sibertin-Blanc, 1991).

Petri nets can provide a sound and suitable approach to model the latter aspect of information systems and are subject to formal specification of executable models of information systems. This will be illustrated in this chapter in the case of an extended entity-relationship model and for the object-oriented approach.

How to model. A structured approach must be used for design and analysis of any information system. For manufacturing systems we adopt the approach proposed in *M**: an information system design and analysis methodology for manufacturing environments (Di Leva *et al.*, 1987; Vernadat *et al.*, 1989). The methodology is structured into three major phases.

Organization analysis (or Requirements Definition Modeling). This phase is concerned with functional analysis of the organization (or a part of the enterprise) in terms of objects, functions and rules (or policies) of the enterprise, based on the analysis of business objectives and constraints as defined by users. It is usually based on high-level models and always involves a functional decomposition principle. The main objective is to build an *organization schema* (made of so-called organization nets) defining the organization functionality, the enterprise entities and the organization behavior. While many other methodologies are using SADT and data-flow diagram (DFD) techniques at this level, we propose to use high-level Petri nets such as predicate/transition nets (PrT-nets) or colored Petri nets as the basic modeling tools for this phase to express more precise and more complete requirements.

Conceptual design. This phase is concerned with the detailed design of the information model which will have to be implemented to support the activities of the physical system and the decision system of the organization. It is concerned with modeling and analysis of the static and dynamic properties of the information system using semantic or object-oriented data models. This is achieved on the basis of data processing requirements (formatted and unformatted requirements) as expressed by users or applications during organization analysis. The objective is to build a *conceptual schema* of the information system, expressed in an implementation independent, computer processable, language subject to formal verifications and consistency checking. We propose to use high-level Petri nets such as the Process and Data Net (PDN) model of *M** for database design (section 5.4) and Updated Petri Nets (UPN) for knowledge-base design (section 5.5).

Implementation design. This phase is concerned with the implementation description of the *internal schema* of the information system to be automated. The internal schema is made of (i) an implemented information model, and of (ii) implemented database transactions performing units of processing on data of the implemented information model. Database transactions can be used by application programs or users of the information system. The implemented information model is made of one or several physical database schemata (data structures, indexes, integrity constraints and access rights) to be implemented on commercial database management systems (DBMSs) used by the manufacturing system. Thus, one objective of this task is to transform the conceptual schema of the previous phase into one or several implementation data models (using the relational model, the hierarchical model, the network model, object-oriented models or even flat-files). Only the implementation of company policies using the Update Dependency Language (UDL) in the context of relational databases is discussed in the book.

5.2.2 Manufacturing information and application systems

Manufacturing information. Manufacturing environments are characterized by the fact that all kinds of data and information are present, from static to very dynamic types of information, from simple to highly structured types of information, from very short to long duration types of transaction processing. Major types of information in manufacturing are (Sartori, 1988; Vernadat, 1984):

- Product information: dealing with product identifiers and product descriptions in terms of drawings, parts lists, NC data, CAD data, engineering data, etc., as well as product versions. They describe what will be produced.
- Process information: dealing with production processes in terms of numerically controlled (NC) programs, work plans, manufacturing processes, activities, operations, etc. They describe how to produce parts and products.
- Resource information: dealing with enterprise facilities (including machines, humans and applications) in terms of abstract and real objects including their production capacity as well as their spatial, temporal and organizational relationships. They describe the production means.
- Planning information: dealing with long-term, medium-term and short-term plans, schedules and inventories. They describe what, when and how much to produce or to buy.
- Production information or shop-floor information: dealing with actual values of process variables in terms of equipment statuses, labor allocation, quantities ordered, quantities produced, quantities stored, options required, etc. They describe the day-to-day operation status and shop-floor progress.
- Administrative information and cost data: dealing with personnel data, supplier's and customer's data, customer orders, procurement orders, cost centers, etc. They describe management information.

Application systems. Manufacturing environments are also characterized by typical functional areas which can be the sources of structural, declarative or procedural knowledge for the information system. The following ones can be mentioned:

- CAD and CAE: computer-aided design and computer-aided engineering are concerned with the definition, preliminary design, detailed design, analysis, testing and documentation of new parts or products based on product specifications.
- Process planning (CAPP): this is concerned with the definition of the manufacturing processes to produce the parts or the products including the selection of the fixtures, the tools and the machines as well as the specification of standard processing times and machinability data. It also covers the generation of NC programs.
- Production planning (MRP II): Manufacturing Resource Planning covers all functions of production planning from master production planning to materials requirements planning, capacity planning, inventory management and production scheduling. It starts from sales forecasts, customer demand and inventory levels and ends up with a list of what

to buy and what to order, in which quantities and for when (procurement lists and work schedules).

- Shop-Floor Control (SFC): this is concerned with the actual execution of the manufacturing orders as defined in the work schedules using available pieces of equipment and labor resources. It is also concerned with data collection about current work-in-progress for feedback to MRP II.

5.2.3 Company objectives – Computer-Integrated Manufacturing

To remain competitive, to increase their competitiveness and to survive in rapid market changes, most companies, from small to large ones, in industrialized countries are striving for the automation of their manufacturing operations. New technologies (CAD, CAE, CAPP, etc.), industrial automation (involving automated cells and Flexible Manufacturing Systems or FMS, robots and automated handling systems such as AGVs), modern management techniques (MRP, Just-In-Time, Kanban, Total Quality Management, etc.), more efficient organization structures (optimized plant layout, flat administrative structures) and enterprise-wide integration, known as CIM for Computer-Integrated Manufacturing, are usually presented as key technologies for success. However, without a clear understanding of company objectives, automation often fails.

Company objectives. Company objectives are concerned with the desired future position of the enterprise in terms of growth, profitability, investments, risk assessment, etc. (Gerelle and Stack, 1988). They are established by top management according to results of an environment analysis (concerning market trends, competitors, emerging new technologies, etc.). They are used to define a company strategy (i.e. to define the nature of products that the enterprise wants to produce, the markets to be targeted, the planned production capacities, etc.).

These high-level company objectives and company strategies must be broken down into sub-objectives and sub-strategies. For instance, company objectives must be decomposed into marketing objectives, manufacturing objectives and management objectives. Manufacturing objectives may concern production productivity, flexibility, quality and adaptability. Company strategies include product strategies, market strategies and manufacturing strategies.

Product strategy concerns decisions about product functionality, product introduction rate, product variety, pricing, quality, etc.

Market strategy concerns decisions about market share, market coverage, customer service, promotion, distribution, etc.

Manufacturing strategy is set according to manufacturing objectives and concerns the manufacturing structure (facility, machinery, tools, materials, utilities, etc.) and the manufacturing infrastructure (knowledge/attitude, rules, organizational structure, etc.). Implementing the manufacturing strategy may require changes in both the manufacturing structure and manufacturing infrastructure.

Computer-Integrated Manufacturing (*CIM*). CIM is generally defined as the automation and integration of the various functions and entities of a manufacturing company through the use of computers and existing standards. Integration can only be achieved if the right information is available at the right place and at the right time. Automation and integration address enterprise functions (business processes and enterprise activities), information (business data and rules) and manufacturing system components (including applications, people and machines). CIM is a manufacturing philosophy, not a technology, aiming at enterprise integration (i.e. system integration, application integration as well as business integration) and contributing to increase enterprise flexibility through the management of change (i.e. new functions can be installed and others changed without major perturbations in the existing system).

Activities of the enterprise covered by CIM encompass product design and engineering, production planning and control and factory automation. They range from customer order entry to product shipment and customer service.

Information is a key ingredient to be managed in such a framework since it is the basis for communications and provides for timely cooperation of components and functions. In theory, information should be stored once and be available at the right place and at the appropriate time. This certainly assumes the availability of suitable data storage systems, computer communications networks and data access protocols, that hardly exist even today.

It is therefore important that information be correctly modeled, its specificities acknowledged and its use carefully planned to avoid unnecessary overload of the system and to optimize its operations. In order to achieve this in a CIM environment, a good functional analysis of the organization has first to be made in order to identify the enterprise processes, activities and the flow of objects. Company policies can then be re-evaluated and established on the basis of a comprehensive enterprise model, which can be widely understood by the organization. Finally, a detailed data-oriented model of the information system complying with the requirements established has to be validated and implemented. This is discussed in this chapter using high-level Petri nets.

5.2.4 Global scenario

Let us consider the following example which will be used as a global scenario throughout this chapter. The manufacturing enterprise ABC produces many complex parts made of steel in small batches using numerically controlled (NC) equipment. Let us assume that we are modeling a part of the enterprise ABC which is organized as follows:

- it makes the design of the products in its Design and Engineering department using CAD and CAPP facilities for producing part data (description and geometry data), product structures (Bills of Materials) and process plans;
- it has a Customer Service department to collect and process customer orders;
- it has a Production Planning department making use of an MRP system to control inventories and to plan materials requirements, to balance the load of work centers and to prepare the manufacturing orders; and
- it has a Manufacturing department where shop-floor control, manufacturing, assembly and shipping activities are performed. Flexible manufacturing cells (FMCs) are used for manufacturing. An FMC is made of machine-tools, measuring machines and handling devices pallets, cutting tools and measuring tools (tools, for short). Each FMC can produce several parts of different types according to predefined work schedules. A part must be mounted on a pallet to be processed. The sequence of machining operations to be applied to a part is given in the form of a process plan.

In the following sections, the modeling of the information flow that reflects this scenario, the model validation and verification techniques, and the model implementation methodologies will be presented.

5.3 HIGH-LEVEL PETRI NETS

5.3.1 Rationale for high-level Petri nets

Petri nets provide a powerful formalism for modeling the dynamic behavior of discrete concurrent systems. Unfortunately, Petri net graphs of basic place/transition models for the representation of complex systems are usually very large and therefore become illegible and difficult to analyze. Information and its transmission in a net is also difficult to model. To construct more compact graphs that include information flow, predicate/transition nets or PrT-nets (Genrich and Lautenbach, 1981; Genrich, 1987) and their improvement called colored Petri nets or CP-nets (Jensen, 1981; 1987; 1990) were introduced. Compared to basic Petri nets, high-level nets are what structured computer programming languages are compared to assembly languages.

The main differences between the two types of high-level Petri nets and the basic Petri net models rely in the facts that (1) in high-level nets tokens have a type (also called *color*) and carry information to represent structured objects, and (2) formal expressions (also called inscriptions) constraining token occurrences, to be used as inputs or outputs of a transition, are attached to the arcs of the graphs. The information can be inspected and modified when a transition fires, thus imposing conditions on transition firing on the basis of token values.

This formalism makes it possible to fold several basic similar subnets into a single, more concise, net. Nets can also be hierarchically organized using so-called *hierarchical CP-nets* (Huber *et al.*, 1990). The conciseness of the resulting model is compensated by more complex inscriptions attached to the arcs of the net. A predicate called *guard* can be attached to the transitions. The guard restricts the set of possible token occurrences (i.e. objects) to be used in the transition, in addition to the constraints implied by the arc inscriptions. The guard is like an input rule which can be used to model preconditions on the execution of the transition or can behave as a filter on possible input token occurrences, by selecting only those verifying a given condition.

It has been shown (Jensen, 1981; 1987) that CP-nets constitute a generalization of PrT-nets and that there is equivalence of both representations in the sense that a formal translation of one formalism to the other exists. PrT-nets are appropriate for formal description and analysis of a system. CP-nets are appropriate for the description and informal explanation of a system. In the remainder of this chapter, we will use PrT-nets, CP-nets and Updated Petri Nets (UPN).

5.3.2 Predicate/transition nets

Definition 5.1. A *multi-set*, over a non-empty and finite set S, is a set which may contain multiple occurrences of elements of S.

Each multi-set m over S is represented by a formal sum:

$$\sum_{s \in S} m(s)s$$

where the non-negative integer $m(s)$ is the number of appearances, called coefficient, of the element s in the multi-set m.

In this chapter, we only consider finite multi-sets. Thus, the formal sum is convergent since m is finite, i.e.

$$\sum_{s \in S} m(s) < \infty$$

The set of all finite multi-sets over the set S will be denoted by S_{MS}.

Example. $\{a, b, d, a\}$ is a finite multi-set over the finite set $\{a, b, c, d\}$. It is represented by the formal sum $2a + b + d$.

Summation, scalar multiplication, comparison and multiplicity of multi-sets over a set S are defined in the following way, for all $m, m_1, m_2 \in S_{MS}$ and $n \in |N$:

(i) $m_1 + m_2 = \sum_{s \in S} (m_1(s) + m_2(s))\, s$ (summation)

(ii) $n \cdot m = \sum_{s \in S} (n \cdot m(s))\, s$ (scalar-multiplication)

(iii) $m_1 \neq m_2 \Leftrightarrow \exists s \in S: [m_1(s) \neq m_2(s)]$ (comparison)

$m_1 \leq m_2 \Leftrightarrow \forall s \in S: [m_1(s) \leq m_2(s)]$

(relations $<$, $>$, $\geq$ can be defined analogously to $\leq$)

(iv) $|m| = \sum_{s \in S} m(s)$ (multiplicity)

When $m_1 \leq m_2$, subtraction can also be defined:

(v) $m_2 - m_1 = \sum_{s \in S} (m_2(s) - m_1(s))\, s$ (subtraction)

Using these notations, PrT-nets, introduced by Genrich and Lautenbach (1981), can now be defined as follows using a formalism close to the one proposed by Jensen (1987).

Definition 5.2. A *predicate/transition net* (*PrT-net*) is defined as a directed graph $N = (\mathbf{P}, \mathbf{T}; \mathbf{F})$ where $\mathbf{P}$ is the set of predicates (or 'first-order' places) represented by circles, $\mathbf{T}$ is the set of transitions represented by boxes and $\mathbf{F}$ is a set of arcs with:

$$\mathbf{P} \cap \mathrm{T} = \varnothing,\ \mathbf{P} \cup \mathbf{T} \neq \varnothing,\ \mathbf{F} \subseteq \mathbf{P} \times \mathbf{T} \cup \mathbf{T} \times \mathbf{P}$$

${}^{\bullet}t = \{p \mid (p, t) \in \mathbf{F}\}$ and $t^{\bullet} = \{p \mid (t, p) \in \mathbf{F}\}$ are called the *preset* and *postset* of $t \in \mathbf{T}$, respectively. The elements of these sets form the set of preconditions and the set of postconditions of transition t, respectively. The graph must verify that:

1. There exists a set V of typed *variables*. Each variable of V has a name v and a type D expressed by the following notation, $v : D$. All variable names in V are distinct.
2. Each place p has attached to it (i) a non-empty set of *token colors* $C(p)$ and (ii) an *initial marking* M_0 which is a function defined on $\mathbf{P}$, such that $M_0(p) \in C(p)_{MS}$ and which is expressed as a formal sum of colors of tokens of p.
3. Each arc has attached to it an *arc expression* EXP (or label) containing a set of free variables $\{v_1 : D_1, v_2 : D_2, \ldots, v_n : D_n\}$ which is a subset of V. Moreover, it is required that the lambda expression $\lambda(v_1, v_2, \ldots, v_n) \cdot \mathrm{EXP}$ defines a function from $D_1 \times D_2 \times \ldots \times D_n$ into $C(p)_{MS}$, where the place p is the source/destination of the arc.
4. Each transition may have attached to it a Boolean expression called the *guard*. Only variables which are involved in the surrounding arc expressions of the transition can be used in the guard expression. The guard restricts the conditions under which the transition can fire. The guard cannot be the constant predicate FALSE.
5. The graph has no isolated places or transitions.

Note. In their original formulation of PrT-nets, Genrich and Lautenbach (1981) define the arc expressions EXP as logical formulas built from a structure Σ consisting of some sorts of individuals together with some operations and relations. The definition of Σ is left open and its grammar can be adapted according to the application domain considered.

For each place p, a capacity $K(p)$ may be defined, where K is a function from **P** to the set of positive integers. It represents the maximum number of occurrences allowed for predicate p. It can be a function of token colors of $C(p)$. By default, the capacity of a place p is assumed to be infinity, if not otherwise stated.

Firing rules for predicate/transition nets. Each element of **T** represents a class of possible changes in the markings of the adjacent predicates. An occurrence (firing) of $t \in \mathbf{T}$ consists of removing/adding copies of items from/to adjacent places according to the expressions labeling the arcs. Firing of a transition *may* occur whenever, for an assignment of variables which satisfies the guard expression of the transition, (1) all input predicates carry enough copies of proper items, and (2) for no output predicate the capacity K is exceeded by adding the respective copies of items. The firing of t changes the marking M into a new marking M'; this fact is denoted by: $M[T\rangle\, M'$.

More formally, a *marking M* for a PrT-net N is a function defined on **P** such that each $M(p)$ is a formal sum of elements of $C(p)$, i.e. $M(p) \in C(p)_{MS}$.

Definition 5.3. A *substitution* $\{v_1/a_1, v_2/a_2, \ldots, v_n/a_n\}$ is an assignment of values $a_i \in D_i$ to variables $v_i : D_i$ of set V under the following rule: if two variables v_i and v_j (with $i \neq j$) have the same type ($v_i : D_i$, $v_j : D_i$) then the two values assigned must be different, i.e. $a_i \neq a_j$.

Definition 5.4. A transition $t \in \mathbf{T}$ is *enabled* by a substitution if the following conditions are verified:

(i) Let $p \in {}^{\bullet}t$ be one of the input places of t. The formal sum of tuples involved in the expression of arc(p, t) consequently to a substitution must be less than or equal to the formal sum $M(p)$ (i.e. the multi-set involved in the arc expression after the substitution must be contained in the multi-set of p before the substitution);
(ii) the substitution must preserve the guard of t to true.

Definition 5.5. The *firing* of an enabled transition t by a substitution transforms the current marking M into a marking M' as follows: let $p \in {}^{\bullet}t$ and $q \in t^{\bullet}$ be an input place and an output place of t, respectively, and let s_p and s_q be the formal sums produced by the expression of arcs(p, t) and (t, p), respectively. The new marking M' is calculated as follows:

$$M'(p) = M(p) - s_p \text{ for each input place } p \text{ of } t$$
$$M'(p) = M(p) + s_q \text{ for each output place } p \text{ of } t$$

Definition 5.6. The *incidence matrix* of a pure ($\mathbf{F} \cap \mathbf{F}^{-1} = \varnothing$) predicate/transition net is a function C from $\mathbf{P} \times \mathbf{T}$ into the set of expressions attached to arcs of **F** such that:

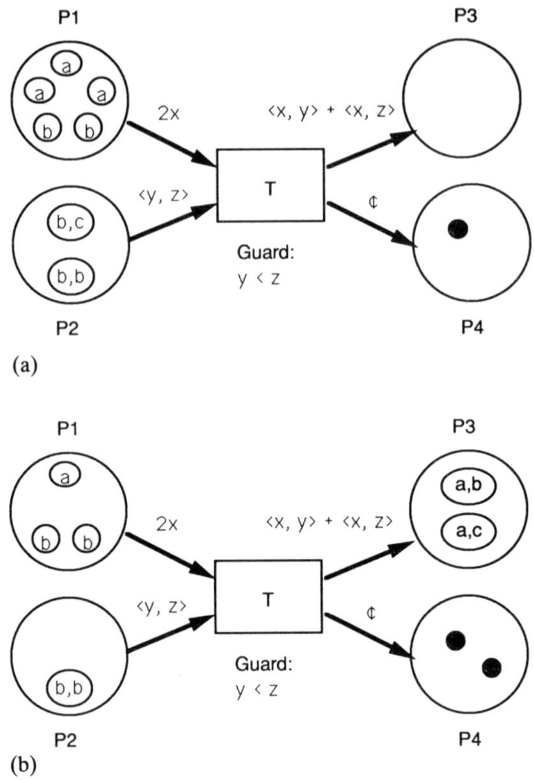

Figure 5.4 PrT-net example: (a) before firing of T; (b) after firing of T.

$$C(p,t) := \begin{cases} -l, & l \text{ is the label of } (p,t) \in \mathbf{F} \\ l, & l \text{ is the label of } (t,p) \in \mathbf{F} \\ 0 & \text{otherwise} \end{cases}$$

The incidence matrix is used to determine invariants of PrT-nets, which is a usual property analyzed on such graphs.

(a) A simple example

The example of Fig. 5.4 shows a net with four predicates P_1, P_2, P_3 and P_4 and one transition T with inscriptions on the arcs and a guard for the transition. Dots in places represent usual Petri net tokens while items in places represent object occurrences (which are allowed to be identical). Figure 5.4 also shows the marking before and after the firing of transition T.

(b) A more realistic example

In an FMC, machines have to be set up before they are used, i.e. the proper tool has to be fixed. Figure 5.5 shows a transition ('Machine Setup')

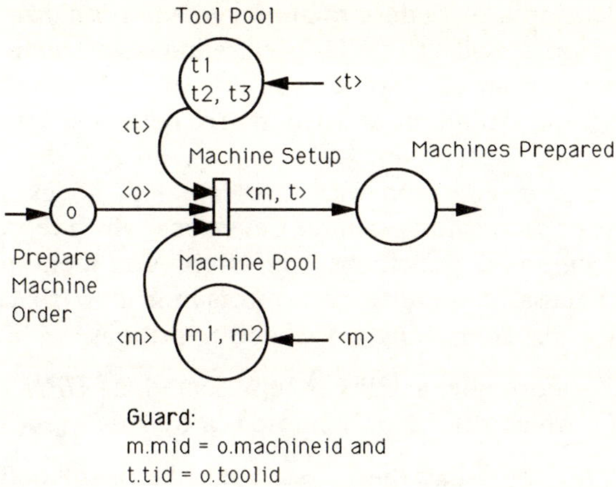

Transitions:

Machine Setup	Prepares machine with proper tool

Predicates:

Prepare Machine Order	Receives machine set-up orders
Machine Pool	Indicates idle machines
Tool Pool	Indicates available machines
Machines Prepared	Indicates machines set up

Marking:

M_0 = {Prepare Machine Order: {o}, Tool Pool: {t1, t2, t3},
Machine Pool: {m1, m2}, Machines Prepared: ∅ }

Figure 5.5 Manufacturing function described by a PrT-net.

and four predicates ('Prepare Machine Order', 'Machine Pool', 'Tool Pool', and 'Machines Prepared') with the initial marking M_0.

When a setup order is received in predicate 'Prepare Machine Order', the transition can fire if the guard is true, i.e. the required machine and the appropriate tool are available. If the transition can fire, a pair $\langle m, t \rangle$ verifying the guard is added in predicate 'Machines Prepared', the occurrence of *m* is removed from predicate 'Machine Pool' and the occurrence of *t* is removed from predicate 'Tool Pool' while the setup order occurrence is removed from predicate 'Prepare Machine Order'.

5.3.3 Updated Petri Nets (UPN)

This section describes the formalism for knowledge representation of business rules and policies modeled by UPN (Lin, 1991). We have extended the primitives of the general colored Petri net (CP-net) descriptions in order to reflect, more closely, the terminology and semantics involved in company policies of manufacturing information systems. These primitives are used to develop a procedure to automate and formalize the interpretation process from a model to a rule-specification language. In the following paragraphs, we present the formal definition of UPN, which is based on both the CP-graph definition and CP-matrix definition given by Jensen (1987).

A UPN is a directed graph with three types of nodes: *places*, which represent facts or predicates; *primitive* transitions, which represent rules or implications; *compound* transitions, which represent metarules (subnets). Enabling and causal conditions and information flow specifications are represented by arcs connecting places and transitions.

Definition 5.7. Formally, a UPN is represented as: $UPN = \langle P, T, C, I^-, I^+, M_0, I_0, MT \rangle$, which can be decomposed in four different parts:

1. P, T, I^-, I^+, M_0 represent the classic color Petri net definition. They identify the part of the information system that provides the conditions for the information control. Only this part of the UPN is used in the validation process. Its entities are defined as follows (Jensen, 1987):

   - $P = \{p_1, \ldots, p_n\}$ denotes the set of *places* (represented graphically as circles).
   - $T = \{t_1, \ldots, t_m\}$ denotes the set of *primitive* transitions (represented graphically as black bars).
   - $P \cap T = \varnothing$ and $P \cup T \neq \varnothing$.
   - C is the color function defined from $P \cup T$ into non-empty sets. It attaches to each place a set of possible token-data and to each transition a set of possible data occurrences.
   - I^- and I^+ are negative and positive incidence functions defined on $P \times T$, such that $I^-(p, t), I^+(p, t) \in [C(t)_{MS} \rightarrow C(p)_{MS}]_L, \forall (p, t) \in P \times T$, where S_{MS} denotes the set of all finite multi-sets over the non-empty set S, $[C(t)_{MS} \rightarrow C(p)_{MS}]_L$ the multi-set extension of $[C(t) \rightarrow C(p)_{MS}]_L$, and $[\ldots]_L$ denotes a set of linear functions.
   - The net has no isolated places or transitions:

   $$\forall p \in P, \exists t \in T : I^-(p, t) \neq 0 \vee I^+(p, t) \neq 0$$

   and

   $$\forall t \in T, \exists p \in P : I^-(p, t) \neq 0 \vee I^+(p, t) \neq 0$$

   - M_0 is the initial marking, a function defined on P, such that:

   $$M_0(p) \in C(p)_{MS}, \forall p \in P$$

2. I_0 is an inhibitor function defined on $P \times T$, such that:

   $$I_0 \in [C(t)_{MS} \rightarrow C(p)_{MS}]_L, \forall (p, t) \in P \times T.$$

3. $MT = \{hm_1, \ldots, hm_l\}$ denotes the set of related transition sets. These are sets of transitions grouped into subnets.

From a rule-based perspective, we have divided the representation of knowledge-base components in the following four groups: data, facts, rules and metarules. In section 5.5, we define the syntax and semantics of the UPN elements representing those components.

5.4 ORGANIZATION ANALYSIS AND CONCEPTUAL DATABASE DESIGN

5.4.1 Organization analysis

This section is concerned with the organization analysis phase of *M** (see section 5.2) for given parts of an enterprise (called environments) from the point of view of information analysis.

(a) Functional departments and activities

From an organizational point of view, any enterprise is made of a certain number of functional departments in which a number of activities are taking place. All departments report to top management (corporate management).

A *department* is an area of responsibilities over a given set of activities (management, operational or support activities) and a set of resources of the enterprise. Departments are usually hierarchically organized into organization units, i.e. sections and services. Typical departments of an enterprise are: general management, finance, marketing, operations management, production, etc.

An *activity* is a task to be done within the enterprise using time and the enterprise's resources in order to contribute to the fulfilment of one or several business objectives under some business constraints. Activities can be organized into business processes. Examples of activities are accounting activities, material requirements planning activities, customer order processing, manufacturing activities such as machining tasks or assembly tasks, etc.

(b) Organization nets

At the organization analysis level (or requirements definition level) of the *M** methodology, subsets of the enterprise, called *environments*, are modeled in terms of *organization nets* which connect *object views* to *functions*. The goal is to obtain an integrated model of a well-defined part of the enterprise and which puts together:

- a functional model (describing the enterprise functionality);
- an information model (describing flows of enterprise objects);
- a behavior model (describing the environment behavior).

An *environment* is a consistent collection of functions, objects and rules of the enterprise achieving one or more business objectives. It is defined by users.

Environments must not be confused with departments. Environments define the scope of the model without paying attention to departmental boundaries or areas of responsibilities. This is an essential concept in order to achieve CIM and to avoid formation of so-called islands of automation.

The organization model of M^* is used to model environments. It is based on high-level Petri nets, especially predicate/transition nets (PrT-nets), for the definition of organization nets. *This is not an executable model.*

Definition 5.8. Formally, an *organization net* is defined as the structure of a PrT-net $N = (O, F; A)$, where O and F are two disjoint sets of places and transitions called object views and functions, respectively, and A is a set of directed arcs connecting object views to functions and vice versa.

An organization net is a causal, descriptive, model made of object views and functions and indicating which kinds of object views are consumed/used/produced by functions.

Object views represent external appearances or states of *enterprise objects* (i.e. physical or information objects) as they are perceived (and described) by users and applications. Enterprise objects represent any physical or abstract entities of relevance to the enterprise environment which can be uniquely identified and have a life cycle. Examples of relevant enterprise objects are: events, persons, orders, products, resources, processes and documents. Enterprise objects are always accessed via one of their object views, one being the physical realization of the object. CIM object views are classified into four major classes in M^*:

- *Messages*: represent real-world events or requests to execute a function as well as preconditions (input messages) and postconditions (output messages) to the execution of functions. Messages can be used to represent timed-events, i.e. start of scheduled actions.
- *Information*: represents information objects used/produced by functions. Four subclasses have been defined: *data* for information items, *files* for structured sets of data items, *forms* for structured documents (statements, reports, etc.) or computer screens and *text* for free-format text.
- *Materials*: represent physical matters used/produced by functions (e.g. raw materials, parts, products, components).
- *Resources*: represent physical means used to carry out functions (e.g. tools, fixtures, machines, people).

Functions represent tasks (or pieces of functionality, i.e. things to be done) of the enterprise with their inputs and outputs (defined as object views). Constraints (either defined as input rules, i.e. predicates, attached to the transitions, i.e. the functions, or as inscriptions attached to the arcs as part of the PrT-net formalism) can be imposed on functions. Furthermore, control objects (defined as object views, usually of information type) can be required by functions. They represent object views which must be known by the function or constrain its execution, but are not modified by the function. They usually provide run-time data (part number, number of parts to be made, etc.) to correctly instantiate function occurrences.

A function at one level can be further described as an organization net at a more detailed level, therefore creating a hierarchical structure of functions. Thus, organization nets are used at the organization analysis level of *M** to define the environments to be analyzed, to identify their functional structure (flow of control and hierarchy of functions) and the flows of information and objects, and to collect formatted and non-formatted requirements for the information system of the CIM system to be designed. Functions described by an organization net (compound functions) are called processes, and functions defined as the leaves of a functional decomposition (atomic functions) are called activities.

Formatted requirements correspond to structured object views, e.g. messages, forms, files, and data screens. *Non-formatted requirements* correspond to textual descriptions of functions, definitions of procedures, and policies supplied by users in verbose form.

Example. In the ABC example of section 5.2.4, shop-floor control activities within the Manufacturing department can consist of two stand-alone functions synchronized by their inputs and triggering events. First, a planning function, which is triggered by a planning request (plan-req) and which produces a validated schedule for shop-floor operations from a production plan given by MRP; and second, an activity control function, which is started by an order (start-ac) and which monitors the shop-floor activity using the validated schedule (Fig. 5.6). The planning function can itself consist of a scheduling activity (scheduling) followed by a schedule analysis activity (analyze schedule) as illustrated in Fig. 5.7. Three other activities have been added: one for updating the shop-floor status in system files (update shop status), another one for updating the shop-floor capacity in system files (update shop capacity) in the case of external events affecting

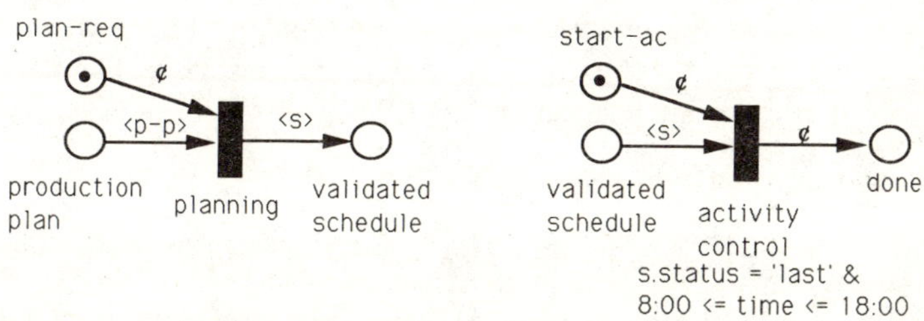

Figure 5.6 Organization net of an activity control function.

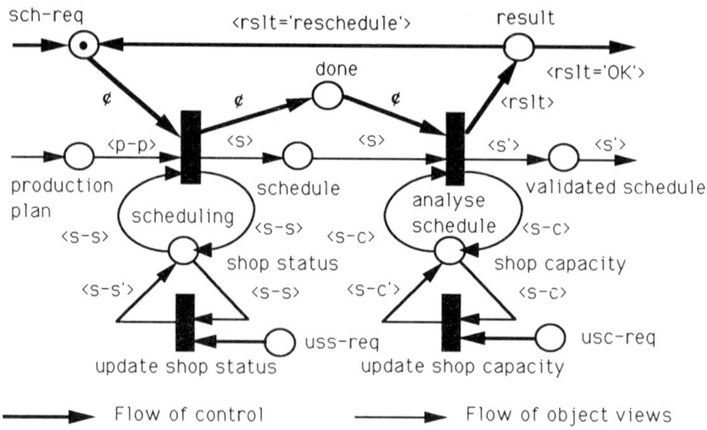

Figure 5.7 Organization net of a planning function.

shop capacity (e.g. machine failures) and another one to decide, according to the result of the 'analyze schedule' activity, whether another schedule must be produced or not using the same production plan (reschedule). Guards can be defined for each transition to model more of the semantics of the application if required (for instance to filter input information).

5.4.2 Conceptual database design and specification

This section concentrates on the conceptual design phase of *M** (see section 5.2) for information systems. The use of high-level Petri nets coupled with advanced data and information models makes it possible to develop an integrated model of the information system embracing the three basic aspects to be modeled: information, functionality and behavior. Such a formalism lends itself to formal specification of the information system being analyzed. This will be first illustrated in the case of an extended entity-relationship model and then for an object-oriented model.

The use of Petri nets in information system methodologies has already been suggested, for instance, for DATAID (Albano *et al.*, 1985) and MERISE (Tardieu *et al.*, 1983). However, this was limited to low-level Petri nets (such as the event/condition nets) and no effort was made to integrate them within a consistent modeling framework as has been done in the *M** methodology using high-level Petri nets (Vernadat *et al.*, 1989) or by Sibertin-Blanc (1991).

(a) Process and Data Net (PDN) model

At the conceptual design level of *M**, the Process and Data Net (PDN) model is used (Di Leva and Giolito, 1989). Its aim is to make easier the specification of the CIM information system on the basis of an extended entity-relationship model for the static properties of data, of PrT nets for the dynamic properties of data, and of process nets for the dynamic behavior of any data processing application. The PDN model is thus made of a data model, a transaction model and a process model.

(i) Data model

Definition 5.9. The entity-relationship data model of the PDN model is defined as a 5-tuple $\langle \mathbf{D}, \mathbf{E}, \mathbf{A}, \mathbf{R}, \mathbf{X}, \mathbf{Q} \rangle$ where:

1. **D** is a finite set of *basic domains*;
2. **E** is a finite set of entity classes (or simply *entities*);
3. **A** is a finite set of *attributes*;
4. **R** is a finite set of relationship classes (or simply *relationships*);
5. **X** is a set of *variables* on entities or relationships;
6. **Q** is a finite set of *operators* on values of attributes, entities and relationships.

(1) The basic domains D_i of **D**, $i = 1, 2, \ldots$, are sets of atomic values (e.g. integers, reals, characters, strings). Compound domains can be constructed on basic domains by means of two mechanisms:

- the Cartesian product $\times$ of domains: e.g. date = day $\times$ month $\times$ year;
- the power set $\Pi(S)$ of a set S: $\Pi(S) = \{S'/S' \subseteq S\}$.

(2) The set of entity classes $E_j \in \mathbf{E}, j = 1, 2, \ldots$, is such that each entity class E_j describes a set of identifiable objects of the real world (or entity occurrences) of interest to the organization and which have identical structure, i.e. common characteristics defined as attributes. Examples of entities are parts, machines and tools, each object can be uniquely identified.

(3) An attribute A_k of an entity class E_i ($k = 1, \ldots, n$, where n is the number of attributes of E_i) is a function which describes a characteristic of the entity defined as follows: $A_k : E_i \rightarrow D_k$.

If $e \in E$ is an entity occurrence, the value of the attribute A_k is denoted by $e \cdot A_k$ where $e \cdot A_k = v_k \in D_k$ if the attribute is simple or $e \cdot A_k = \{v_k\} \in \Pi(D_k)$ if the attribute is repeating, i.e. it can take several values.

An internal *identifier I* (or key) of the entity E, based on attribute $A_i : E \rightarrow D_i$, is a bijection between the co-domain D'_i of D_i ($D'_i \subseteq D_i$) of the attribute and the entity E, i.e. $I : D'_i \rightarrow E$.

The equality relation between two objects $e_1, e_2 \in E$ can be stated by:

$$e_1 = e_2 \text{ iff } I^{-1}(e_1) = I^{-1}(e_2) \quad \text{and} \quad e_1 \cdot A_k = e_2 \cdot A_k, \ \forall A_k \in E$$

Identifiers of entities may also be defined by means of groups of attributes by:

$$I : D'_1 \times D'_2 \times \ldots \times D'_n \rightarrow E$$

where D'_i, $i = 1, n$, are the co-domains of attributes of E involved in the identifier.

Some entities can also be identified by means of other entities on which they are dependent (*external identifiers*). Such entities are called *weak entities* because they depend on the existence (E) or the identification (ID) of other (external) entities.

(4) A relationship $R \in \mathbf{R}$ between entities $E_1, E_2, \ldots, E_n$ describes the set of relationship occurrences (or links) linking objects $e_1 \in E_1$, $e_2 \in E_2, \ldots, e_n \in E_n$, i.e. $R = \{\langle e_1, e_2, \ldots, e_n \rangle\} \mid E_1 \times E_2 \times \ldots \times E_n$.

The equality relation between two relationship occurrences can be stated by:

$$\langle e_1, e_2, \ldots, e_n \rangle = \langle e'_1, e'_2 \ldots, e'_n \rangle \text{ iff } e_i = e'_i,\ \forall i = -1, 2, \ldots, n.$$

The identifier of a relationship $\langle e_1, e_2, \ldots, e_n \rangle \in R$ is defined by the values of identifiers of objects involved in the link, i.e. $\langle I^{-1}(e_1), I^{-1}(e_2), \ldots, I^{-1}(e_n) \rangle$.

Like entities, relationships may have attributes A_k other than attributes involved in their identifier. They are defined by: $A_k : R_i \rightarrow D_k$.

(5) A variable $x \in \mathbf{X}$ relatively to entity E or to relationship R will be denoted by $x \mid E$ or $x \mid R$, respectively.

(6) Operators on values of domains are usual operators (such as addition, multiplication, division and comparision, for integers and reals, concatenation, comparision and sub-string, for character strings, etc.). Operators on entities include *insert*, *update* and *delete* to respectively add, modify and remove occurrences of an entity class. Operators on relationships include *connect* and *disconnect* to respectively add and remove occurrences in a relationship. A formal definition of the operators for entities and relationships is provided by Table 5.1.

Table 5.1 Operators for entities and relationships

Operator	*Equivalent formula*
insert e $(e \cdot id \leftarrow v,\ e \cdot a_l \leftarrow v_1, \ldots)$	$e \in E$ and $I^{-1}(e) = v$ and $e \leftarrow a_1 = v_1$ and ...
delete e	$e \notin E$
$e' \leftarrow$ update e $(e \cdot a \leftarrow v)$	$I^{-1}(e') = I^{-1}(e)$ and $e' \cdot a = v$ and $e' \cdot b = e \cdot b$ and ...
connect $R(e_1, e_2)$	$\langle e_1, e_2 \rangle \in R$
disconnect $R(e_1, e_2)$	$\langle e_1, e_2 \rangle \notin R$

Note: e and e' are variables defined on E, id is the identifier of entity E, e_1 and e_2 are variables defined on E_1 and E_2, respectively and $v, v_1, \ldots$ are values or expressions.

The update operator, denoted by $e' \leftarrow$ update $e(e$ fi$a \leftarrow v, \ldots)$, means that the object to be updated e is first deleted and that another object e' is created in the database with value v for attribute a and with previous values of e for the other attributes including the identifier of e (which can never be changed by an update operator). If the identifier of an entity must be changed, this means that a new occurrence must be created and that the old one must be destroyed. Thus, in this case two operators must be used: first a *delete* then followed by an *append*.

(i) Abstraction hierarchies The entity-relationship model used is an extended model in the sense that it makes use of two *abstraction hierarchies* to capture more of the semantics of data (Smith and Smith, 1977). An abstraction hierarchy makes it possible to represent some entities in terms of more general entities. The hierarchies are the subset hierarchies and the partition hierarchy.

Subset hierarchy: an entity S is a subset of an entity E, or E is a generalization of an entity S, denoted by $S \Rightarrow E$, if each occurrence of S is an occurrence of E.

Partition hierarchy: entities $P_1, P_2, \ldots, P_n$ form a partition of entity E, denoted by $E \Leftarrow \{P_1, P_2, \ldots, P_n\}$, if each occurrence of E is also an occurrence and only one of either P_1 or P_2 or ... or P_n.

Remark: abstraction hierarchies define a partial order on the set of entities. These two data abstraction mechanisms are special cases of the generalization mechanism, or 'is-a' link (Smith and Smith, 1977), associated with the property inheritance principle.

(ii) Integrity constraints A number of *integrity constraints*, i.e. rules constraining static properties of data, can be defined in this model. Among these we have:

1. *Constraints on entities*: entities are uniquely identified by at least one identifier.
2. *Constraints on relationships*: they concern the mapping functionalities of relationships with the related entities and are defined by a pair of positive integers, called cardinalities and noted (c_1, c_2). c_1 is called the *minimum cardinality*. It represents the minimum number of times (possibly 0) that an occurrence of the entity can be involved in occurrences of the relationship. c_2 is called the *maximum cardinality*. It represents the maximum number of times (minimum 1) that occurrences of the entity can be involved in occurrences of the relationship. If this number is unknown, the symbol 'n' is used.

   Remark: the cardinality of a weak entity in a relationship with the entity on which it depends is always $(1, 1)$.
3. *Constraints on attributes*: an attribute can be *simple* (i.e. its domain is atomic) or *compound* (i.e. defined over the Cartesian product of several domains) or else *repeating* (i.e. it can take several values). Furthermore, an attribute can be *total* (i.e. its value is always mandatory) or *partial* (i.e. its value may be unknown).
4. *Constraints on hierarchies*: either in generalization $E \Leftarrow S$ and in partition $E \Leftarrow \{P_1, P_2, \ldots\}$ hierarchies, attributes and relationships of entity E are attributes and relationships of sub-entities, but not vice versa

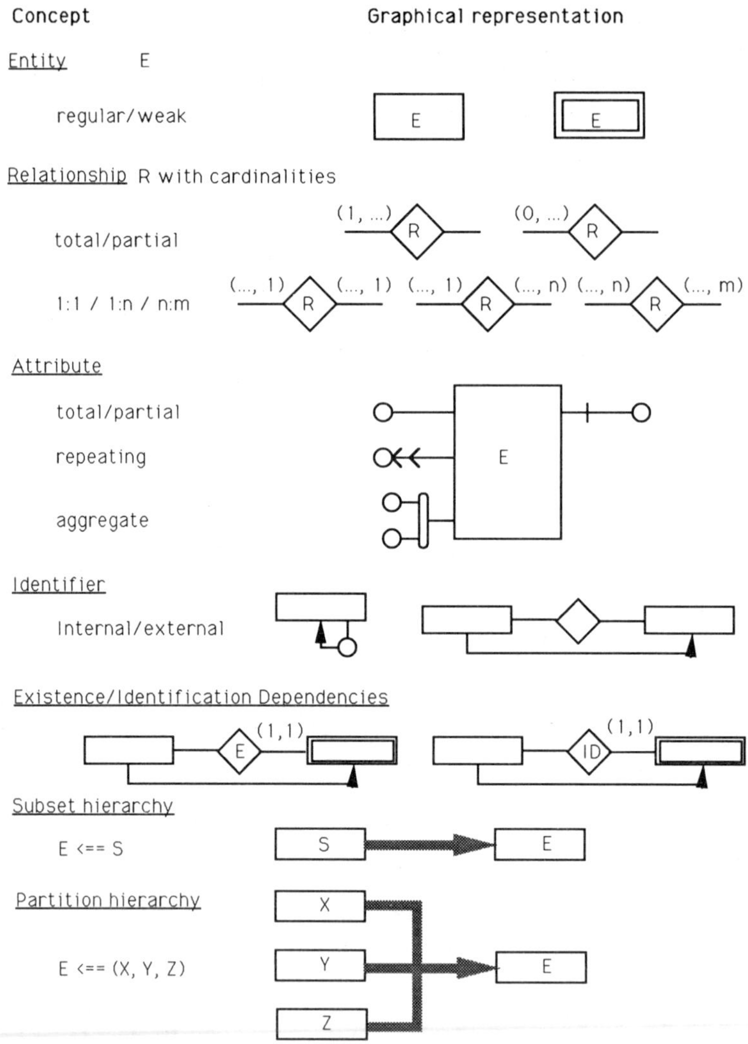

Figure 5.8 Graphical formalism for the EER model.

(inheritance principle). Usually, sub-entities have additional attributes that define particular features of these entities with respect to *E*.

Figure 5.8 provides a graphical formalism for the extended entity-relationship (EER) model.

Definition 5.10. Formally, a database schema *S* can be specified as a finite set of logical records and a finite set of abstraction hierarchies, respectively defined by:

ENTITY (attribute_1, attribute_2, . . .)
  KEY: (attribute_α, attribute_β, . . .) [WEAK ON entity _1]
RELATION (entity_1: [role_1] card_1, entity_2: [role_2] card_2, . . .)
  [ATTRIBUTES: (attribute_1, attribute_2, . . .)]
$E \Leftarrow S$ or $E \Leftarrow \{P_1, P_2, \ldots\}$

where:

1. ENTITY represents the name of an entity class followed by its list of attributes (attribute_1, attribute_2, . . .) and the definition of its identifier (KEY) made of one or more attributes of the entity class (attribute_α, attribute_β, . . .). If it is a weak entity class the optional clause WEAK ON is used where entity_1 is the name of another entity class;
2. RELATION represents the name of a relationship class followed by the list of names of entity classes (entity_1, entity_2, . . .) and respective cardinalities (card_1, card_2, . . .) involved in the relationship class and optionally the list of attributes of the relationship class (attribute_1, attribute_2, . . .). Roles (role_1, role_2) are only used to label relationship arcs of unary relationship, i.e. defined on only one entity class;
3. E, S, P_1, P_2, . . . are names of entity classes used in the definition of abstraction hierarchies.

Example. Figure 5.9 gives a simplified EER schema for a flexible manufacturing cell. The manufacturing system produces lots of parts (entities LOT and PART) which must be mounted on pallets by means of fixtures (entity FIXTURE). Any part belongs to a part-type (entity PART TYPE) and thus is connected to entity PART via the relationship PP. The same pallet and fixtures can be used for different part-types and each part can be mounted on different possible pallets. Possible relationships between pallets and part-types are indicated by the relationship FP, while actual relationships between pallets and parts are given by the relationship PF. Each part has an identifier (*p#*), a type (same as for PART TYPE), a status (stat), the values of which can be {'raw', 'ready', 'mounted', 'in-process', 'worked', 'completed'} and a location (loc), the values of which are the input buffer IM, the output buffer OM and the in-process buffers of the cell. This allows for tracking the storage location of each part. In this example, storage locations are not defined as entities (this could have been done if more detailed information about them was required). The sequence of operations (entity OPERATION) to be executed in a part routing (entity PLAN) is linked to the routing by the relationship PO. The main routing of a part is linked to the entity PART by SP. PLAN is a weak entity of entity PART since a process plan only exists relatively to a part definition. Alternate routings for the same part can be recursively obtained from relationship SS. SS.succ (succ stands for successor) and SS.pred (pred stands for predecessor) indicates the role played by the routing in

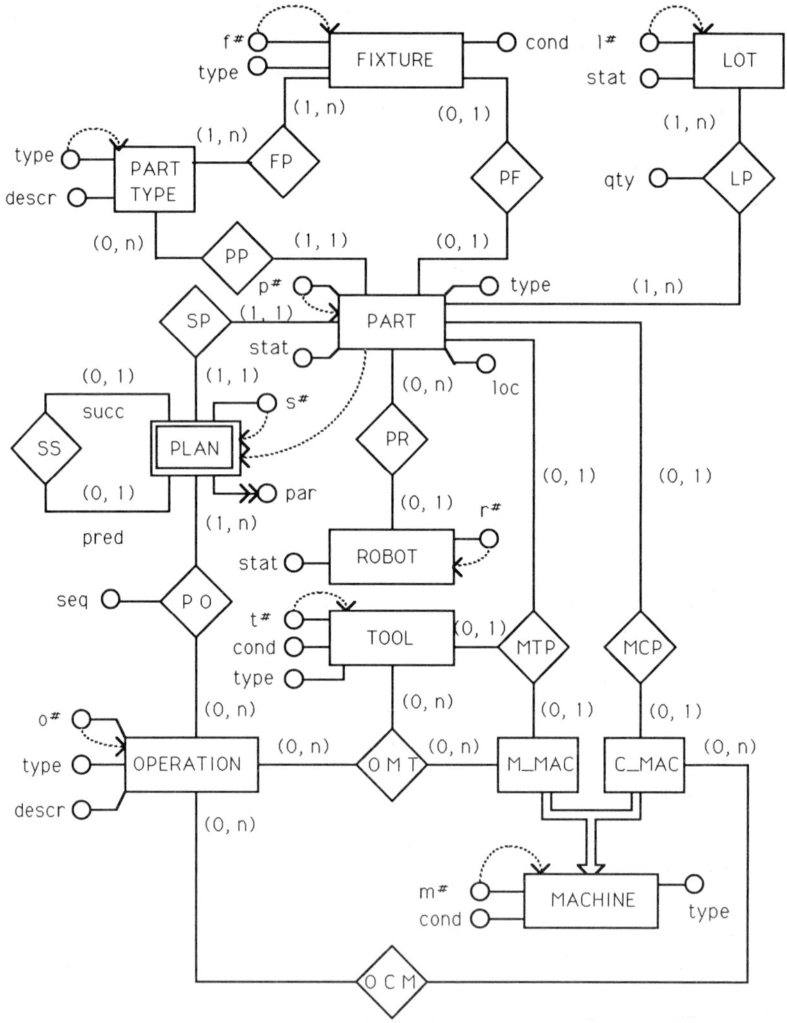

Figure 5.9 EER schema for an FMC.

SS. A routing consists of an increasing sequential number (*s*#) and of a set of parameters (par). An operation has an identifier (*o*#), a type, the values of which can be {'mfg' for manufacturing, 'asy' for assembly, 'cnt' for quality control} and a description (descr). Each operation is linked to a tool (entity TOOL) and to a machining machine (entity M_MAC) by the relationship OMT if it is a machining operation. It is linked to a measuring machine (C_MAC) by relationship OCM if it is a quality control or measuring operation. Machining machines and measuring machines are generalized by the concept of machine (entity MACHINE). Machines are described by a number (*m*#), a type and a condition (cond), the values of which are {'available', 'working', 'down'} depending if they are idle or working. Similarly, tools are described by a number (*t*#), a type and a condition (cond), the values of which are {'available', 'in-use', 'worn-out'}. Relationships MTP and MCP model the link between a part and a machine depending if it is a machining machine or a measuring machine. Finally, parts on pallets are handled in the cell by robots (entity ROBOT). Relationship PR between parts and robots indicates that one part can be handled by one robot at a time. The formal definition of the database schema depicted by Fig. 5.9 follows:

Entities:
LOT (*l*#, stat) KEY: (*l*#)
FIXTURE(*f*#, type, cond) KEY: (*f*)
PART TYPE (type, descr) KEY: (type)
PART (*p*#, type, stat, loc) KEY: (*p*#)
PLAN (*s*#, par) KEY: (*s*#, *p*#) WEAK ON PART
OPERATION (*o*#, type, descr) KEY: (*o*#)
TOOL (*t*#, cond, type) KEY: (*t*#)
ROBOT (*r*#, stat) KEY: (*r*#)
MACHINE (*m*#, type, cond) KEY: (*m*#)

Relationships:
LP (LOT: $(1, n)$, PART: $(1, n)$) ATTRIBUTES: (qty)
PF (FIXTURE: $(0, 1)$, PART: $(0, 1)$)
FP (FIXTURE: $(1, n)$, PART TYPE: $(1, n)$)
PP (PART TYPE: $(0, n)$, PART: $(1, 1)$)
SP (PART: $(0, 1)$, PLAN: $(1, 1)$)
SS (PLAN: succ $(0, 1)$, PLAN: pred $(0, 1)$)
PO (PLAN: $(1, n)$, OPERATION: $(0, n)$) ATTRIBUTES: (seq)
PR (PART: $(0, n)$, ROBOT: $(0, 1)$)
OMT (OPERATION: $(0, n)$, TOOL: $(0, n)$, M_MAC: $(0, n)$)
MTP (M_MAC: $(0, 1)$, TOOL: $(0, 1)$, PART: $(0, 1)$)
MCP (C_MAC: $(0, 1)$, PART: $(0, 1)$)
OCM (C_MAC: $(0, n)$, OPERATION: $(0, n)$)

Abstraction hierarchies:
MACHINE ⇐ {M_MAC, C_MAC}

(ii) Transaction model

The transaction model makes use of a transaction definition language to formally specify database transactions to be executed on the database schema. The language used in *M** is based on the EER model and makes use of the operators on entities and relationships (i.e. insert, delete, update, connect, disconnect). The language makes it possible to define *views* on the database schema and to specify data manipulations in the form of *transactions*.

(i) The concept of view

Definition 5.11. A view W on a database DB, defined by its schema S, is a restriction (or a subset) of the database defined by a subschema S' defined over S.

A view W on the database DB can be expressed by:[†]

view W (⟨⟨db_parameters⟩⟩)
range ⟨data_segments⟩ [, ⟨data_segments⟩]
where ⟨selection⟩

where:

1. W is the name of the view.
2. Parameters ⟨db_parameters⟩ are defined as:

   ⟨variable⟩ [, ⟨variable⟩]|⟨db_element⟩

   where ⟨db_element⟩ represents an entity or a relationship of the database schema S and ⟨variable⟩ is the name of a variable, the values of which are occurrences of this entity or this relationship.
3. The optional clause range allows for the description of relationships involved in the view and is made of data segments (⟨⟨data_segments⟩⟩) which define access paths to data in the database schema. A data segment has the following form:

   $$\pi_h\langle\mu_i\rangle\pi_k \ldots \langle\mu_m\rangle\pi_n$$

   in which $\pi_h, \pi_k, \ldots, \pi_n$ are entity terms and $\mu_i, \ldots, \mu_m$ relationship terms. An entity term (respectively, a relationship term) can be a variable specified in ⟨db_parameters⟩ or can be a new variable $z|Z$ defined on element Z of the database schema. A sequence of terms of the form $\pi_h\langle\mu_i\rangle\pi_k$, called a *link predicate*, means that an occurrence $e_h \in E_h$ (represented by the variable corresponding to term π_h) must be connected by a relationship $r_i \in R_i$ (represented by μ_i) to an occurrence $e_k \in E_k$ (represented by π_k). Using a tuple notation, we have:

   $$r_i = \langle e_h, e_k\rangle \in R_i \subseteq E_h \times E_k$$

   Occurrences of a data segment are obtained by concatenating entity occurrences which satisfy the link predicate chain. Different data segments can be connected by means of common variables, i.e. variables defined in a term of a data segment and used in another data segment. For instance, in the flexible cell example, the following range clause:

   range m|M_MAC⟨o|OMT⟩y|OPERATION, ⟨o⟩t|TOOL

   describes the set of machines with the associated tools and the operations which can be executed on them. Notice that the example makes

[†] ⟨.⟩ means non-terminal symbol and [.] means zero or many occurrences of the clause.

Table 5.2 Possible occurrences

m	*y*	*t*	*o*
⟨521, available, m_1⟩	⟨3, sawing, mfg⟩	⟨20, available, t_1⟩	⟨521, 3, 52⟩
⟨015, available, m_2⟩	⟨5, milling, mfg⟩	⟨21, available, t_2⟩	⟨015, 5, 21⟩
⟨215, working, m_3⟩	⟨7, drilling, mfg⟩	⟨24, in-use, t_3⟩	⟨215, 7, 24⟩

use of a ternary relationship. A possible set of occurrences at a given instant for this range clause is given by Table 5.2 where $o \ldots \langle I^{-1}(m), I^{-1}(y), I^{-1}(t)\rangle$.

4. The selection predicate (⟨selection⟩) in the 'where' clause specifies a logical formula on variables introduced in the view parameters or in the range clause. Occurrences of the view are obtained by retaining only occurrences of the 'range' clause which satisfy the selection predicate. The set of occurrences of a view *W* provides the *view extension* $M(W)$.

For instance, let us consider the following views *IM*, *MM* and P_1 defined on the data schema of Fig. 5.9 and on the data of Table 5.2:

IM: 'Select all parts which are stored in the input buffer *IM*'

view *IM* (*p*|PART) where *p*.loc = '*IM*'

MM: 'Select tools and machining machines for all machining operations ready to be executed'

view *MM* (*m*|M_MAC, *y*|OPERATION, *t*|TOOL)
range *m*⟨*o*|OMT⟩*y*, ⟨*o*⟩*t*
where *m*.cond = 'available' and *t*.cond = 'available'
and *y*.type= 'mfg'

Note: The 'range' clause could also have been written: *y*⟨*o*|OMT⟩*m*, ⟨*o*⟩*t*

P_1: 'Select all parts which have been processed on machines and are stored in buffer P_1'

view P_1 (*m*|M_MAC, *t*|TOOL, *p*|PART)
range *m*⟨*x*|MTP⟩, ⟨*x*⟩*p*
where *p*.loc = 'P_1'

The relationship MTP keeps track of all parts processed by machines with a certain tool until it is purged.

The *context* $\Xi(W)$ of a view *W* is the conjunction of the link predicates and the selection predicate. Variables which do not appear as view parameters must be existentially quantified in $\Xi(W)$.

Therefore, the context defines the actual data which are involved in a view. For instance, the context of the view *MM* is given by the formula:

$\Xi(MM) \ldots (\exists o)\ o \in$ OMT and $\langle m, y, t\rangle = o$ and m.cond = 'available' and t.cond = 'available' and y.type = 'mfg'

Table 5.3 Tuples

m	*y*	*t*
⟨521, available, m_1⟩	⟨3, sawing, mfg⟩	⟨20, available, t_1⟩
⟨015, available, m_2⟩	⟨5, milling, mfg⟩	⟨21, available, t_2⟩

The purpose of the view mechanism is twofold. On one hand, it is used to define the intention (or definition) of the part of the database (or database subschema) to be accessed. On the other hand, it can be used to build the extension $M(W)$ of view W by binding a set of database objects to the set of variables which appear in its context so that the formula $\Xi(W)$ is satisfied.

For instance, considering the range clause of *MM* and assuming that the extension of the database at a given instant is the one given by Table 5.2, then the extension of *MM* becomes $M(MM)$ and only contains tuples of Table 5.3.

(ii) The concept of message

Definition 5.12. A message can be defined as a view V with no parameters ⟨db_parameters⟩, no 'range' clause and no 'where' clause, i.e. as a 0-argument predicate.

It is used to model events, i.e. requests to do something or to provide indications that something has happened (e.g. signals).

(iii) The concept of transaction The transaction system in the PDN model is based on the following considerations:

1. Transactions receive data as input views and produce data as output views. Data manipulations are expressed as a sequence of actions using a set of basic database operators: *append*, *delete*, *update* (for entities) and *connect* and *disconnect* (for relationships) as introduced above;
2. Transactions can use local data (i.e. data not explicitly stored in the database). Data are then specified by means of local views which may use local variables in the range clause of their data segments;
3. Data specified by input and output views and by local variables of a transaction can be checked before the execution of the transaction using an input rule (or guard);
4. A 0-argument predicate represents a real-world condition (i.e. a place in the usual condition/event Petri nets). Since the PDN model introduces some form of procedurality in the system behavior description, it has been decided to model enterprise events by *messages* as 0-argument predicates, i.e. text as character strings.

Definition 5.13. A *transaction T* on a database or on a subset of a database can be defined by:

transaction *T* [accept ⟨i_messages⟩, ⟨i_views⟩;
return ⟨o_messages⟩, ⟨o_views⟩]:
range ⟨data_segments⟩; [⟨local_views⟩]
where ⟨input_rule⟩
{⟨transaction_block⟩}

where:

1. *T* is the transaction name.
2. The 'accept' clause specifies the input messages ⟨i_messages⟩ and the input views ⟨i_views⟩ of *T*. The 'return' clause specifies the output messages ⟨o_messages⟩ and the output views ⟨o_views⟩ of *T*. A view is specified by a term of the form $P(p_1, \ldots, p_n)$ in which *P* is the name of the view and $p_1, \ldots, p_n$ is a list of variables corresponding to parameters of *P*. Variables specified in the input/output views are used in the transaction as well as those which might have been introduced in the range clause or in the definition of local views (local_views).
3. The 'range' clause specifies the relationships on data which are involved in the transaction. The syntax is the same as for views. Use of local views (local_views) may be required. They are defined as:

   Wi (⟨db_parameters⟩) $\Leftarrow$ {⟨data_segments⟩ where . . .};

   in which *Wi* is the name of the view and ⟨db_parameters⟩, ⟨data_segments⟩ and the 'where' clause have the same meaning as defined for global views.
4. In the 'where' clause of the transaction, the input rule ⟨input_rule⟩ specifies a formula which must be verified by the data which are involved in the transaction. Such a rule can be used to express conditions on local views using set-theory operators (contained_in, contains, difference, member_of, equals).
5. The transaction body ⟨transaction_block⟩ consists of a sequence of operations on data variables specified in input/output views. It describes the processing to be done on database data using the view definitions. Control structures which can be used include sequential, conditional, iterative and case structures. They are expressed using operators on entities, relationships and sets.

For example, let us consider the following transaction using views *IM*, *MM* and P_1 defined previously.

*W*1: 'If machine 521 and the required tool are available, then perform the pending operation for this machine':

transaction *W*1 [accept *IM* (*p*), *MM* (*m*, *y*, *t*); return P_1 (*m*, *t*, *p*)]:
range *p*⟨SP⟩*s*|PLAN⟨PO⟩*y*
where *m.m*# = 521

{update $m(m.\text{cond} \leftarrow$ 'working');
update $t(t.\text{cond} \leftarrow$ 'working');
update $p(p.\text{loc} \leftarrow$ 'P_1');
connect MTP (m, t, p)}

The context of a transaction can be defined in the same way as for the context of a view as follows:

The *context* $\Xi(T)$ of a transaction T is obtained from the conjunction of input view contexts, the link predicates which are mentioned in the 'range' clause of T and the selection formula. Variables which do not appear as parameters of input views must be quantified existentially in $\Xi(T)$.

The context of a transaction characterizes actual data which are involved in the transaction. For instance, the context of transaction W_1 is given by:

$$\Xi(W1) \equiv \Xi\,(IM) \quad \text{and} \quad \Xi\,(MM)$$

and

$$(\exists s)\ s \in \text{PLAN} \quad \text{and} \quad \langle p, s\rangle \in SP \text{ and } \langle s, y\rangle \in PO \quad \text{and} \quad m \cdot m\# = 521$$

(iii) The process model

The two previous sections make it possible to describe the static properties of data (using an EER model) and the dynamic properties of data (using the concepts of views and transactions). The process model presented in this section is used to specify the dynamics of the data processing system being analyzed, i.e. to specify how the transactions and the views are related to one another and used in a given application system. In fact the process model is used to specify at the conceptual design level the internal behavior of activities and functions defined at the organization analysis level. The model can also serve as a validation tool of data schemata previously obtained by simulating the execution of the transactions in their application context.

In the PDN model, the behavior of a physical system is described from an information point of view by means of *process nets*.

Definition 5.14. A process net relative to a database DB is defined by the following components:

1. A directed graph $(\mathbf{P}, \mathbf{T}; \mathbf{F})$, which describes the control structure of the organization being analyzed. $\mathbf{P}$ is a set of places (representing information elements), $\mathbf{T}$ is a set of transactions on the database DB (representing actions on the real-world system) and $\mathbf{F}$ is a set of arcs linking each transaction T to its input and output places, i.e.:

   $$\mathbf{F} \subseteq \mathbf{P} \times \mathbf{T} \cup \mathbf{T} \times \mathbf{P} \quad \text{with} \quad \mathbf{P} \cap \mathbf{T} = \varnothing \quad \text{and} \quad \mathbf{P} \cup \mathbf{T} \neq \varnothing$$

   $\cdot\mathbf{T} = \{P | (P, T) \in \mathbf{F}\}$ and $\mathbf{T}\cdot = \{P | (T, P) \in \mathbf{F}\}$ are called set of input places and set of output places of transition $T \in \mathbf{T}$, respectively. $\cdot\mathbf{T}$ and $\mathbf{T}\cdot$ are usually not disjoint. The set of places **P** is made of two disjoint sets $\mathbf{P} = \mathbf{V} \cup \mathbf{M}$, where **V** is a set of views on database *DB* and **M** is a set of *messages*.

2. Each transaction T represents a simple or complex, but indivisible, action of the organization functionality. The execution of a transaction usually implies changes of the state of adjacent views of transaction T for the associated database.
3. Given a transaction T, a *substitution* of T is a set of occurrences extracted from the input views and input messages. A substitution of T is admissible (i.e. the transaction T can be fired) if the link predicates of the range clause and the input rule of T are satisfied; in other words, if the context $\Xi(T)$ is satisfied.
4. The *marking* M of the process net is given by a set of extensions $M(P_i)$ of the views P_i of the net. The initial marking M_0 of the net is directly deduced from the initial content of the database. It must be noticed that an extension of a view is always a set and not a multi-set because occurrences of entities and relationships are uniquely identified in a database.
5. A firing rule: if the transition T can be fired, i.e. if the marking M is such that:

   - each extension of input views of T contains at least one tuple,
   - the set of input tuples of T contains at least one admissible substitution for T,

   then transaction T can be executed. During its execution, data manipulation operations imply modifications on the database and, consequently, the related view extensions are updated according to the following rules:

   - input tuples which appear in an admissible substitution are removed from the extension of input views;
   - relevant output tuples are added to the extension of output views;
   - input messages cease to hold and output messages become valid.

In fact, messages model events of the real-world used to trigger a process net, i.e. a data processing procedure. They also model state places as used in basic Petri nets.

(iv) Example

As an example, let us consider the dynamic description of the FMC system as illustrated by the process net of Fig. 5.10. Parts $\langle p \rangle$ stored in the input buffer R and ready to be machined are mounted on pallets with proper fixtures $\langle f \rangle$ given by view F (transaction Mount). A part is processed (transaction Work) according to a sequence of specific operations given by its process plan and performed by a machine $\langle m \rangle$ using the proper tool $\langle t \rangle$. Available machines $\langle m \rangle$ and tools $\langle t \rangle$ for a given operation are represented by the place U. At the end of each operation $\langle y \rangle$, the part must be prepared for the next operation. The transaction Prep returns the machine

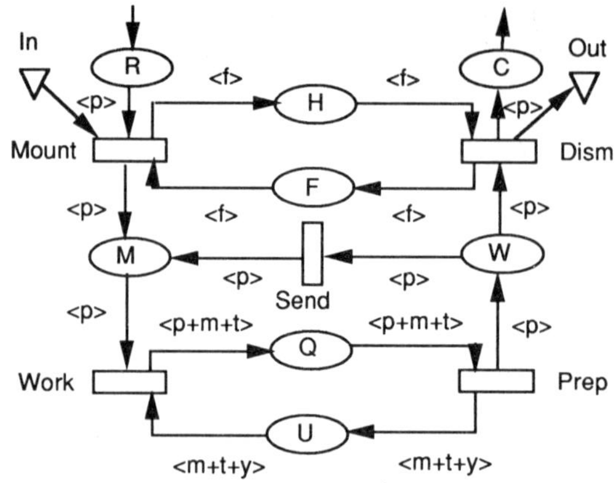

Figure 5.10 Process net for part processing in an FMC.

used in the previous operation and selects the next operation for a part $\langle p \rangle$ which is in-process. If all the operations on the part have been done, the part is dismounted and is stored in an output buffer *C* (transaction Dism); otherwise, a transaction Send sends the part back into the cell for further processing in buffer *M*.

view *F*(*f*|FIXTURE) where *f*.cond='available'
view *H*(*f*|FIXTURE) where *f*.cond='used'
view *R*(*p*|PART) where *p*.stat='ready'
view *M*(*p*|PART) where *p*.stat='mounted'

message In {'start process'}

transaction Mount [accept In, *R*(*p*), *F*(*f*); return *M*(*p*), *H*(*f*)]:
range *p*⟨FP⟩*pt*|PART_TYPE ⟨PP⟩*f*
{update *p*.stat←'mounted', *f*.cond←'used'; connect *PF*(*p*, *f*)}

view *U*(*m*|M_MAC, *y*|OPERATION, *t*|TOOL)
range *p*⟨*r*|OMT⟩*y*, ⟨*r*⟩*t*
where *m*.cond='available' and *t*.cond='available'

view *Q*(*p*|PART, *m* | M_MAC, *t*|TOOL)
range *p*⟨*r*|MTP⟩*m*, ⟨*r*⟩*t*

transaction Work [accept *M*(*p*), *U*(*m*, *y*, *t*); return *Q*(*p*, *m*, *t*)];
range *p*⟨SP⟩*s*|PLAN⟨PO⟩*y*
{update *m*.cond←'working', *t*.cond←'in-use'; connect MTP(*p*, *m*, *t*)}

view *W*(*p*|PART) where *p*.stat='in-process'
view *C*(*p*|PART) where *p*.stat='worked'

transaction Prep [accept *Q*(*p*, *m*, *t*); return *W*(*p*), *U*(*m*, *y*, *t*)]:
range *p*⟨SP⟩*s*|PLAN⟨PO⟩*y*; *s*⟨SS.succ⟩*s*'|PLAN

{disconnect MTP(p, m, t), SP(p, s); connect SP(p, s');
update m.cond←'available', t.cond←'available', p.stat←'in-process'}

transaction Send [accept $W(p)$; return $M(p)$]:
range p⟨SP⟩s|PLAN
{update p.stat←'mounted'}

message Out {'process completed'}

transaction Dism [accept $W(p)$, $H(f)$; return Out, $F(f)$, $C(p)$]:
range f⟨PF⟩p
{disconnect PF(p, f); update p.stat←'worked', f.cond←'available'}

Thanks to the formalism provided by Petri nets, advantages of a model such as the PDN model are: (1) it allows a nice and integrated extension of the entity-relationship model to model the dynamics of the information system, and (2) it lends itself to formal specifications of the information system which, because of the increasing complexity of such systems, is becoming an important factor in CIM system design.

(b) *Petri nets with objects – the O-PDN model*

Object orientation is the new trend in information system analysis. Therefore, a new version of M^* has recently been developed to deal with objects and called M^*-Object.

While the PDN model is made of three tightly-coupled submodels (data model, transaction model and process model), the goal of the Process and Data Net with Objects (O-PDN) model of M^*-Object is to provide a unique model to cover the same features using object orientation.

The O-PDN model is made of an object-oriented (OO) semantic data model (which replaces the EER model in the PDN model) and a process model defined as a PrT-net. The OO model is used to specify the structure and behavior of objects (derived from the definition of object views and from formatted and non-formatted requirements) whereas the process model is used to specify the application behavior, i.e. the dynamic behavior of objects and functions identified at the organization analysis level.

(i) The data model

Definition 5.15. The object-oriented data model used in M^*-Object is defined by the 5-tuple ⟨**D**, **E**, **A**, **X**, **O**⟩, where:

1. **D** is a set of basic domains;
2. **E** is a set of classes of objects;
3. **A** is a set of attributes A_{ki}, $i = 1, 2, \ldots$;
4. **X** is a set of variables on the classes of **E**;
5. **O** is a set of operators applicable to objects.

(1) Basic domains include integers, reals, characters, strings, as well as record structures, enumerated data types and range data types.

(2) Each class $E_i \in \mathbf{E}$ describes a set of similar objects having the same structure (i.e. the same *attributes*) and the same behavior (i.e. the same operators or *methods*). Objects are occurrences of classes. Each object is identified by a unique object identifier (OID) and belongs to one class only. Object classes can be organized into class hierarchies. There are two types of class hierarchies:

- The *generalization hierarchy* allows a class (the subclass) to be defined as a specialization of another existing class (the superclass). The subclass inherits attributes and methods from the superclass and may add specific attributes and methods. It defines the 'is-a' semantic link.
- The *aggregation hierarchy* allows a class (the composite class) to be specified as a composition of existing classes (the component classes). It defines the 'part-of' semantic link.

A class identifier of a class E_i is a set of one or more attributes which uniquely identify each occurrence of the class (if this set includes an identifier of another class related to E_i, E_i is said to be a dependent class).

(3) Attributes A_{ki} are characteristics of classes of **E**. An attribute can be classified as:

- A member attribute, i.e. a structural property of an object class (the price of a machine, operation time, etc.). It is defined as a mapping $A_{ki} : E_i \rightarrow D_k$ where E_i is a class and D_k is a domain of **D**.
- A relationship attribute, i.e. a link property between an object class E_i and another class E_j defined as a mapping $A_{ki} : E_i \rightarrow E_j$ (e.g. a machine 'uses' a tool). The reverse link can be defined from E_j to E_i. The value of the attribute is the value of the identifier(s) of the linked object(s).

The properties of an attribute A_{ki} can be further specified by means of two natural numbers (min, max) which mean that an object $e \in E_i$ from which A_{ki} is an attribute must take a minimum of min and a maximum of max values; if A_{ki} is a relational attribute pointing to class E_j, this means that e must be related to a minimum of min and a maximum of max objects in E_j, at all times.

(4) Variables on the classes of **E** are declared as follows:

$$\text{use } p|P$$

meaning that variable $p \in \mathbf{X}$ can represent any object of class $P \in \mathbf{E}$.

(5) Operators applicable to objects are defined by means of methods on the classes of **E**. Methods are procedures or routines applicable to attributes of the classes. They define the dynamic part of the objects of the class. They are specified in the implementation part of the object class.

Furthermore, predefined methods are also available in the model to select and manipulate objects (Create, Update, Delete, Select, Search), to

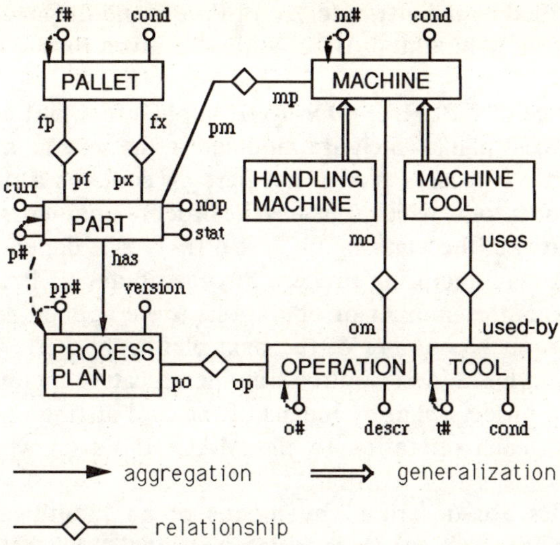

Figure 5.11 Database schema for an FMC.

specify objects in a generalization hierarchy (Specialize, Generalize) and to add (remove) objects into (from) a composite object (Add, Remove). Referring to the example in Fig. 5.11, one can write:

use *m*|MACHINE; *n*|MACHINE-TOOL; *t*|TOOL; *f*|PALLET;
  p|PART; *k*|PROCESS-PLAN
m.Create (*m*#:'14', cond: 'available', . . .); . . .
  «create a machine *m* whose number is . . . »
n.Specialize (*m*; uses:*t*); . . .
  «machine 14 is specialized as a machine-tool which uses a tool *t*»
p.Select (*p*#='101'); . . .
  «select the part whose number is . . . »
f.Select (*f*#'F47'); . . .
  «select the pallet whose number is . . . »
p.Update (*px* : *f*, . . .); . . . «*f* becomes the pallet on which *p* is mounted»

Figure 5.11 gives an illustration of the OO data model for the FMC example. It can be interpreted in the following way:

- The system manages PARTs which must be mounted on a PALLET to be processed by a MACHINE. Each part is described by means of a 'stat'(us) ('ready', 'mounted', 'working', 'worked', 'completed') and a part number '*p*#'.
- A pallet is described by means of a 'cond'(ition) ('free', 'held') and a pallet number '*f*#'. Parts can be mounted on different pallets (specified by the attribute '*pf*'), and a pallet can be used for different parts (specified by the attribute '*fp*'). The pallet which is actually in use for a given

part is specified by the attribute '*px*' of PART and the machine on which a part is actually mounted to be worked is given by the attribute '*pm*' of PART.

- The sequence of OPERATIONs to be applied to a part is described by a set of process plans which are modeled in the schema as components of that part. A process plan for a part *p#* contains a progressive sequence number '*pp#*' which specifies the order of operations to be applied on that part, i.e. the class PROCESS-PLAN is a dependent class. To control the manufacturing process, two attributes of PART are used: 'nop' contains the number of operations to be applied and 'curr' contains the sequence number of the next plan to be performed.
- An operation has a 'descr'(iption), an identification number '*o#*', and is related to a process plan by means of the '*op*' attribute. The attribute '*om*' relates each operation to the MACHINEs on which it can be performed.
- MACHINEs are described by means of an identification number '*m#*' and a 'cond'(ition) ('available', 'working'). A machine can be a HANDLING-MACHINE or a MACHINE-TOOL. In the latter case, it can use a set of TOOLs specified by the attribute 'uses'.

In Fig. 5.11 only some attributes are reported. For instance, a more detailed specification of the class PROCESS-PLAN is given by:

> class PART composed of (has: PROCESS-PLAN) identifier *p#*
>   member attributes *p#*: integer; . . .
> class PROCESS-PLAN identifier (has, *pp#*)
>   member attributes
>   *pp#*: integer; rev_date: date;
>   descr: string; planner: string; version: string;
>   status: ('created', 'held', 'released', 'obsolete');
>   relationship attributes
>   *po* $(1, n)$: OPERATION inverse of *op*
>   methods Create (*p*|PART, . . .); Initialize (. . .); Release (. . .);
>   Plan (. . .); Review (. . .); Purge (. . .);

where 'descr' is a plan description, 'planner' is the process planner name, 'version' the version number, and 'rev_date' the date of revision. The aggregation 'has' connects a part to the set of process plans which specifies the operations to be applied on it.

The collection of all object class definitions with their implementation parts defines the object-oriented *database schema.*

(ii) Database views and processes

(i) View

Definition 5.16. A view on a database schema is defined by a statement of the following form:

view ⟨view-name⟩ (⟨target-list⟩)
use ⟨variable-declaration-list⟩
where ⟨predicate⟩

where

1. ⟨view-name⟩ is the name of the view.
2. ⟨target-list⟩ has the following form: $p_h \mid E_h, p_k \mid E_k, \ldots$ and introduces a set of parameters $p_h, p_k \ldots$ on the classes $E_h, E_k, \ldots$ It specifies what must be retrieved as a set of tuples $\langle e_h, e_k, \ldots \rangle$ where $e_h \in E_h$, $e_k \in E_k$, and so on.
3. The ⟨variable-declaration-list⟩ has the form $v_m \mid E_m, v_n \mid E_n, \ldots$ which can be used to introduce a set of local variables on the classes $E_m, E_n, \ldots$ that must be used to construct the query result.
4. The predicate in the 'where' clause is called a *selective path* because it is, in general, a conjunction of *restriction predicates* (which are used to select relevant objects) and a *path predicate* which specifies a navigation in the database and assumes the form: $p_h \langle r_{hk} \rangle p_k \langle \ldots \rangle v_m \langle r_{mn} \rangle v_n \ldots$, where r_{hk} is the name of a link which connects E_h with E_k, and so on; a link can be a relationship attribute or an arc of a hierarchy (the name can be omitted if there is just one link between the classes); the path predicate is satisfied with the substitution $\{p_h/e_h, p_k/e_k, \ldots, v_m/e_m, v_n/e_n\}$ if, and only if, the object $e_h \in E_h$ is connected by the link r_{hk} to the object $e_k \in E_k$, and so on. A tuple $\langle e_h, e_k, \ldots, e_m, e_n \rangle$ which satisfies the selective path is called an *occurrence* of the path.

The view occurrences are obtained by discarding from the path occurrences all the variables that are not specified as view parameters; the set of occurrences of a view is called the *view extension*.

For instance, referring to Fig. 5.11, the query: *W* 'Select the tools which can be used to work the part 27' can be expressed as (remember that a machine-tool inherits all the properties of a machine):

view *W*(*t*|TOOL)
use *p*|PART; *pp* | PROCESS-PLAN; *o*|OPERATION;
  mt|MACHINE-TOOL
where *p.p*#=27 and *p*⟨has⟩*pp*⟨po⟩*o*⟨om⟩*mt*⟨uses⟩*t*

Note. The tuple values returned by views depend upon time.

(ii) Messages A routine models a manipulation on data under the request of an event. It specifies a change in the information system state.

Definition 5.17. An event is described by means of a *message* which is issued at a given instant in time. It may have some parameters and can be declared as:

message ⟨message-name⟩ (⟨parameter-list⟩)

where ⟨message-name⟩ is the name of the message and ⟨parameter-list⟩ is used to exchange data between the routine and the external world.

(iii) Routines

Definition 5.18. The routine mechanism is used to define pieces of functionalities to be executed on objects of the CIM system. A routine can receive and send messages and views (called places) as inputs and outputs and do some processing (the routine-block). They have the following form:

```
routine ⟨routine-name⟩
[accept ⟨input-place-list⟩; return ⟨output-place-list⟩]
use ⟨local-variable-list⟩; ⟨local-view-list⟩
guard ⟨predicate⟩
exec ⟨routine-block⟩ end
```

where:

1. A place in ⟨input-place-list⟩ and ⟨output-place-list⟩ is specified by a term of the form $P(p_1, \ldots, p_n)$ where P is the place name and $p_1, \ldots, p_n$ is a list of parameters corresponding to the formal parameters in the place declaration (parameters refer to values if P is a message, to database objects if P is a view).
2. An input message describes an event that must be active for the routine execution, while an output message describes an event that becomes active after the routine execution.
3. Input views (in the 'accept' clause) describe the data that are relevant for the routine execution (only changing data must be specified), while output views (in the 'return' clause) describe the data that have been changed after the execution. A parameter which appears in the input and output list means that the *same* object which is received in input is also given back as an output (some of its properties can be changed by the routine).
4. With the 'use' clause it is possible to introduce local variables (as in the view definition) as well as local views, which describe database objects that will not be changed.
5. The predicate in the 'guard' clause can be used to specify conditions on the input and local data by means of path and restriction predicates.
6. The 'routine-block' consists of a sequence of operators which manipulate database objects; control structures (including *sequence*, *conditional*, *iterative*, *parallel* and *rendezvous* structures) can be used to specify the program flow.

For instance, mounting a part (with a given part number and ready to be worked) on a pallet (which must be free) can be described in the following way:

message In (*pn*: integer)
  (* the part with the part number '*pn*' must be worked *)
view *R*(*p*|PART) where *p*.stat = 'ready'
  (* parts ready to be worked *)
view *F*(*f*|PALLET) where *f*.cond = 'free' (* free pallets *)
view *M*(*p*|PART) where *p*.stat = 'mounted' (* mounted parts *)
view *H*(*f*|PALLET) where *f*.cond = 'held' (* held pallets *)

routine Mount [accept In(num), *R*(*p*), *F*(*f*) return *M*(*p*), *H*(*f*)]
guard *p.p*#=num and *p*⟨*pf*⟩*f*
exec *p*. Update (stat: 'mounted', *px*: *f*, curr: 1);
  f.Update (cond: 'held', *fx*: *p*)
end Mount

(iii) The process model

The process model of the O-PDN model of *M**-Object is used to define pieces of the CIM system behavior, called *processes*, to specify the enterprise functions identified at the organization analysis level. It is based on the concepts of *views*, *messages* and *routines* as just introduced. It defines causal relationships between routines.

Definition 5.19. The process model of *M**-Object is defined as a PrT-net by:

1. A directed net (**P**, **T**; **F**), which represents the control structure of the system. **P** is a set of places such that $\mathbf{P} = \mathbf{V} \cup \mathbf{M}$, where **V** is a set of views and **M** is a set of messages, **T** is a set of routines on the database and **F** is a set of arcs connecting each routine to its input and output places ($\mathbf{F} \subseteq \mathbf{P} \times \mathbf{T} \cup \mathbf{T} \times \mathbf{P}$, $\mathbf{P} \cap \mathbf{T} = \varnothing$, $\mathbf{P} \cup \mathbf{T} \neq \varnothing$).
2. An object-oriented (OO) database schema to specify system objects.
3. The marking Δ of the process net, which consists of the collection of the extensions $\Delta(V_i)$ of its views $V_i \in \mathbf{V}$ and the condition (active or inactive) of its messages $M_i \in \mathbf{M}$.
4. A set of routines **T**, which represent how the system reacts to a given marking Δ. A marking Δ enables a routine $T \in \mathbf{T}$ if all the input messages of T are active, all the output messages of T are inactive, and there exists an input substitution which satisfies its guard predicate.
5. Firing rules such that if the routine T is enabled, it can be *executed*, i.e. statements of the 'exec' clause are performed on the input substitution so that input tuples are removed from the input views, output tuples are added to the output views, input messages become inactive (they no longer hold) and output messages become active (they begin to hold).

As an example, let us consider again the dynamic description of the FMC system as illustrated by the process net of Fig. 5.12 (which is similar to the one of Fig. 5.10 but refers to the data schema of Fig. 5.11). The activities in the FMC system, besides the Mount routine described above, are:

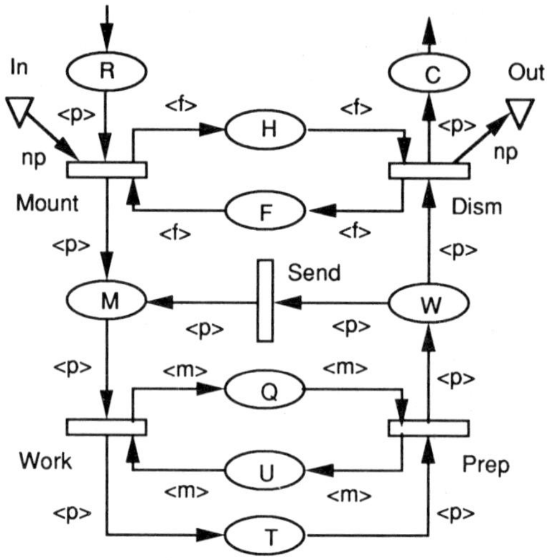

Figure 5.12 Process net for a flexible manufacturing cell.

- A part is processed via the execution (routine Work) of a sequence of specific operations, performed by a machine. The actual number of the plan to be executed is stored in the attribute 'curr' of PART.
- At the end of each operation, the part must be prepared for the next operation; the routine Prep returns the machine used in the previous operation and selects the next plan by incrementing the attribute 'curr'.
- If all the operations on the part have been completed, the part is dismounted and is stored into an output buffer (routine Dism); otherwise, a routine Send sends the part back into the cell for further processing.

Buffers containing parts at different stages of manufacturing (parts 'ready' to be worked, 'mounted' on pallets, 'working', 'worked' and 'completed') are represented by means of the views *R*, *M*, *T*, *W* and *C*, respectively, views *F* and *U* represent 'free' pallets and 'available' machines, while views *H* and *Q* represent 'held' pallets and 'working' machines respectively. The message Out gives the number of the part whose manufacturing process has been completed. In this case, the specification of the system is based on the following definitions of views, messages and routines.

message In (*np*: integer)
message Out (*np*: integer)

view *R*(*p*|PART) where *p*.stat = 'ready'
view *M*(*p*|PART) where *p*.stat = 'mounted'
view *T*(*p*|PART) where *p*.stat = 'working'
view *W*(*p*|PART) where *p*.stat = 'worked'
ciew *C*(*p*|PART) where *p*.stat = 'completed'
view *F*(*f*|PALLET) where *f*.cond = 'free'

view *H*(*f*|PALLET) where *f*.cond = 'held'
view *U*(*m*|MACHINE) where *m*.cond = 'available'
view *Q*(*m*|MACHINE) where *m*.cond = 'working'

routine Mount [accept In(num), *R*(*p*), *F*(*f*) return *M*(*p*), *H*(*f*)]
guard *p*.*p*#=num and *p*⟨*pf*⟩*f*
exec *p*.Update (stat: 'mounted', *px*: *f*, curr: 1);
  f.Update (cond: 'held', *fx*: *p*) end Mount

routine Work [accept *M*(*p*), *U*(*m*) return *T*(*p*), *Q*(*m*)]
use *s*|PROCESS-PLAN, *y* | OPERATION
guard *p*⟨has⟩*s*⟨po⟩*y*⟨om⟩*m* and *p*.curr = *s*.*pp*#
exec *m*.Update (cond: 'working', mp: *p*);
  p.Update (stat: 'working', pm: *m*) end Work

routine Prep [accept *Q*(*m*), *T*(*p*) return *U*(*m*), *W*(*p*)]
guard *p*⟨pm⟩*m*
exec *m*.Update (cond: 'available', mp: nil);
  p.Update (stat: 'worked', pm: nil, curr: curr + 1) end Prep

routine Send [accept *W*(*p*) return *M*(*p*)]
guard *p*.curr ≤ *p*.nop
exec *p*.Update (stat: 'mounted') end Send

routine Dism [accept *W*(*p*), *H*(*f*) return *C*(*p*), *F*(*f*), Out(num)]
guard *p*.curr > *p*.nop and *p*⟨px⟩*f*
exec *p*.Update (stat: 'completed', px: nil);
  f.Update (cond: 'free', fx: nil) end Dism

The purpose of this section was to show that a similar framework can be used either for the object-oriented data models and for the entity relationship data models for modeling the behavior of information systems.

5.5 MODELING, ANALYSIS AND IMPLEMENTATION OF COMPANY POLICIES

5.5.1 Overview of the knowledge-base design methodology

The design and maintenance of a knowledge-based system (KBS) to control the functional relationships and information flow is briefly described here. Our general design methodology for it is depicted in Fig. 5.13. It starts from user-defined rule specifications reflecting a specific company policy, which is then modeled using a special set of colored Petri nets – UPN (Updated Petri Nets) – and a hierarchical modeling methodology, discussed in section 5.5.3. The next step is to convert the UPN model into general Petri nets (GPN) for validation purposes and feed the results back to the user to resolve (i) conflicting company rules, and (ii) errors

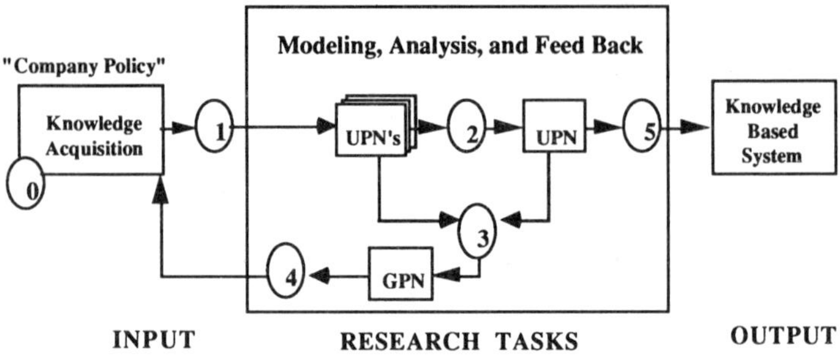

Figure 5.13 Knowledge-base design methodology.

introduced during the modeling phase. This process is described in section 5.5.4. After the model has been validated, a parser translates the UPN model into a rule-specification language described in section 5.5.5. The end result is a software package that controls the dataflow and accessibility between several databases. In short, the input is a specific company policy, expressed as a set of logical and procedural rules, and the output is an Artificial Intelligence (AI) production system for controlling operations, accessibility, precedence and updates of data within the manufacturing applications involved. An overview of the major design phases of the system is presented below.

(a) Knowledge acquisition (company policy)

The design of the model is based on the information flow established between all the manufacturing applications, namely CAD/CAPP/MRP II and SFC. As mentioned above, the expert rules embedded in the knowledge base are extracted from company expertise, policies and procedures, which can be obtained through a number of individual interviews and group meetings with experts from all manufacturing application systems to be integrated and from managers responsible for making company policy. Therefore, substantial effort may be required for gathering all expert rules to form the knowledge base. However, since we are here to develop and demonstrate a generic design methodology, the prototype presented here only includes limited rules extracted from our own industrial experience.

(b) Scenarios of the proposed CIM system

The development of a knowledge-based system usually starts from designing a set of abstract rules for the initiation and maintenance of specific entities within the system. A set of scenarios under each entity which represent these abstract rules is listed below.

(i) Part data

- Adding New Product Parts in CAD
- New Product Part Revisions in CAD
- New Non-Product Parts in CAPP
- Adding New Non-Product Part Revisions in CAPP
- Adding New Non-Product Parts in MRP II
- Adding New Non-Product Revisions in MRP II
- Making Parts Obsolete
- Deleting Parts.

(ii) Product structures (bills of material)

- Adding Component Relationships in CAD (for products)
- Adding Component Relationships in CAPP (for tools, jigs, fixtures, etc.)
- Deleting Component Relationships
- Substituting Components in Relationships
- Changing the Required Quantity of a Component
- Copying Relationships from One Assembly to Another.

(iii) Work centers

- Establishing New Work Centers in MRP II
- Modifying Work Centers in MRP II, CAPP and SFC
- Deleting Work Centers in MRP II.

(iv) Process plans

- Establishing New Process Plans in CAPP
- Modifying Process Plans in CAPP
- Deleting Process Plans in CAPP.

(v) Manufacturing orders

- Adding Orders in MRP II
- Modifying Orders in MRP II
- Modifying Order BOM/Routing in SFC
- Updating Job Status in SFC (including actual material issued and time taken)
- Deleting Orders from MRP II.

(c) Status codes

The flow of information within the system is controlled using a set of status codes assigned to each set of entity data within each functional area. The status codes are designed to provide for triggering the right action while controlling the sequence of various part and process design and manufacturing related activities. The following are the status codes used in the system.

- *Working*: the 'working' status is given to CAD part data related to designs that have not yet been finalized. In a similar fashion, it is given to process plans in CAPP, work centers in CAPP and shop orders in SFC. In the case of work centers, it is intended to signify that some work center data is still missing in CAPP.
- *Released*: the 'released' status is indicative of an entity becoming active in the system. It is applied to CAD and MRP II part revision data; CAPP, SFC and MRP II work center data; CAPP and MRP II process plan data; SFC and MRP II order data.
- *Hold*: the 'hold' status is normally given to an entity, when it is being reviewed for possible revision or replacement, and the entity should not be used while on hold. For example, it can be given to a work center in the case of an extended breakdown. It is used for CAD and MRP II part revision data; CAPP, SFC and MRP II work center data; CAPP and MRP II process plan data; SFC and MRP II order data.
- *Obsolete*: data related to entities that are no longer considered active, are given the 'obsolete' status. This code is used by CAD part revisions and CAPP routings. However, MRP II and SFC handle obsolescence with the use of effectivity start and effectivity end dates. Therefore, they do not require this status code.

(d) Modeling of the domain knowledge

The hierarchical modeling approach facilitates the modeling task. It incorporates: (a) a top-down stepwise refinement technique for the modeling of each scenario from an abstract and aggregate level to a detailed level, and (b) a synthesis technique for synthesizing separate nets, which represent different scenarios of the system, to form a coherent net. The approach necessitates the development of new Petri net modeling entities, which include two types of transitions presented in section 5.3.3; one to represent primitive operations, and the other to represent compound operations which can be further exploded into subnets. The design process for each scenario (each set of functional relations) starts from an abstract net with both primitive and compound transitions, and continues by exploding each compound transition until no compound transition exists. This improved modeling capability enables the structuring of Petri nets in a progressive manner and facilitates the transformation of the 'company policy' of Fig. 5.13 to a formal model.

(e) Knowledge verification

Several analysis techniques for Petri nets, including reachability trees and net invariants, can be used (Peterson, 1981; Jensen, 1986). Net invariants, which represent mutually exclusive conditions within the 'company policy', can reveal logical conflicts in the specification of the original rules and, possibly, errors introduced during the modeling process. Reachability trees can be used to detect any deadlock or inconsistencies in the model. Programs for computerizing these analysis methods have been developed and applied extensively. Some reduction rules (Lee and Favrel, 1985) can also be investigated to reduce the complexity of nets for further analysis.

However, these analysis techniques were developed for generalized Petri nets (GPN), but not applied to colored Petri nets (CPN) in which a great diversity of linear functions are associated to the arcs. Therefore, unlike analysis algorithms for GPN that use integer matrices, analysis algorithms for CPN need to manipulate matrices composed by linear functions. This fact introduces a high complexity in the development and execution of these algorithms. An alternative approach is to *convert* UPN into GPN before they can be analyzed. The unfolding algorithm is presented in section 5.5.4.

(f) Implementation

We have adopted a fairly new concept in systems integration, known as database interoperability. It is being realized through the development of the Update Dependency Language (UDL) in the Department of Computer Science at the University of Maryland (Mark and Roussopoulos, 1987). Database interoperability can be described as the concatenation of the schemata of each of the databases of the application systems, along with a rule set constructed for each separate database, called update and retrieval dependencies. These update and retrieval dependencies control inter-database consistency through inter-database operation calls. We propose the use of Update Dependencies as a special rule specification language, to be used for the implementation of our knowledge-based system. The algorithm of automatic translation between UPN and UDL is presented in section 5.5.5. It reduces dramatically the implementation effort. The details of the Update Dependency Language are also presented in the same section.

5.5.2 Establishment of company policies – a typical scenario

(a) Data specification

The basic work center data fields supported by the majority of MRP II, CAPP and SFC systems are listed below. Initially the data fields established by MRP II are described, as the majority of these are supported by all three application systems. Subsequently, the remaining CAPP instantiated data fields and SFC instantiated data fields will be described.

- *ID Number*: this is a unique identifying number assigned to each work center (WC) in the system. MRP II is the sole center for assigning the ID Numbers to new work centers, and these numbers identify the work center in CAPP and SFC as well.
- *Description*: this field is filled in by MRP II users and contains information as to the type of work center identified by that particular ID Number. This record is also maintained in CAPP and SFC.
- *Department*: this represents the department to which the work center is assigned. MRP II assigns work centers to various departments, such as the machine shop, assembly line, and quality inspection. CAPP and SFC include this field in their records.
- *Capacity*: this field represents the number of hours a particular work center is available for use in a given day. This information is established and maintained in MRP II and transferred to SFC for scheduling purposes.
- *Resource Code*: this is a code which is inserted into the work center record, allocating a particular measure to the resource being utilized. The measure could be man-hours, machine-hours, etc. This is done for costing purposes.
- *Rate Code*: the Rate Code associates a particular dollar rate per hour, with the specified Resource Code.
- *Dispatch Horizon*: the Dispatch Horizon defines how many working days into the future are to be scanned, so as to provide workload for a work center.
- *Effectivity Start Date*: this represents the date on which a particular work center becomes active in the system. This date is determined by MRP II users, based on the introduction of the work center in the system, and the time required to set it up, so that it becomes fully operational. CAPP and SFC users maintain this information as well, in order to determine whether the new work center will be ready in time to be utilized by a particular routing. (In fact, upon release of a work order to SFC, SFC should check the effectivity start date of work centers involved to make sure they are currently active.)
- *Effectivity End Date*: this represents the date on which the work center is expected to become inactive in the system, due to obsolescence or any other reason. This information is not required to be entered when establishing a new work center. It can be entered at a later time.
- *Status Code*: status codes are used to control the state of a work center in the system at any given time. MRP II, CAPP and SFC users can change the status of a particular work center, meaning that they can release, place on hold or delete a work center subject to the rules of the knowledge base.
- *Work Center Load Profile*: this field shows the capacity loads on a specific work center scheduled by SFC within the dispatch horizon of both MRP II and SFC.

- *Work Center State*: this field shows the physical state, either 'operational' or 'broken', of a work center used on the shop floor. Both SFC and MRP II use this field to monitor the availability of all work centers and resolve problems related to machine breakdowns.

In addition to the fields described above, CAPP systems normally maintain additional technical performance data in their work center files, which are essential for the generation of process plans and the selection of work centers for specific operations by CAPP, and are accessible by SFC for routing selection, work order scheduling and resource allocation.

(b) Policy specification

Work centers are originated in the system primarily through the MRP II module. MRP II users are responsible for establishing as well as phasing out work centers in the system and maintaining work center data in MRP II. Because CAPP requires detailed work center information for generating process plans, its work center files incorporate both the work center information maintained in MRP II and other detailed technical information. Similarly, SFC also needs detailed work center information as in MRP II and CAPP. Additional work center information in SFC includes the state of a work center, to be able to schedule operations of work orders generated by MRP II. Once again, there is a great deal of similarity between the sets of data maintained by each application system.

Establishing new work centers in the system (Add): ADD via MRP II
1. MRP II user enters the basic data (WC ID, Description, Department) to create the Work Center Record with *h* status in MRP II:

check WC ID does not exist in MRP II

2. MRP II user enters additional information (Capacity, Resource Code, Rate Code, Dispatch Horizon and Effectivity Start Date) through a modification transaction. MRP II user then releases the WC. If this additional information was not entered, the system will prompt the user during releasing of the WC in MRP II.

check WC ID exists in MRP II
WC with *h* status exists in MRP II
All the necessary data fields are filled
WC record does not exist in CAPP
update WC record status changed from *h* to *r* in MRP II
Skeletal WC Record automatically created in CAPP with *w* status

3. CAPP user enters additional information (Horse Power, Speed Range, Feed Range, Work Envelope, Accuracy, Tool Change Time, Feed Change

Time, Speed Change Time, Table Rotation Time, Tool Adjusting Time and Rapid Traverse Rate) through a modification transaction. CAPP user then releases the WC. If this additional information was not entered, the system will prompt the user during releasing of the WC in CAPP.

check WC ID exists in CAPP
WC with *w* status exists in CAPP
All the necessary data fields are filled
WC record does not exist in SFC
update WC record status changed from *w* to *r* in CAPP
WC Record automatically created in SFC with *h* status

4. SFC user releases the WC with the work center state as *av* for available

check WC ID exists
WC with *r* status exists in MRP II and CAPP
WC with *h* status exists in SFC
update WC record status changed from *h* to *r* in SFC,
and state changed from *na* to *av* in SFC and MRP II

(c) *Detailed policy specification*

MRP II users have the sole authority to establish new work centers. This is because MRP II is the execution function in most manufacturing organizations and is used for all types of purchasing and procurement activities. To establish a new work center in the system, MRP II users must provide the following basic information, in addition to assigning the identification number.

- WC ID
- Description
- Department

Subject to the condition that this WC ID does not already exist in the system, this work center record is established in MRP II with its status set to hold. MRP II users then finalize all the other work center details needed in MRP II module as follows:

- Capacity
- Resource Code
- Rate Code
- Dispatch Horizon
- Effectivity Start Date.

Then the MRP II user releases the work center.

In addition to the data fields mentioned, effectivity end date, status code, work center state and work center load profile are also part of the work center record in MRP II. The status code is not a user input, but is automatically updated from *h* for hold to *w* for released by the release transaction on the work center. The work center state is maintained by SFC users and is not provided at this stage. The work center load profile will be entered and maintained by SFC and updated automatically in MRP II after the work center is allocated for job scheduling.

Invoking the work center release transaction in MRP II triggers a set of consistency checks, which are as follows: the WC ID provided must exist in MRP II with hold status; all the required data fields should have been filled and any data fields left out by users are requested at this stage with the help of system-generated prompts. If all these checks are satisfied, the system changes the work center status code from 'hold' to 'released' and a skeletal work center record is automatically created in the work center file in CAPP, with its status set to 'working' as well as a work center record in the work center file in SFC with its status set to 'hold'. These work center records contain all the common information between MRP II and CAPP and between MRP II and SFC.

CAPP users then input the detailed technical information regarding that work center. The following fields are required to be completed in CAPP, before the work center can be given a released status and made effective and ready to be used in process plans. If a particular data field does not apply to the specific work center, then 'inapplicable' will be entered automatically. However, no field can be left blank.

- Horse Power
- Speed Range
- Feed Range
- Work Envelope
- Accuracy
- Tool Change Time
- Feed Change Time
- Speed Change Time
- Table Rotation Time
- Tool Adjusting Time
- Rapid Traverse Time.

Similar to the MRP II, invoking the work center release transaction in CAPP triggers a set of consistency checks, which are as follows: the WC ID is checked to ensure that there exists in CAPP a work center file with status set to 'working'; all the required data fields should have been filled and any data fields left out by users are requested at this stage with the help of system–generated prompts. Upon the satisfaction of these checks, the work center gets a released status in CAPP and the common data fields between CAPP and SFC are automatically copied from CAPP to SFC.

This scenario ends by releasing the work center in the SFC module. Invoking the work center release transaction in SFC triggers a consistency

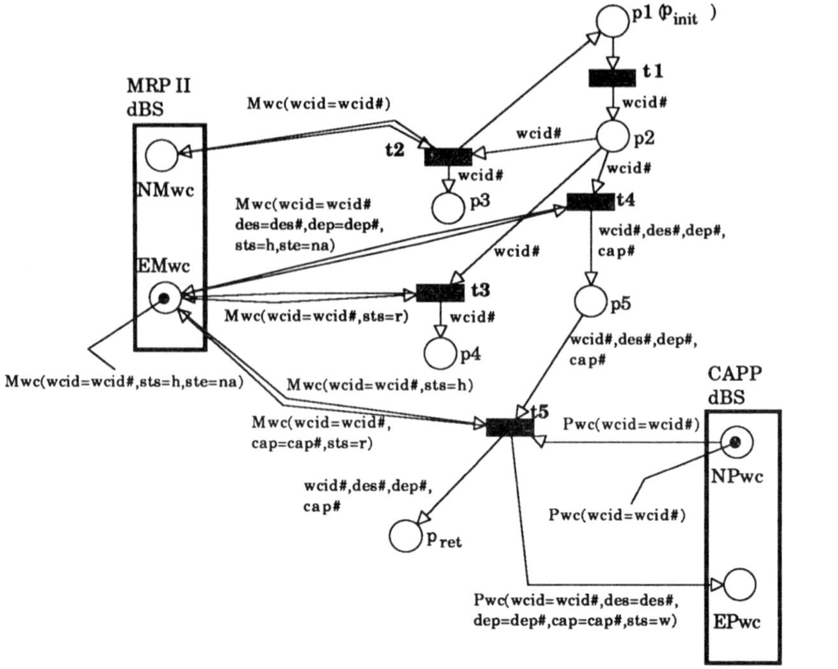

t1: request and read wcid
t2: write error message and restart
t3: write error message
t4: request other information
t5: update work center record in MRP II dBase with sts=r, and additional data, insert a work center record in CAPP dBase

p1(p_init): user starts the transaction
p2: wcid is provided
p3: work center ID does not exist in MRP II
p4: work center already has 'r' status in MRP II
p5: all the necessary data is provided
p_ret: return of the procedure call
EMwc: existence of work center in MRP II dBase
NMwc: non-existence of work center in MRP II dBase
EPwc: existence of work center in CAPP dBase
NPwc: non-existence of work center in CAPP dBase

Figure 5.14 Subnet of the work center creation scenario 'Release of a work center in MRP II' with initial marking.

check: the WC ID is checked to ensure that it exists in the SFC work center file with status set to 'hold'; the WC status in both MRP II and CAPP are set as *r* to ensure all the necessary information has been provided; the current date is past the effectivity start date. The work center state is then automatically set to *av* in SFC and MRP II as being available. Upon the satisfaction of these checks, the work center gets a released status in SFC and the work center state is updated in MRP II.

5.5.3 Model design of the knowledge-based system

Let us focus on the example scenario, *creation of a work center via MRP II* and, particularly, in one of its sub-scenarios, the *release of a work center in MRP II* as shown in Fig. 5.14 with its initial marking, which is used throughout this section (Lin, 1991). The specification of the example was expressed in a natural language (section 5.5.2) and then represented here using UPN primitives.

(a) Data

The user needs to refer to atomic data and establish relations between different data by structuring information into composed data objects, called database relations or records. In order to: (a) provide a more adequate representation to facilitate the translation of the user specifications, (b) give a clearer graphical representation, and (c) make the validation process easier, UPNs allow for the specification of the following classes of information (in general we refer to any of these data classes by the generic name 'data' or 'color'):

Atomic data is individual information with a lexical representation. Each atomic data is represented by an identifier, a color, which is attached to a specific token, the syntax of which is an alphanumerical sequence. As an example, we know that a part in CAD can be in four different statuses: w (working), r (released), h (hold) or o (obsolete). We can describe them as four constant symbols (w, r, h, o), which can be grouped to form the status color set: $DSTS = \{w, r, h, o\}$.

Atomic data can belong to one of the following two classes: *fixed*, if the color set is completely known in advance, such as the set status which was described above $DSTS = \{w, r, h, o\}$ and *non-fixed*, if not all of the color set components are completely known in advance, such as the information about a part identification number in CAD, which can be an alphanumerical sequence (e.g. *C157635*) or the description of a part record, which can be any string (e.g. 'Fine-pitch involute spur gear').

This last example raises another interesting point related to non-fixed data. There is some information, such as a part identification number, that is normally used to guide the information control; generally, preconditions of a rule can be related to a specific part identification number, but normally specifications do not care about the specific instances of its description; in other words, the description is associated to a specific part and it will be flowing with it in the system, but no precondition check for a specific description is necessary. Thus, *non-conditional* data refer to information, which is not important for the control of the system and their flow is controlled by *conditional* data.

Data structure is the classical mechanism to aggregate related data. Therefore, instances of data structures can be seen as ordered tuples $\langle s\_1, s\_2, \ldots, s\_n \rangle$. Different tuples can be identified by their relation name, $\langle R \rangle$, and different elements of a tuple can also have textual identifications $A_1, \ldots, A_m$. This means that every tuple will be represented as $\langle R \rangle$ $(A_1, \ldots, A_m)$, where $\langle R \rangle$ is the database relation name and $A_1, \ldots, A_m$ are textual identifications for its attributes. A specific element of a tuple can be accessed in two ways; either by referring its specific position within that tuple ('_, _, . . . , v_i, . . . , _') or by referring to its attribute name ('$A_i = v_i$').

To illustrate the model adopted for data representation, let us consider the specifications related to a work center record in MRP II. This record is represented in UPN by a data structure named *Mwc* (where *M* identifies the MRP II database and *wc* identifies the work center record). Its composition, characteristics of its attributes, and its representation are shown in Table 5.4.

(b) Facts

Fundamental in describing an information system are the facts within it, i.e. structural and declarative knowledge. Each fact declares a piece of information about some data or data structure in the system. To be able to model the access and exchange of information in a consistent and generic way, we need to define specific semantics to describe facts. This becomes more important if we consider a modular modeling methodology where the user must be allowed to create different scenarios, which can be linked together later, based upon shared knowledge.

Facts in UPNs are represented by *places*. Places are one of three kinds of nodes in a UPN and are graphically represented as circles. The fact asserted by one place is determined by the place name and its content.

This content defines the marking of the place; we will refer to the marking of place *p* by *M*(*p*). For example, the fact that a token of color *A* is in place *inprocess* can be interpreted as: '*A is in process*' or *inprocess*(*A*) in logic syntax.

Places *NMwc* and *NPwc* in Fig. 5.14 represent the non-existence of a record in MRP II and CAPP databases, respectively. On the other hand, places *EMwc* and *EPwc* represent the existence of a record. The interpretation associated with each of the rest of the places is provided in Fig. 5.14.

Due to the modular approach adopted in this methodology, UPNs allow the splitting of the whole system into separate scenarios represented by subnets. Each subnet can be designed and verified separately from each other and then the user can follow a stepwise process for merging different subnets. Regarding facts, places can be of two different types:

- *Local scope*: when a place is used to represent a local fact (relevant only to one specific scenario or subnet), which will not affect the state of the other scenarios or subnets. They are typically used to represent intermediate state facts within the decision process. Examples of local places are p_1, p_2, p_3, p_4, p_5 in Fig. 5.14.
- *Global scope*: when a place is used to represent a fact relevant to (accessible by) different scenarios or subnets. Global places must be referred to by the same name in all of their occurrences. Typical examples of global places are facts about database state information, such as places

Table 5.4 Data structures and color sets of work centers

Attribute	*Data class*	*Color set*	*DB data type*	*Description*
wcid	non-fixed	WCID	identification	identification number
des	non-fixed	DES	text	description
dep	non-fixed	DEP	text	department
cap	non-fixed	CAP	integer	capacity
sts	fixed	MSTS	{*h*, *r*} (hold, released)	work center status code
ste	fixed	MSTE	{*na*, *av*} (not avail, avail)	work center status code
res	non-fixed	RES	text	resource code
esd	non-fixed	RSD	date	effectivity start date

Complete data structure for work center in MRP II
Mwc(*wcid, des, dep, cap, sts, ste, res, esd*)

Attribute	*Data class*	*Color set*	*DB data type*	*Description*
wcid	non-fixed	WCID	identification	identification number
des	non-fixed	DES	text	description
dep	non-fixed	DEP	text	department
cap	non-fixed	CAP	integer	capacity
sts	fixed	PSTS	{*w*, *h*, *r*} (working, hold, released)	work center status code

Complete data structure for work center in CAPP
Pwc(wcid, des, dep, cap, sts)

EMwc and *NMwc* (in the MRP II database) and *EPwc* and *NPwc* (in the CAPP database) shown in Fig. 5.14.

Regarding the user interface, there are two special types of places to specify the interchange of information with the user (these places are of local scope):

- *Input places* represent facts whose initialization is generated by an external action (i.e. by a request from the user through the interface). An example of an input place is the place p_1 shown in Fig. 5.14, which needs the introduction of a token for the release process to start.
- *Output places* are used to represent situations where some information must exit (i.e. some information to be displayed). An example of an output place is the place p_4 shown in Fig. 5.14; the value provided in *wcid#* will be displayed.

The concept of *input* and *output* places is similar to that of *source* and *sink* places (Beck and Krogh, 1986) and, in addition, provides a mechanism to specify terminal facts during consistency and completeness verifications of the knowledge model.

(c) *Rule specification*

Another important feature in modeling information systems is the representation of information flow specifications, i.e. procedural knowledge and information dynamics. Here, we are considering domains where the user specifies information flow policies using 'if then' rules. These rules are expressed in UPN by means of *transitions*. Any transition t has a color set, $C(t)$, associated with it. The net in Fig. 5.14 shows five transitions (t_1, t_2, t_3, t_4, t_5) and their associated interpretations.

In this section, we explain how several components of the rules (preconditions, postconditions, variables, etc.) are represented and used in UPN.

(i) Variables

UPNs characterize references to generic data using variables (in the sense that a relation can hold for different data occurrences). There exists a set V of typed variables. Each variable $v : D$ has a name v and an associated color set D. They are represented graphically by names ending with the symbol '#'. Examples of variables are *wcid#*, *des#*, *dep#* and *cap#* as shown in Fig. 5.14.

(ii) Arcs

Arcs are relations which connect facts and actions to form rules and constitute the rule preconditions and postconditions. Arcs also identify the flow of information. Formally, each arc has attached to it an arc expression, *exp*, containing a set of variables or constants $\{v_1 : D_1, v_2 : D_2, \ldots, v_n : D_n\}$, where D_i identifies the color set of the variable or constant v_1.

The lambda expression, $\lambda(v_1, v_2, \ldots, v_n)$fiexp is defined as a mapping from $D_1 \times D_2 \times \ldots \times D_n$ into $C(p)$, where the place p is the source/destination of the arc.

UPNs use variables as in typical rule-based systems, where they are used to specify data sets within the enabling arcs of a rule and as a mechanism to transfer data from the preconditions to the postconditions. For example, looking at transition t_4 in Fig. 5.14, the information about variables *des#* and *dep#* are needed for place p_5, which will be transferred from place *EMwc*, as shown in their arc expressions.

In general in colored Petri nets, upon the firing of a transition, the enabled colors are taken out from the incoming places. We have introduced a special case in UPN in situations where the place represents a database-related fact and the color represents a record. In case of firing, all the information from that record is removed from the place, while only a subset of its attributes may be involved for the update of information. The information of those unspecified attributes must also be included in the color set of the transition, in order to transfer it from one database to the other or back to itself. For example, looking at transition t_5 in Fig. 5.14, the attributes *des*, *dep* and *ste* of the MRP II work center table are not shown in the arc expressions for the update of the work center record (pair of arrows from and to the place *EMwc*). However, these attributes have to be included in the colored set of transition t_5 as shown in Fig. 5.14. One of these variables is used for each record attribute, which is not specified in the expressions of arcs connected to that transition. They are identified by names ending with the symbol '0'.

Each transition can have an associated predicate, *pred*, to impose additional constraints in the enabling of the transition. This predicate is a Boolean expression which can only contain those variables which are already in the expressions of all arcs connected to the transition. To avoid degenerated transitions, the predicate must differ from the constant predicate *false*. The predicate is supposed to be *true* by default. Simple predicates can be grouped by means of the following logical operators: $\neg$ (not), $\vee$ (or) and $\wedge$ (and).

The color set, $C(t)$, of a transition, t, $\forall t \in T$, is determined by (Jensen, 1987):

$$C(t) = \{(d_1, d_2, \ldots, d_n) \in D_1 \times D_2 \times \ldots \times D_n \mid (\lambda(v_1, v_2, \ldots, v_n)\cdot \text{pred})(d_1, d_2, \ldots, d_n)\}$$

where:

- *pred* is the predicate attached to t; and
- $V(t) = \{v_1 : D_1, v_2 : D_2, \ldots, v_n : D_n\}$ is the set of all variables appearing in the expressions of all arcs connected to the transition.

As an example, let us look at Fig. 5.15(a) which focuses on transition t_5 from the net shown in Fig. 5.14. The color sets for the associated places are as follows (the simple color sets are specified in Table 5.4):

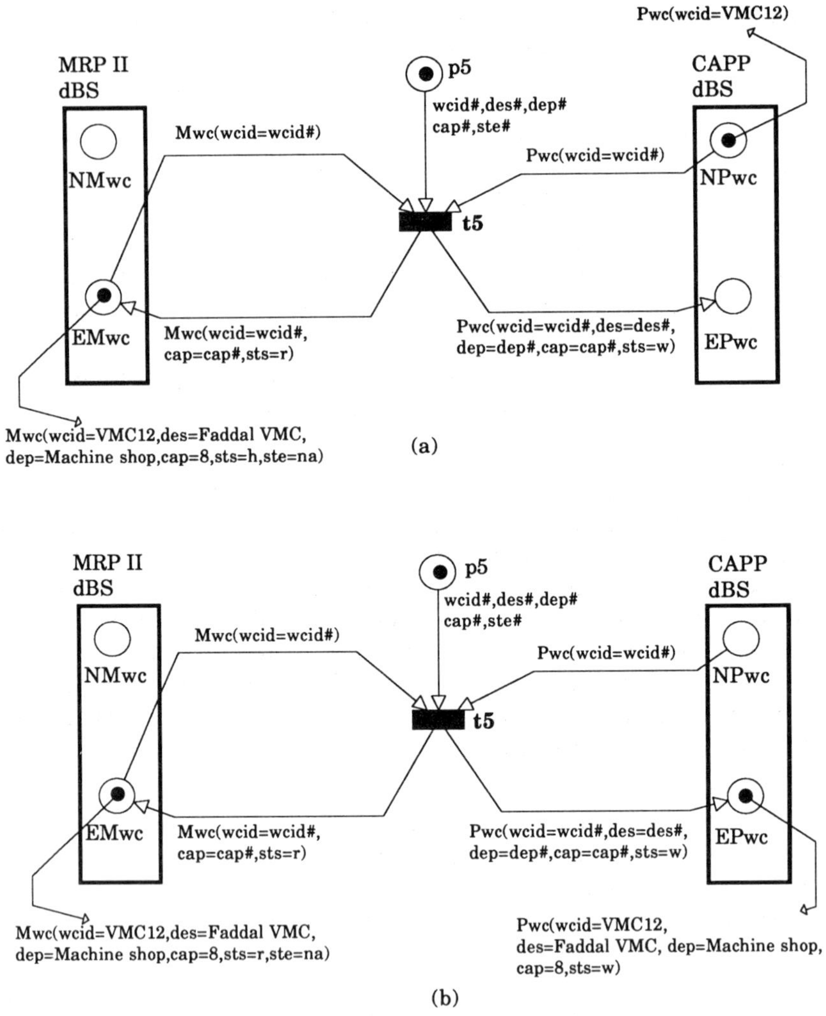

Figure 5.15 UPN representation of the rule 'Release work center from MRPII'.

$$C(EMwc) = MWC = WCID \times DES \times DEP \times CAP \times MSTS \times MSTE \times RES \times ESD$$
$$C(NPwc) = WCID$$
$$C(EPwc) = PWC = WCID \times DES \times DEP \times PSTS$$
$$C(p_5) = WDDCS = WCID \times DES \times DEP \times CAP$$

There are four variables in the arcs connected to t_5 (*wcid#*, *des#*, *dep#* and *cap#*). On the other hand, t_5 has an incoming arc from place *EMwc*, a database place for the record *Mwc* (*wcid*, *des*, *dep*, *cap*, *sts*, *ste*, *res*, *esd*) and there are five attributes which do not need to be specified in the arcs to/from *EMwc*: *des*, *dep*, *ste*, *res* and *esd*. The total set of variables and the color set for the transition t_5 are:

$$V(t_5) = (wcid\# : WCID, des0 : DES, dep0 : DEP, cap\# : CAP, mste0 : MSTE, res0 : RES, esd0 : ESD, des\# : DES, dep\# : DEP)$$
$$C(t_5) = MWCSDD = WCID \times DES \times DEP \times CAP \times MSTE \times RES \times ESD \times DES \times DEP$$

Functions in I^- and I^+ are defined in terms of lambda expressions with the form $f(c) = \lambda(V)exp(c)$, where $c \in C(t)$ and *exp* is the expression associated to the arc. For transition t_5, V and $c \in MWCSDD$ can be represented as follows:

$$V = (wcid\#, des0, dep0, cap\#, ste0, res0, esd0, des\#, dep\#)$$
$$c = (wcid, deso, depo, cap, steo, reso, esdo, des, dep)$$

UPNs provide different types of arcs.

Enabling arcs are directed arcs, which connect a place/fact with a transition/action and define a precondition for the transition. They indicate the data that must be in a place for a transition to be enabled and must be removed from that place on firing. Given $C(t)$ and $V(t)$ defined as above for some transition $t \in T$, if a single arc from a given $p \in P$ exists, with an arc expression *exp*, $I^-(p, t)$ is then defined to be the function $\lambda(v_1, v_2, \ldots, v_n) \cdot$ exp (mapping between $C(t)$ and $C(p)$). If no arc from p to t exists, $I^-(p, t)$ is the zero-function. For example, the following are the enabling arcs for transition t_5 in Fig. 5.15(a):

- $I^-(p_5, t_5) : exp = [wcid\#, des\#, dep\#, cap\#]$
  $\lambda(V) \cdot exp \in [MWCSDD_{MS} \rightarrow WDDCS_{MS}]_L$ such that:
  $\lambda(V) \cdot exp(c) = (wcid, des, dep, cap)$
- $I^-(EMwc, t_5) : exp = [wcid = wcid\#]$,
  $\lambda(V) \cdot exp \in [MWCSDD_{MS} \rightarrow MWC_{MS}]_L$ such that:
  $\lambda(V) \cdot exp(c) = (wcid, deso, depo, -, -, msteo, reso, esdo)$
- $I^-(NPwc, t_5) : exp = [wcid\#]$,
  $\lambda(V) \cdot exp \in [MWCSDD_{MS} \rightarrow WCID_{MS}]_L$ such that:
  $\lambda(V) \cdot exp(c) = wcid$

Causal arcs are directed arcs which connect a transition with a place and define a postcondition for the rule. Causal arcs describe modifications to be performed in the state of the net when the transition is fired; and, more concretely, they indicate which data must be added to a place after the transition has been fired. I^+ is defined in the same way as I^-, but by means of the arcs from transitions to places. For example, these are the causal arcs for transition t_5 in Fig. 5.15(a):

- $I^+(EMwc, t_5) : exp =$ \B\BC\[(\A\VS6($wcid = wcid\#, des = des\#, dep = dep\#, cap = cap\#, sts = r$)),
  $\lambda(V) \cdot exp \in [MWCSDD_{MS} \rightarrow MWC_{MS}]_L$ such that:
  $\lambda(V) \cdot exp(c) = (wcid, deso, depo, cap, r, msteo, reso, esdo)$

- $I^{+}(EPwc, t_5) : exp = [wcid = wcid\#, des = des\#, dep = dep\#, sts = w]$, $\lambda(V) \cdot exp \in [MWCSDD_{MS} \rightarrow PWC_{MS}]_L$ such that: $\lambda(V) \cdot exp(c) = (wcid, despo, depo, cap, w)$

Checking arcs indicate which data must be present in a place in order to enable a transition, but no data are removed upon firing. They are represented by an enabling and a causal arc, with a unique arc expression. These are generally used to indicate database checkings. For example, transition t_2 and place *EMwc* in Fig. 5.14 are connected by a checking arc with the associated arc expression: $Mwc(wcid = wcid\#)$. This is a precondition for t_2 to be enabled, but after the transition has been fired, no data are removed from or added to *EMwc*. Arcs between t_2 and *NMwc* and arcs between t_4 and *EMwc* are also checking arcs.

Inhibitor arcs indicate which data must not be present in a place in order to enable a transition. It is graphically represented as an enabling arc ending in a small circle.

It is also possible for a transition to request information from the user interface. This is done whenever an arc expression requires some information that has not been provided by the variables in the enabling or checking arcs.

(iii) Transition enabling and firing

The dynamic behavior of the net is provided by transition enabling and firing. A transition $t \in T$ is said to be enabled with respect to a color $c \in C(t)$ if the current marking M is such that:

$$M(p) \geq I^{-}(p,t)(c) \quad \text{and} \quad M(p) < I_o(p,t)(c), \ \forall p \in P$$

An enabled transition may be chosen to fire. The firing of a transition t with respect to a color c consists of removing $I^{-}(p,t)(c)$ colors from each of its input places and adding $I^{+}(p,t)(c)$ colors to each of its output places. We may think of a firing as an event which may take place if certain conditions are satisfied. Each firing creates a new set of conditions and the total number of colors in the net may change after each firing.

In order to illustrate the previous concepts, we focus on transition t_5 in Fig. 5.15(a). For this explanation, let us consider the following initial marking:

$$
\begin{aligned}
M_0(NMwc) &= \emptyset \\
M_0(EMwc) &= (vmc12, \text{`}Fadal\ VMC\text{'}, \text{`}machine\ shop\text{'}, null, h, na, m12, \\
&\quad null) \\
&= Mwc(wcid = vmc12, des = \text{`}Fadal\ VMC\text{'}, dep = \text{`}machine \\
&\quad shop\text{'}, cap \\
&= null, sts = h, ste = na, res = m12, esd = null) \\
M_0(NPwc) &= vmc12 \\
M_0(EPwc) &= \emptyset \\
M_0(p_5) &= (vmc12, \text{`}Fadal\ VMC\text{'}, \text{`}machine\ shop\text{'}, 8)
\end{aligned}
$$

It is clear that transition t_5 is enabled with respect to the following specific color (note that the entire color set for transition t_5 is, $C(t_5) = MWCSDD$):

c = (*vmc12*, '*Fadal VMC*', '*machine shop*', *null*, *na*, *m12*, *null*, '*Fadal VMC*', '*machine shop*')

Because *c* satisfies the following enabling rules, t_5 is then enabled.

$M_0(NMwc) \geq \lambda(V)exp(c) = \varnothing$
$M_0(EMwc) \geq \lambda(V)exp(c) =$ *vmc12*, '*Fadal VMC*', '*machine shop*', *null*, *h*, *na*, *m12*, *null*
$M_0(NPwc \geq n(V)exp(c) =$ *vmc12*
$M_0(EPwc) \geq n(V)exp(c) = \varnothing$
$M_0(p_5) \geq \lambda(V)exp(c) =$ (*vmc12*, '*Fadal VMC*', '*machine shop*', *8*)

where:

exp represents the corresponding arc expression,
V is as defined previously in this section

The firing of transition t_5, with respect to color *c*, implies the removal of the tokens corresponding to the arc expressions from the input places and the addition of the tokens defined by the causal arcs to the output places:

$M_0(NMwc) = \lambda(V)exp(c) = \varnothing$
$M_0(EMwc) = \lambda(V)exp(c) =$ (*vmc12*, '*Fadal VMC*', '*machine shop*', *null*, *r*, *na*, *m12*, *null*)
$M_0(NPwc) = \lambda(V)exp(c) = \varnothing$
$M_0(EPwc) = \lambda(V)exp(c) =$ (*vmc12*, '*Fadal VMC*', '*machine shop*', *8*, *w*)
$M_0(p_5) = \lambda(V)exp(c) = \varnothing$

The final marking after transition t_5 fires is the following (Fig. 5.15(b)):

$M_0(NMwc) = \varnothing$
$M_0(EMwc) =$ (*vmc12*, '*Fadal VMC*', '*machine shop*', *null*, *r*, *na*, *m12*, *null*)
$M_0(NPwc) = \varnothing$
$M_0(EPwc) =$ (*vmc12*, '*Fadal VMC*', '*machine shop*', *8*, *w*)
$M_0(p_5) = \varnothing$

(d) Modeling approach

Generally speaking, any 'company policy' starts from the specification of general global rules, which describe aggregate operations for a given entity within the system. These rules are then further refined into more detailed specifications on a step-by-step basis until no aggregate operations are left. Following a similar concept, a hierarchical modeling method using UPN has been developed, which allows the system designer to start from abstract global nets and continue with successive refinements until the desired degree of detail has been reached. In addition to the refinement of rules within each scenario, it is necessary to synthesize all scenarios for all entities to form a coherent net representing the company-wide policy in the system.

Some work in hierarchical representation using Petri nets has been done for various applications (Valette, 1979; Suzuki and Murata, 1983; Narahari and Viswanadham, 1985; Jeng and DiCesare, 1990). The hierarchical modeling methodology facilitates the modeling task. It is discussed in detail in the following paragraphs.

(i) Top-down stepwise refinement technique

The top-down stepwise refinement technique has been developed for the modeling of each scenario from an abstract and aggregate level to a detailed level. This approach necessitates the development of new Petri net modeling entities which include two types of transitions as mentioned in the previous section; one is *primitive* transitions representing primitive rules, and the other is *compound* transitions representing metarules, which can be further refined into subnets. The connections are represented by calls from one *compound* transition of the net at the abstract level to the subnets at the more detailed level. Three techniques for constructing *compound* transitions, representing metarules, are presented below.

(i) Metarule specification Metarules provide a higher level representation mechanism. They establish relations between rules. In UPN formalism, a metarule is represented by a subnet. Metarules increase the efficiency of the rule control process. Grouping related rules together saves a substantial amount of time in searching for appropriate sets of rules, which are due to be executed together and may be invoked many times. In other words, a metarule reflects a restricted set of rules that are bound to be called upon together and repeatedly. Metarules are mainly used in UPN as a mechanism to define subnets, which allow for the structural composition of the rule-specification knowledge.

(ii) Horizontal composition of rules: hmrules Rules at the same level of abstraction can be connected to form subnets. This horizontal composition allows the aggregation of rules under specific criteria. Horizontal compositions are established by means of what we call 'hmrules'. An hmrule, hm_a, specifies a relation in a subnet with a set of transitions $\{t_1, t_2, \ldots, t_m\}$, where $m \geq 1$ and t_i is defined at the level of abstraction a , $\forall t_i \in hm_a$. A subnet, defined by metarule hm_a, is composed by the set of transitions and the places that are interconnected.

Hmrules are generally used to identify various scenarios used to refine a transition at a lower level of abstraction. This aggregation is very useful at the implementation stage, because a complete subnet is translated into one procedure, which avoids the use of local variables and makes the code more efficient. For example, if the complete subnet were implemented by

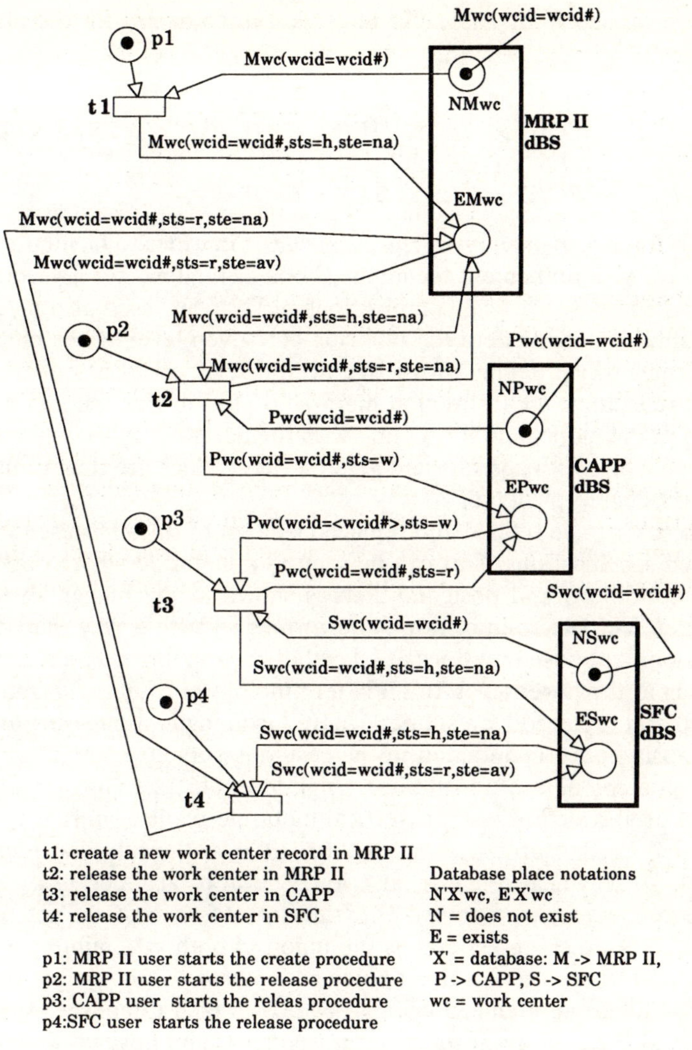

Figure 5.16 UPN graph of the scenario 'Creation of a work center via MRP II' at an abstract level with initial marking.

several procedures, additional local variables would be required to pass the state of one procedure to the other in order to complete the execution of the complete subnet without user interruption.

An example of a horizontal composition is one of the subnets for the work center creation scenario, related to the release of a work center in MRP II, shown in Fig. 5.14. The five rules related to that scenario are grouped together, represented by transition t_2 in Fig. 5.16 with its initial marking. The formal representation of that subnet is specified by its incidence functions as shown in Tables 5.5 and 5.6. The color sets for the places are such that:

$$E = \{\varepsilon\}$$
$$MWCSC = WCID \times DES \times DEP \times CAP \times MSTE \times RES \times ESD \times DES \times DEP \times CAP$$
$$WDDC = WCID \times DES \times DEP \times CAP$$

(iii) Vertical composition of rules: vmrules In order to facilitate the top-down stepwise refinement technique, UPN also allows for a vertical composition of rules. Vertical compositions in UPN are used as a mechanism to establish relations between one rule at a given level of abstraction and other rules which define knowledge at a lower level of abstraction. As a result, rules form an abstraction hierarchy. This abstraction facilitates the design of the model and it is in line with the natural way a company policy is defined: first with abstract and general rules, which are then refined with a higher degree of detail.

One rule can be refined and replaced by a set of rules representing more detailed specifications. This can be seen as a metarule where only part of its preconditions and postconditions is shown at the higher level of abstraction. Vertical composition is performed in such a way that rule preconditions and postconditions at a level of abstraction are preserved when working at the lower level. In UPN terminology, a *compound* transition is refined and replaced by a subnet, where incoming and outgoing arcs from the transition are maintained in the resulting subnet.

Given a *compound* transition, t_a (high-level abstraction transition), and a level of abstraction $i - 1$, a vertical metarule (called 'vmrule') $vm^t a_{i-1}$ is identified by the tuple $\langle t_a, hm_i \rangle$, where hm_i is an hmrule that identifies a subnet, *UPN*, which refines t_a at a lower level of abstraction *i*.

The refinement of a *compound* transition of an abstract net produces a new UPN which, in general, is the union of both nets minus the refined *compound* transition. A subnet being refined from a *compound* transition is formed with an attached calling protocol, which establishes the link of the net at the abstract level with the subnet at the next lower level. The *compound* transition is replaced by the subnet with a calling net, which contains one initiation transition (t_{init}), one waiting place (p_{wait}) and one returning transition (t_{ret}). In addition, p_{init} is the starting place of the subnet and p_{ret} is the ending place of it. One arc connecting t_{init} to p_{init} in the subnet and another arc connecting p_{ret} to t_{ret} in the subnet are added to link the calling protocol to the subnet. The calling protocol and the subnet are shown in Fig. 5.17.

It is also possible to explode a *compound* transition to several metarules. Let UPN_1 be a net representing specifications at a high abstract level and let $UPN_{2_1}, \ldots, UPN_{2_n}$ be the subnets representing the metarules $hm_{2_1}, \ldots, hm_{2_n}$, from $\langle t_a, hm_{2_1}, \ldots, hm_{2_n} \rangle$, that refine the compound transition $t_a \in T_1$. The new *UPN* net is defined as follows:

Table 5.5 Positive incidence matrix

		t_1	t_2	t_3	t_4	t_5
		WCID	*WCID*	*WCID*	*MWCS*	*MWCS*
NM_{wc}	WCID	0	id	0	0	0
EM_{wc}	MWC	0	0	$dbm^{mwc}\{wcid = P_1, sts = r\}$	$dbm^{mwc}\{wcid = P_1, des = P_2$ $dep = P_3, sts = h\}$	$dbm^{mwc}\{wcid = P_1, des = P_2$ $dep = P_3, cap = P_4, sts = r\}$
NP_{wc}	WCID	0	0	0	0	0
EP_{wc}	PWC	0	0	0	0	$dbm^{mwc}\{wcid = P_1, des = P_2$ $dep = P_3, cap = P_4, sts = w\}$
p_1	E	0	abs	0	0	0
p_2	WCID	id	0	0	0	0
p_3	WCID	0	id	0	0	0
p_4	WCID	0	0	id	0	0
p_5	WCID	0	0	0	$\{P_1, P_2, P_3, P_4\}$	0

Table 5.6 Negative incidence matrix

		t_1	t_2	t_3	t_4	t_5
		WCID	*WCID*	*WCID*	*MWCS*	*MWCS*
NM_{wc}	WCID	0	id	0	0	0
EM_{wc}	MWC	0	0	$dbm^{mwc}\{wcid = P_1, sts = r\}$	$dbm^{mwc}\{wcid = P_1, des = P_2$ $dep = P_3, sts = h\}$	$dbm^{mwc}\{wcid = P_1\}$
Np_{wc}	WCID	0	0	0	0	$\{P_1\}$
EP_{wc}	PWC	0	0	0	0	0
p_1	E	abs	0	0	0	0
p_2	WCID	0	id	id	$\{P_1\}0$	0
p_3	WCID	0	0	0	0	0
p_4	WCID	0	0	0	0	0
p_5	WCID	0	0	0	0	$\{P_1, P_2, P_3, P_4\}$

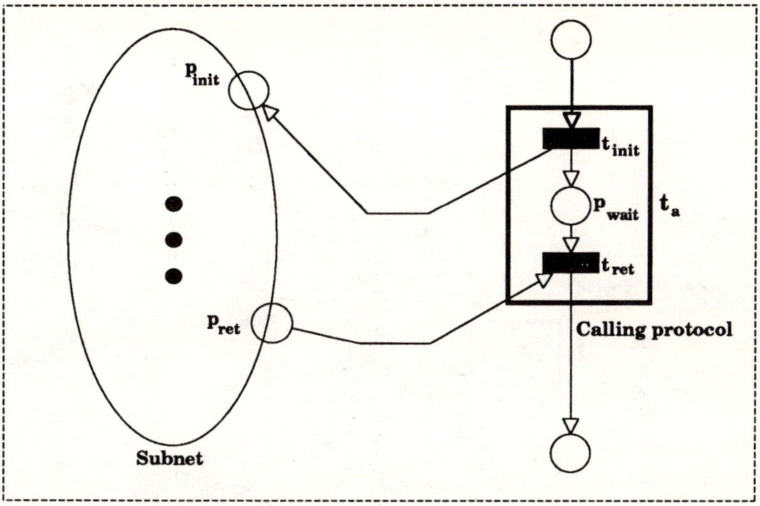

Figure 5.17 Example of a subnet and a calling protocol.

1. $P = P_1 \cup P_{2_1} \cup \ldots \cup P_{2_n} \cup P_{call}$, where $P_1, P_{2_1}, \ldots, P_{2_n}$ denote the sets of places of $UPN_1, UPN_{2_1}, \ldots, UPN_{2_n}$, respectively, and P_{call} denotes the set of places of the *UPN* for the calling protocol.
2. $T = \{T_1 - \{t_a\}\} \cup T_{2_1} \cup \ldots \cup T_{2_n} \cup T_{call}$, where $T_1, T_{2_1}, \ldots, T_{2_n}$ denote the sets of transitions of $UPN_1, UPN_{2_1}, \ldots, UPN_{2_n}$, respectively, and T_{call} denotes the set of transitions of the UPN for the calling protocol.
3. All the incoming and outgoing arcs are preserved, except the ones related to transition t_a:

$$I^-(t,p) = I_1^-(t,p),\ \forall p \in P_1,\ \forall t \in \{T_{1}\text{-}\{t_a\}\}$$
$$I^-(t,p) = I_{2_i}^-(t,p),\ \forall p \in P_{2_i},\ \forall t \in T_{2_i},\ 1 \le i \le n$$
$$I^-(t,p) = I_{call}^-(t,p),\ \forall p \in P_{call},\ \forall t \in T_{call}$$

Similarly for $I^+(t,p)$ and $I_o(t,p)$. In addition,

$$I^-(t_{init}, p) = I_1^-(t_a, p),\ \forall p \in P_1$$
$$I^+(t\_\{ret\}, p) = I_1^+(t_a, p),\ \forall p \in P_1$$

$$I^+(t_{init}, p_{initi}) = \cup\ \{I_{2_i}^-(t, p_{init}) \mid t \in T_{2_i},\ 1 \le i \le n\}$$
$$I^+(t_{ret}, p_{reti}) = \cup\ \{I_{2_i}(t, p_{ret}) \mid t \in T_{2_i},\ 1 \le i \le n\}$$

$I^-(t,p) = \varnothing,\ \forall p \in P_{2_i},\ \forall t \in T_{2_i},$ where $I_{2_i}^-(t,p) \cap I_1^-(t_a,p),$
and $\$\ I_{2_i}^-(t, p_{ret}) \ne \varnothing$
$I^+(t,p) = \varnothing,\ \forall p \in P_{2_i},\ \forall t \in T_{2_i},$ where $I_{2_i}^+(t,p) \cap I_1^+(t_a,p),$
and $I_{2_i}^+(t, p_{ret}) \ne \varnothing$

4. $M(p) = M_1(p) + M_{2_i}(p),\ \forall p \in P_1 \cap P_{2_i},\ 1 \le i \le n$
$M(p) = M_1(p),\ \forall p \in \{P_1 - (P_1 \cap P_{2_i}) - \ldots - (P_1 \cap P_{2_n})\}$
$M(p) = M_{2_i}(p),\ \forall p \in \{P_{2_i} - (P_1 \cap P_{2_i})\},\ 1 \le i \le n$

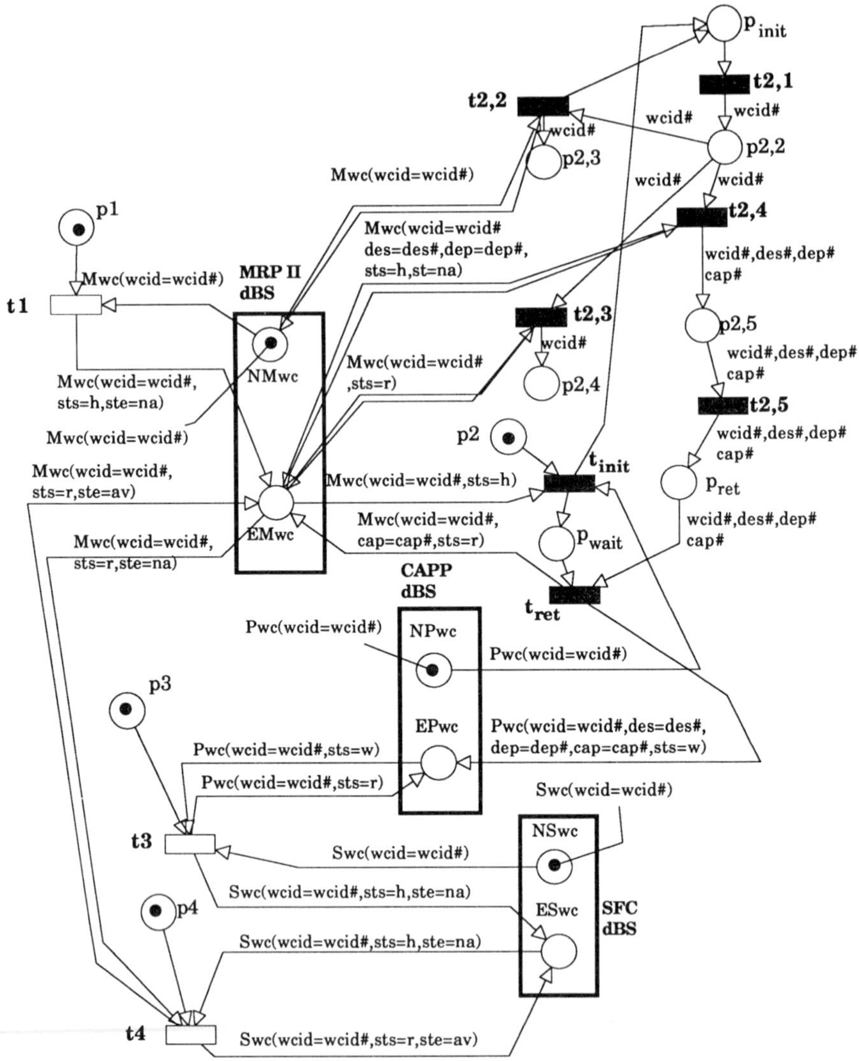

Figure 5.18 Partially refined UPN of the scenario 'Creation of a work center via MRP II' with initial marking.

An example of the top-down refinement technique is the refinement of the rule 'Release of a work center in MRP II' (Fig. 5.14), which is the *compound* transition t_2 in the abstract scenario 'Creation of a work center via MRP II' (Fig. 5.16). During the refinement procedure a calling protocol is used, and the refined UPN is shown in Fig. 5.18 with initial marking. Preconditions and postconditions represented by transition t_2 are preserved (arcs to/from *EMwc*, *NPwc* and *Epwc*).

(ii) Synthesis technique

It is necessary to synthesize related scenarios to build the company-wide policy represented by one single net. There have been some synthesis techniques presented (Suzuki and Murata, 1983; Narahari and Viswanadham, 1985; Jeng and DiCesare, 1990), based on their application domain. In our work, we take advantage of the features of UPN, such as global places that reflect the state of the databases involved. In addition, we take advantage of standard modification procedures embedded in the database management systems associated with each application system. The synthesis of UPN is achieved through the following mechanism.

(i) Global places (representing database states) Due to the representation of database states in UPN, every scenario involves checkings, updates and retrievals in some of the databases of the system. Therefore, connections from and to global places, which represent database states, exist in every UPN. These global places provide the connectivities between all scenarios. For example, the scenario 'Creation of a work center via MRP II' (Fig. 5.16) and 'Removal of a work center via MRP II' (Fig. 5.19) are synthesized through the work center global places of the databases in MRP II, CAPP and SFC. The synthesized net is shown in Fig. 5.20.

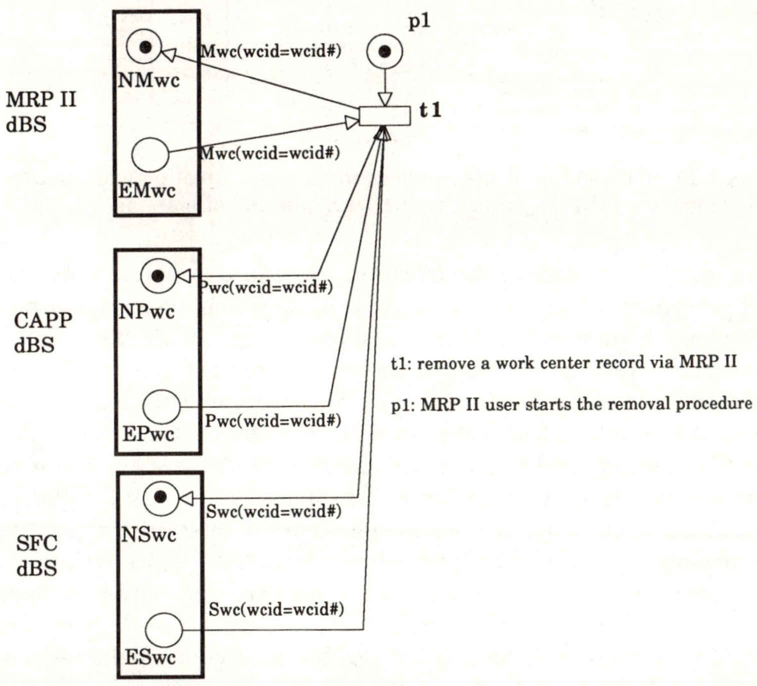

Figure 5.19 UPN graph of the scenario 'Removal of a work center via MRP II' at an abstract level with initial marking.

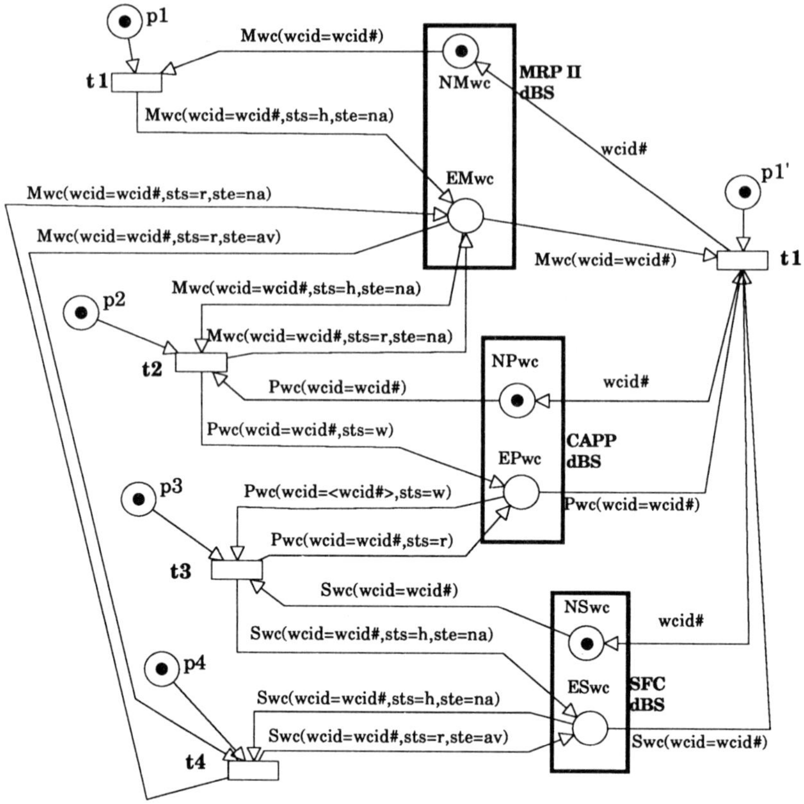

Figure 5.20 UPN graph of the synthesized scenario 'Creation and removal of a work center via MRP II' at an abstract level with initial marking.

Alternatively, standard modification procedures (representing the default procedures maintained in database management systems) can be used to synthesize subnets together. As shown in Fig. 5.21, there are three procedural levels in a typical database management system. The lowest level is the physical procedural level, which represents the actual database changes. The procedures at this level are represented here by *ins*, *del* and *upd*. The second level is the modification procedural level, which represents the modification procedures with built-in rules in the DBMS. The procedures at this level are represented here by *insert*, *delete* and *update*. The highest level is the application procedural level, which represents the user-defined procedures (*create*, *release*, *remove*, etc.) based on company policy and expert rules, that are the focal points of this work. Note that the modification procedural level could also be modeled using the same methodology and techniques presented in this chapter. In that case, instead of the global places defined in the previous section, these modification procedures would become the links between application procedures, which

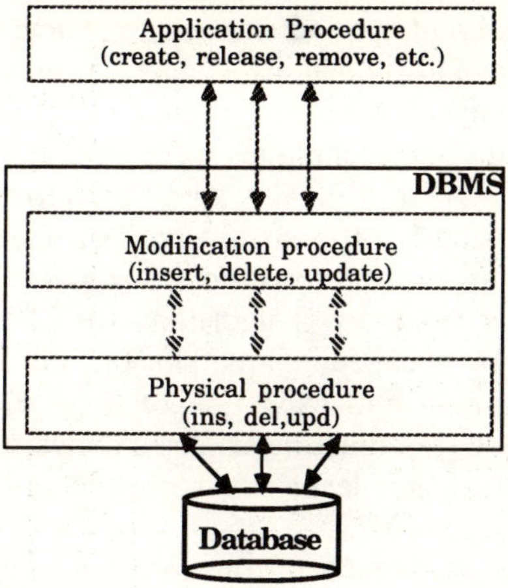

Figure 5.21 Procedural levels in Database Management Systems.

have been developed in various scenarios. The reasoning is that all application procedures call for the standard modification procedures embedded in the DBMS in order to make changes in the databases involved.

In order to avoid 're-inventing the wheel', we decided not to model the database management systems of the manufacturing application systems involved using UPN, but to define the database states as global variables and to interface the application procedures (company policy) through those global places representing database states, as explained above.

5.5.4 Model validation and specification

The major objective of creating a KBS using Petri nets is the ability of validating the KBS mathematically and systematically. The *completeness* (dead-end rules, unfirable rules), *consistency* (redundant rules, subsumed rules, under-constrained rules) and *conflicts*, are the major issues in knowledge-base verification (Nguyen *et al.*, 1987; Lopez *et al.*, 1990). The incidence matrices of Petri nets representing the rule base can be used to perform some of these verification checks, which can be complemented by the user with the aid of specific domain knowledge. Several other analysis techniques for Petri nets, including reachability trees and net invariants, are also used (Peterson, 1981; Martinez and Silva, 1982; Jensen, 1986). The net invariants, which represent mutually exclusive conditions within the 'company policy', can reveal logical conflicts in the specification of the original rules and possibly errors introduced during the modeling process. The reachability tree can be used to detect any deadlocks or inconsistencies in the model. The programs for computerizing these analysis methods have been developed and applied extensively. Some reduction rules (Lee and Favrel, 1985) have also been investigated to reduce the complexity of nets prior to the analysis phase.

However, these analysis techniques were initially developed for generalized Petri nets (GPN) and do not apply to colored Petri nets (CPN), which are characterized by a great diversity of linear functions that are associated to their arcs. Therefore, unlike analysis algorithms for GPN that use integer matrices, analysis algorithms for CPN need to manipulate matrices composed by linear functions. This fact introduces high complexities in the development and execution of these algorithms. As a consequence, a necessary step must be taken to convert the high-level net of the UPN model into a low-level net expressed in terms of generalized Petri nets (GPN) (Lin, 1991).

Based on the above discussion, the following two major aspects are considered.

Structural verification focuses on the correctness of the knowledge-base structure, which mainly depends on the KBS representation formalism used. In our case, it is the structure of Petri nets. With a formal representation of the KBS, it is possible to verify the KBS structure mathematically. The structural verification does not depend on the domain which the KBS is applied to, or the rule-specification language used for the implementation. Thus, it is generic for all KBS using the same representation formalism. The following properties can be tested using structural methods (Nguyen *et al.*, 1987):

- *Completeness*: the goal is to detect possible gaps in the rule base that have been overlooked in the modeling process.
- *Consistency*: the goal is to perform a static analysis of the logical semantics of the rule structure.

Domain knowledge verification focuses on assuring a proper behavior with respect to the domain of the system model. It depends on the functionalities of the company policy and can be very different from one application to the other. Therefore, it cannot be performed fully automatically; it needs user input to determine the correctness of the functionalities of the system and the compliance with the company policy.

Petri nets have traditionally provided the mathematical background to carry out this kind of verification. Generic analysis methods and properties of Petri nets are shown in the following section, which provide the basis of verifying the KBS.

The staged approach to our KBS verification system, shown in Fig. 5.22, consists of three major parts:

1. generic analysis methods and properties of Petri nets;
2. structural verification;
3. domain knowledge verification.

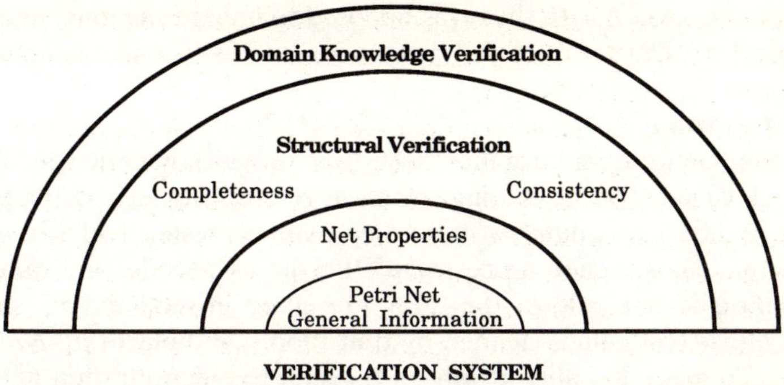

Figure 5.22 Staged model verification.

The following sections describe first the generic analysis methods and properties of Petri nets, which can be used for verification. We then present the verification methodology and finally some examples for demonstration purposes.

(a) Conversion of UPN to generalized Petri nets

We call this conversion process: '*abstraction and unfolding*'. The first step is to perform an abstraction on the identification number of a database entity, which if unfolded will create a number of duplicated nets, one for each possible value of its identification number. For example, unfolding the *wcid* of the work center record in MRP II will create a large number of nets, as many as the number of work centers which can be inserted into the database. In reality, we need only one net to represent the company policy for any work center. More specifically, we retain only one net to model the control of information flow in the system for one 'generic' work center and we can use it for the verification of the KBS. The functionality of the model does not differ for different work centers. It is specified with respect to any work center in the system. Thus, the potential variety of hundreds, if not more, work centers in a typical company can be reduced to one 'generic' work center net, without harming the generality of the model. The second step in the net conversion process is to apply the formal colored Petri net unfolding algorithm on a UPN, which consists of the information that is aggregated in the color sets of its places and transitions and the functions of its arcs. The result is a new net with more places, transitions and arcs. The unfolding of the color sets in places and transitions is done using only the conditional data fields (e.g. *sts* and *ste* of the work center record in MRP II). The rest of the data fields in the database record, which do not contribute to the control of the information flow in the system and are only used for carrying supplementary information, do not need to be unfolded (examples include work center description and work center capacity). The unfolded arcs of the net have only integer numbers associated with them (as opposed to linear functions) and tokens are colorless. The result of this conversion process is an analyzable GPN.

(i) Abstraction

In a manufacturing environment, the user provides generic specifications for work centers, parts, routings, without referencing any particular work center number, part number or routing number, respectively. In order to illustrate the abstraction process of a UPN, let us consider an example, the abstraction of one rule of the UPN presented in section 5.5.3 shown in Fig. 5.23(a). This rule is defined by transition t_1 and places p_1, *NMwc* and *EMwc*. To make the abstraction on this net, the identification number of the work center record is removed from all the color sets of places and transitions and from the functions of the arcs. The same is done for all supplementary information related to a specific work center. As a result, only the color sets of the *conditional* data (e.g. *wcid*, *sts* and *ste* – see section 5.5.3) are considered, because these are the ones which are used to control the information flow in the net. The new net is shown in Fig. 5.23(b), where *wcid* has been removed from the color sets of places and transitions and from the functions of arcs. The color set of place *NMwc* was changed from *WCID* to unity and the color set of place *EMwc* was changed from $WCID \times STS \times STE$ to $STS \times STE$. Similarly, the color set of transition t_1 was changed from $WCID \times \{set = h\} \times \{ste = na\}$ to $\{sts = h\} \times \{ste = na\}$. The function of the arc between *NMwc* and t_1 was changed from *Mwc* (*wcid* = *wcid#*) to unity, whereas the function of the arc between p_1 and t_1 needs no change. Note that p_1 reflects the intention of a user to enter a work center record to the system and as such it contains an uncolored token. Finally, the function of the arc between t_1 and *EMwc* was changed from $Mwc(wcid = wcid\#, sts = h, ste = na)$ to $Mwc(sts = h, ste = na)$. This new net preserves the behavior of the model for a 'generic' work center. Therefore, a token flowing in the net represents a 'generic' work center and reflects only all other color sets in the work center data structure except *wcid*.

The incidence matrices of the net are changed from:

I^- =

	t_1
NMwc	$wcid = wcid\#$
EMwc	0
p_1	1

I^+ =

	t_1
NMwc	0
EMwc	$wcid = wcid\#, sts = h, ste = na$
p_1	0

to

I^- =

	t_1
NMwc	1
EMwc	0
p_1	1

I^+ =

	t_1
NMwc	0
EMwc	$sts = h, ste = na$
p_1	0

Note that I^- and I^+ still contain functions, instead of just 0 and 1.

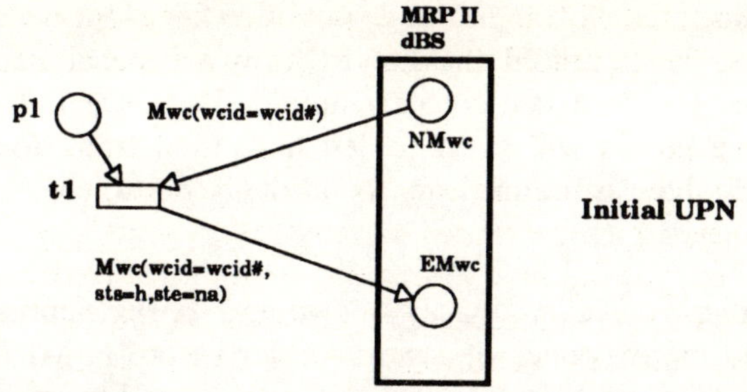

(a)

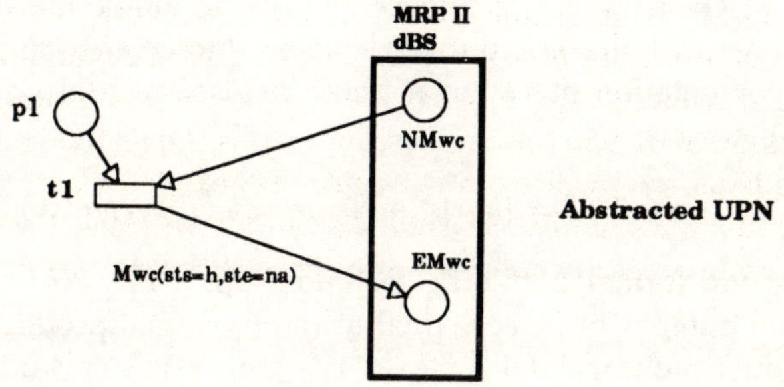

(b)

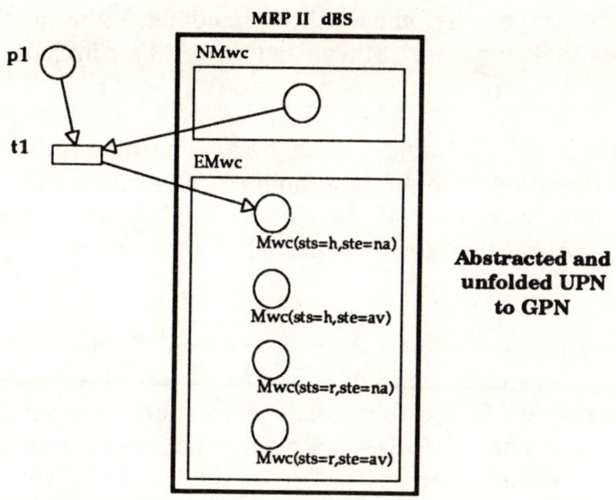

(c)

Figure 5.23 Conversion of a UPN rule to GPN.

(ii) Unfolding

The next step is to unfold the net to form a general Petri net, which is analyzable. To construct the new unfolded net, we follow a stepwise process where places will be unfolded first, then transitions and finally arcs. We will show in detail how to unfold UPN to GPN by using the example of Fig. 5.23(b).

(α) Unfolding of places Places *NMwc* and *EMwc* represent the non-existence and the existence of a work center record in MRP II database, respectively. The unfolding of place *NMwc* is straightforward because its color set is already equal to unity. Therefore, *NMwc* will be unfolded to just one place representing the non-existence of a 'generic' work center. On the other hand, place *EMwc* contains two color sets of the work center record in MRP II, namely *sts* and *ste*. In order to unfold the values of these fields in a net with uncolored tokens, we need to create one place for each possible combination of values in those fields, which are defined in two separate data sets:

$$MwcSTS = \{h, r\} \quad \text{and} \quad MwcSTE = \{na, av\}$$

Following the formal unfolding technique for CPN, *EMwc* will be unfolded to a number of places equal to the Cartesian product of the color sets of those conditional fields (i.e. $\{h, r\} \times \{na, av\}$) as shown in Fig. 5.23(c). Each of those four different places represents one possible combination of values in the above data sets.

Note. Although three places ($EMwc(sts = h, ste = av)$, $EMwc(sts = r, ste = av)$ and $EMwc(sts = r, ste = na)$) are not connected to any transitions, they are kept temporarily, in case they are connected to other transitions in the subnet that contains t_1, otherwise they will be eliminated from these subnets.

(β) Unfolding of transitions Each transition will be unfolded into a number of transitions equal to the number of possible color sets of that transition. In this case, the color set of transition t_1 contains only one color, $sts = h, ste = na$ and therefore it is unfolded to one transition, as shown in Fig. 5.23(c).

(γ) Unfolding of arcs Each arc will be unfolded into a set of arcs with functions equal to unity and linked to places, which are specified in the initial arc functions. In the above example, the arc between *NMwc* and t_1 has a function of unity and does not need to be unfolded; the same is true for the arc between p_1 and t_1. The arc between t_1 and *EMwc* with an arc function $Mwc(sts = h, ste = na)$ is unfolded into one arc connecting the unfolded transition to the place of $\{sts = h, ste = na\}$. The fact that this arc ends at that place and not any other one is determined by the initial function of this arc.

Following the unfolding algorithm, the incidence matrices of the net are changed from

$$I^- = \begin{array}{r|l} & t_1 \\ \hline NMwc & 1 \\ EMwc & 0 \\ p_1 & 1 \end{array} \qquad I^+ = \begin{array}{r|l} & t_1 \\ \hline NMwc & 0 \\ EMwc & sts = h,\ ste = na \\ p_1 & 0 \end{array}$$

to

$$I^- = \begin{array}{l|c} & t_1 \\ \hline NMwc & 1 \\ EMwc(sts = h, ste = na) & 0 \\ EMwc(sts = h, ste = av) & 0 \\ EMwc(sts = r, ste = na) & 0 \\ EMwc(sts = r, ste = av) & 0 \\ p_1 & 1 \end{array} \qquad I^+ = \begin{array}{l|c} & t_1 \\ \hline NMwc & 0 \\ EMwc(sts = h, ste = na) & 1 \\ EMwc(sts = h, ste = av) & 0 \\ EMwc(sts = r, ste = na) & 0 \\ EMwc(sts = r, ste = av) & 0 \\ p_1 & 0 \end{array}$$

Note now that all elements of the incidence matrices are equal to 0 or 1. The completely unfolded net, shown in Fig. 5.23(c), is now represented as a general Petri net and follows the firing rules of general Petri nets. For example, t_1 is enabled by at least one colorless token residing in each of its input places. Therefore, existing analysis methods for general Petri nets can be applied for the verification of our rule base.

(b) Checking completeness

Completeness verification involves the revelation of unreachable places or unfirable transitions in the rule base.

(i) Unfirable transitions A transition, which will never fire from any possible state of the system, is an unfirable transition. It happens usually because of over-constrained preconditions. This can be detected by checking the reachability tree of the Petri net representing the KBS with a finite set of possible initial markings determined by the model designer.

(ii) Unreachable places A place, which will never be marked throughout the life cycle of the system, is an unreachable place. This is due to a modeling error, unless that place corresponds to an input place, i.e. an interface of the system with its surroundings. An unreachable place is one which does not have any input arcs and is not an *input* place (see section 5.5.3). Usually, unreachable places are those whose input transitions are all unfirable. This can be detected from the incidence matrices of the Petri net representing the KBS as follows:

For $PN = \langle P, T, I^+, I^-, M_0 \rangle$, if $I^+(pi, t) = 0, \forall t \in T$ and $p_i \notin P_{input}$

p_i is then an unreachable place, where P_{input} is the set of *input* places of *PN*.

(iii) Dead-end places A place, which will never be unmarked once it is marked, is a dead-end place. This is due to a modeling error, unless that place corresponds to an output place, meaning an interface with the user. A dead-end place is then a place which does not have output arcs and is not an output place. This can be detected from the incidence matrix of the Petri net representing the KBS. In addition, places whose output transitions are all unfirable, are also considered as dead-end places.

For $PN = \langle P, T, I^+, I^-, M_0 \rangle$, if $I^-(p_i, t) = 0, \forall t \in T$ and $p_i \notin P_{output}$

p_i is then a dead-end place, where P_{output} is the set of output places of *PN*.

(c) Checking consistency

One of the most critical issues in the design of a knowledge base is the consistency check, which eliminates conflicts and redundancies in it. Some of the properties related to the consistency of rules are listed and discussed below.

Some simple cases of such problems can be detected by using incidence matrices of Petri nets representing KBSs, based on an idea initially suggested by the CHECK system. It is used for verification by matching facts and rules in KBSs (Nguyen *et al.*, 1987). The advantage of using Petri nets, as opposed to dealing directly with the facts and rules in a knowledge base, is that the relations between facts and rules are now represented explicitly in the structure of a Petri net. This eliminates the effort for searching and matching facts and rules in some kind of a rule specification language.

(i) Redundant rules Rules are redundant if they succeed from the same preconditions and have the same results, e.g.:

{If *a* then *b*, If *a* then *b*

An example of a simple redundant rule in a Petri net form is shown in Fig. 5.24.

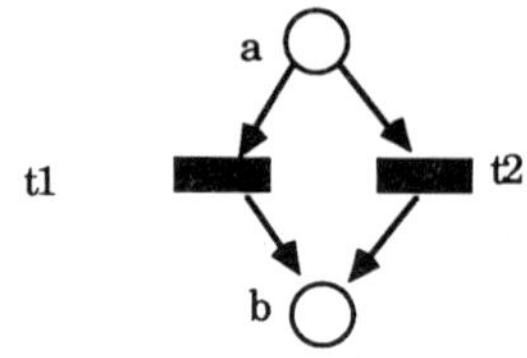

Figure 5.24 Simple redundant rules.

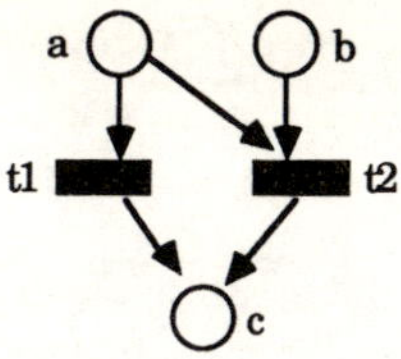

Figure 5.25 Subsumed rules.

If two transitions have the same set of input places and the same set of output places, they are then redundant. Redundant rules can be detected in an incidence matrix where transitions t_x, t_y have identical input place and output place sets: $i = 0$, $\forall_i \in \{I^-(t_x) - I^-(t_y)\}$ and $\forall_i \in \{I^+(t_x) - I^+(t_y)\}$. The incidence matrices shown below represent the net in Fig. 5.24.

$I^- =$

	t_1	t_2
a	1	1
b	0	0

$I^+ =$

	t_1	t_2
a	0	0
b	1	1

By computing $I^-(t_1) - I^-(t_2) = (0, 0)$ and $I^+(t_1) - I^+(t_2) = (0, 0)$, we conclude that t_1 and t_2 are redundant.

(ii) Subsumed rules Two rules are subsumed if they have the same results but one contains constraints additional to those in the other, e.g.:

{If *a* then *c*, If *a* and *b* then *c*

This happens if two transitions have the same output places and the input places of one transition are a subset of the input places of the other one. This can be represented by the Petri net graph in Fig. 5.25.

Subsumed rules can be detected from the incidence matrices where transitions t_x, t_y have identical output place sets and the input place set of t_y is a subset of the input place set of $t_x : i \geq 0$, $\forall i \in \{I^-(t_x) - I^-(t_y)\}$ and $\Sigma\, i \neq 0$ and $i = 0$, $\forall_i \in \{I^+(t_x) - I^+(t_y)\}$. The incidence matrices shown below represent the net in Fig. 5.25.

$I^- =$

	t_1	t_2
a	1	1
b	0	1
c	0	0

$I^+ =$

	t_1	t_2
a	0	0
b	0	0
c	1	1

By computing $I^-(t_1) - I^-(t_2) = (0, -1, 0)$ and $I^+(t_1) - I^+(t_2) = (0, 0, 0)$, we conclude that t_2 is subsumed to t_1.

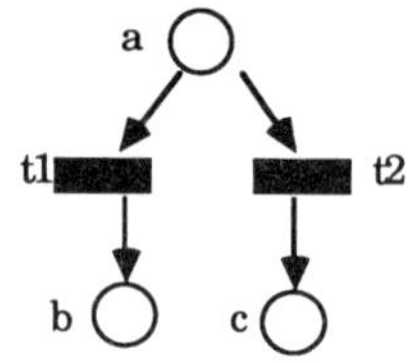

Figure 5.26 Free-choice rules.

(d) Domain knowledge verification

Assuming the correctness of the rule-based system structure, the system domain verification amounts to assuring that the model of the domain embodied in the rule base complies with the semantics of the real world (Lopez *et al.*, 1990).

(i) Data constraints Each of the data records in the database, such as work center records, part records and routing records, cannot be at different statuses at the same time; potential statuses of a work center record are mutually exclusive. For example, a work center cannot be at *hold* and *released* statuses at the same time.

These mutually exclusive properties can be partially detected by the place invariants (*s*-invariants), if the total number of the tokens in the invariant is 1. From this set of *s*-invariants, mutually exclusive conditions can be checked against the logical content of the places of the model. For example, a net involving a work center record should have an *s*-invariant, which consists of all the possible statuses of a work center.

(ii) Free choice rules These are rules succeeding from the same preconditions but their results are not contradictory, e.g.:

{If *a* then *b*, If *a* then *c*, *b* and *c* are not conflicting

This can be represented by the Petri net graph in Fig. 5.26. If two transitions, with the same input places and before more constraints are introduced, result in non-contradictory output places, these transitions are choice-free by nature. This is not allowed within our current domain knowledge, which is restricted to single paths for a specific marking of the net. This is called a *decision-free* net. The decision making is made within the application systems (CAD, CAPP, MRP II and SFC).

(iii) Conflicting rules Rules are conflicting if they succeed from the same preconditions but have conflicting results, based on specific domain knowledge, e.g.:

{If *a* then *b*, If *a* then *c*, *b* and *c* are conflicting

This can be represented by the Petri net graph in Fig. 5.27. If two transitions, with the same input places, and before more constraints are introduced,

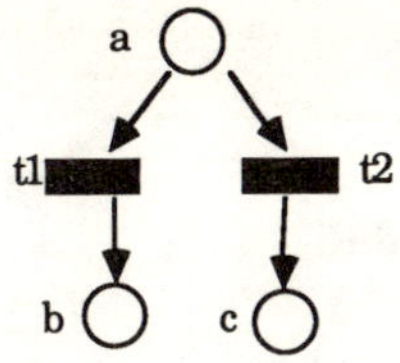

Figure 5.27 Conflicting rules.

result in a contradictory output, these transitions are conflicting by nature. Some of these conflicting rules can be detected by using the incidence matrices and comparing the output place sets with the user specifications. Therefore, we look for transitions t_x, t_y that have identical input place sets and unequal output place sets: $i = 0$, $\forall_i \in \{I^-(t_x) - I^-(t_y)\}$ and $\exists_i \neq 0$, $\forall_i \in \{I^+(t_x) - I^+(t_y)\}$. The incidence matrices shown below represent the net in Fig. 5.27.

$$I^- = \begin{array}{c|cc} & t_1 & t_2 \\ \hline a & 1 & 1 \\ b & 0 & 0 \\ c & 0 & 0 \end{array} \qquad I^+ = \begin{array}{c|cc} & t_1 & t_2 \\ \hline a & 0 & 0 \\ b & 1 & 0 \\ c & 0 & 1 \end{array}$$

By computing $I^-(t_1) - I^-(t_2) = (0, 0, 0)$ and $I^+(t_1) - I^+(t_2) = (0, -1, 1)$, we conclude that t_1 and t_2 might be conflicting with each other; the user should look at the logical content of the output place sets to confirm the conflict.

(i) Structural verification of complex rules

The knowledge-base verification cases discussed above, which have also been studied in various research projects (Nguyen *et al.*, 1987; Lopez *et al.*, 1990) can only handle the verification of simple rules but not complex cases as the examples described below.

A more complicated case of redundant rules involves one primitive transition and one or more compound transitions. For example, let us consider two subnets N_1 and N_2 that have the same set of input places which are not output places of any transition in those nets and the same set of output places which are not input places of any transition in those nets. These two nets are then redundant. See Fig. 5.24 where t_1 becomes N_1 and t_2 becomes N_2.

Our experience in modeling a knowledge base using Petri nets indicates that it is possible to verify some complicated rules by applying restricted reduction rules. There has been some work done in Petri net reduction methods (Lee and Favrel, 1985).

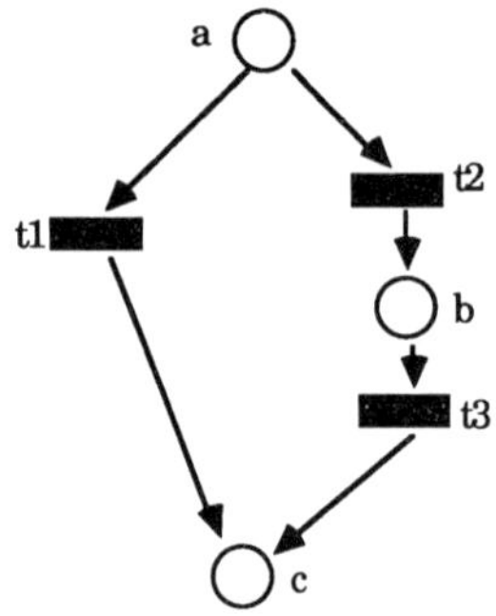

Figure 5.28 Complicated redundant rules.

The procedure to apply these reduction rules on the Petri nets which represent our knowledge-base system is described below with an example shown in Fig. 5.28:

1. Find a place which has only one output transition: place b.
2. The output transition of this place should not have any other input place:

$$I^{-}(a, t_3) = 0, I^{-}(b, t_3) = 1, I^{-}(c, t_3) = 0$$

3. Then, this place can be eliminated and its output transition can be merged into its input transition ($t_{2,3}$). the output places (c) of the merged transition will then be connected to the merged transition ($t_{2,3}$). The reduced net is shown in Fig. 5.29, which can then be checked for the redundancy of rules t_1 and $t_{2,3}$.

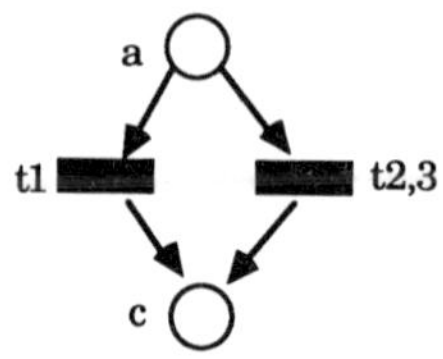

Figure 5.29 Reduced net.

(e) Example

The model verification procedures described in the previous sections are applied in the example shown in Fig. 5.14. The abstracted and unfolded net of Fig. 5.14 is shown is Fig. 5.30 and attention should be paid to the way the database places have been unfolded; any places that do not connect to any transitions within the net have been eliminated. For example,

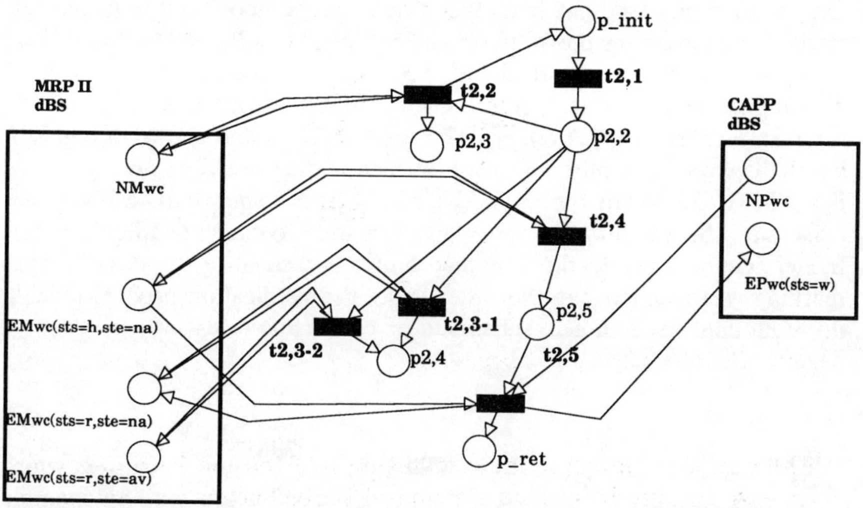

Figure 5.30 Unfolded UPN of the subnet 'Release of a work center in MRP II'.

the database place *EMwc* was abstracted and unfolded into four places: *EMwc*(*sts* = *h*, *ste* = *na*), *EMwc*(*sts* = *h*, *ste* = *av*), *EMwc*(*sts* = *r*, *ste* = *na*) and *EMwc*(*sts* = *r*, *ste* = *av*). Of those, only three places *EMwc*(*sts* = *h*, *ste* = *na*), *EMwc*(*sts* = *r*, *ste* = *na*), *EMwc*(*sts* = *r*, *ste* = *av*) have been retained and used in the net. Similarly, the database place *EPwc* was abstracted and unfolded into three places: *EPwc*(*sts* = *w*), *EPwc*(*sts* = *h*) and *EPwc*(*sts* = *r*) and only one place *EPwc*(*sts* = *w*) has been retained and used in the net. The net invariants, which represent mutually exclusive conditions within the 'company policy' can reveal logical conflicts in the specification of the original rules and possibly errors introduced during the modeling process. For the above example, there are 10 invariants found as shown below and they are all correct having been checked against the company policy. For example: invariant number 7 represents that a work center cannot be at *hold* and *released* statuses at the same time.

1. $NMwc$
2. $EMwc(sts = r, ste = av)$
3. $NPwc,\ p_{ret}$
4. $EMwc(sts = h, ste = na),\ p_{ret}$
5. $p_{init},\ p_{2,2},\ p_{2,4},\ p_{2,5},\ p_{ret}$
6. $NPwc,\ EMwc(sts = r, ste = na)$
7. $EMwc(sts = h, ste = na),\ EMwc(sts = r, ste = na)$
8. $p_{init},\ p_{2,2},\ p_{2,4},\ p_{2,5},\ EMwc(sts = r, ste = na)$
9. $NPwc,\ EPwc(sts = w)$
10. $p_{init},\ p_{2,2},\ p_{2,4},\ p_{2,5},\ EPwc(sts = w)$

One reachability tree has been generated for the above net with the following initial marking: one token in each of the places P_{init}, *EMwc*(*sts* = *h*, *ste* = *na*) and *NPwc* as shown in Fig. 5.31. There should be only one final marking in the tree: one token in each of the following places, *EMwc*(*sts* = *r*, *ste* = *na*), *EPwc* (*sts* = *w*), p_{ret}. The reachability tree is shown in Fig. 5.32 which displays the number of tokens in each of the places (p_{init}, $p_{2,2}$, $p_{2,2}$, $p_{2,4}$, $p_{2,5}$, *NMwc*, *EMwc*(*sts* = *h*, *ste* = *na*), *EMwc*(*sts* = *r*, *ste* = *na*), *EMwc*(*sts* = *r*, *ste* = *av*), *NPwc* and *EPwc*(*sts* = *w*)). It is used to detect deadlocks in the model. There is no deadlock detected in this case since the desired final marking was reached. On the other hand, the verification process reveals no redundant, free-choice, subsumed or conflicting rules.

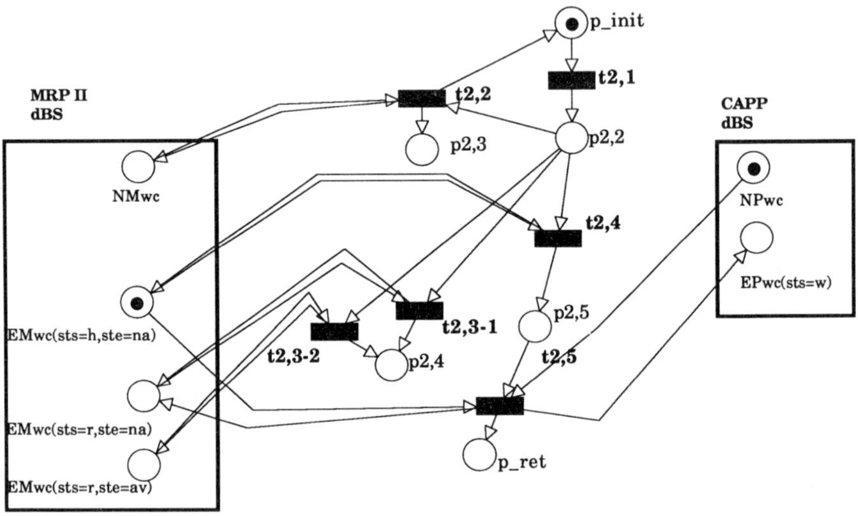

Figure 5.31 Marked PN of the subnet 'Release of a work center in MRP II'.

(1,0,0,0,0,0,0,1,0,0,1,0)

↓ t2,1

(0,1,0,0,0,0,0,1,0,0,1,0)

↓ t2,4

(0,0,0,0,1,0,0,1,0,0,1,0)

↓ t2,5

(0,0,0,0,0,1,0,0,1,0,0,1)

Figure 5.32 Reachability tree generated for the net in Fig. 5.31.

5.5.5 The use of interpreters and implementation using database technology

The emphasis in this section is placed on the automatic translation of the structural representation (UPN) into a rule-specification language, which facilitates the implementation stage and reduces the design cycle of frequently changing rule-based systems (Lin, 1991).

A rule-specification language is needed for the implementation of the system. There exists a variety of programming languages and software development tools: LISP, PROLOG, PASCAL and C for general-purpose programming functions; OPS-5 for performing simulation, KEE for knowledge engineering, LOTOS (Language for Temporal Ordering Specification) developed by the ISO for specifying data communication protocols, services and CIM system architectures (Biemans and Blonk, 1986); SAM-* (Su, 1986) by the National Institute of Standards and Technology (NIST) in its Automated Manufacturing Research Facility (AMRF) project and *M** for modeling data and activities in a manufacturing environment. More recent research has focused on object-oriented programming and database management systems, which facilitate the development of new applications and improve system performance. ROSE developed by the Rensselaer Polytechnic Institute (Hardwick and Spooner, 1989; Spooner *et al.*, 1990) and KRON (Knowledge Representation Oriented Nets) by the University of Zaragoza (Muro *et al.*, 1989) are some examples. The Update Dependency Language (UDL) (Mark and Roussopoulos, 1987) was selected for our implementation due to the similarity of features between it and the UPN and due to the advantage that UDL is designed especially for rule specifications and data updates. It consists of a rule set constructed for each separate database with its update and retrieval dependencies to control inter-database consistency through inter-database operation calls.

This section is structured as follows. To provide the reader with a better understanding of the features of the UDL, we chose to outline its basic syntax and semantics in the second following subsection. The third subsection presents the correspondence between UPN and UDL. The fourth subsection details the implementation strategy and the translation procedure. The fifth subsection provides examples of the automatic translation between UPN models and UDL code, based on an example of a rule specification in the CAD/CAPP/MRP II/SFC integrated system.

(a) The Update Dependency Language: syntax and semantics

The Update Dependency Language (UDL) is a means to specify and control the semantics of a database under update. A set of update dependency procedures gives a declarative operational specification of an update of a relation in terms of a set of alternative sequences of implied updates of the relation, and possibly of other relations, and specifies the conditions under which the implied updates must succeed for the original one to succeed.

The syntax and semantics of the language are formally presented in the following subsections. The rest of this chapter, in addition to the translation algorithm from UPN to UDL, provides a number of examples of how the scenario used throughout this chapter is translated into the formalism presented here.

(b) UDL syntax

For each relation and view defined in a relational database, the database designer defines procedures for the three database modifications: insertion, deletion and update. In addition, a set of application procedures for each relation may be defined or, as is the case in this work, automatically generated by the translation from UPN to UDL.

Definition 5.20. *Procedures* have the following form:

$$\begin{aligned} OR(A_1 = V_1, \ldots, A_n = V_n[; A_1 = W_1, \ldots, A_n = W_n]) \\ \rightarrow C_1, O_{1,1}, \ldots, O_{1,n_1} \\ \rightarrow \ldots \\ \rightarrow C_m, O_{m,1}, \ldots, O_{m,n_m} \end{aligned}$$

where [ ] indicates an optional element.

A procedure is uniquely identified by its operation type O and the name R of the base relation or view for which it is defined. The type of a modification procedure is either *insert*, *delete* or *update*; the type of an application procedure is a *user-defined* name. The formal parameter list, required for all procedures, binds the values of relation R's attributes A_i to the variables V_i, $1 \leq i \leq n$. The *replacement* parameter list, used only in update procedures, binds the replacement values for relation R's attributes A_i to the variables W_i, $1 \leq i \leq n$.

As an example, an application procedure named `release` is applied on the work center relation in the MRP II database and involves two modification procedures: `insert` and `update`. The example of releasing a work center record in MRP II, is shown in Fig. 5.14 and discussed in detail below.

The *body* of a procedure consists of a set of procedure alternatives, each with the elements:

- a condition C_i, $1 \leq i \leq m$, on the database state; and,
- a sequence of procedure invocations O_i, $1, \ldots, O_{i,n_i}$, $1 \leq i \leq m$.

Conditions are safe expressions formed through conjunction and negation of the following atoms (parentheses are used to alter the default precedence of operators):

- *Tuple existence tests* with the form, $R(A_1 = V_1, \ldots, A_k = V_k)$, where R is the name of any base relation or view defined in the database, A_i, $1 \leq i \leq k$, are attribute names of R and V_i, $1 \leq i \leq k$, are constants or variables. The relation, *Mwc*, used in the above example, represents the work center record in MRP II database and it contains the following attributes: *wcid*, *des*, *dep*, *cap*, *sts*, *ste*, *res*, *esd*. A tuple existence test evaluates to true if there exists at least one tuple in relation (or view) R, such that, for every instantiated variable V_i, the value of attribute A_i is equal to the value of V_i. A test of the existence of a work center record in MRP II with work center identification number `Wcid`, would have the following form:

  ```
  Mwc(wcid = Wcid, des = Des, dep = Dep, cap = Cap)
  ```

  Every uninstantiated variable V_j, in this example `Des`, `Dep` and `Cap`, will be instantiated as a result of the evaluation. The instantiated variables act as selection values and the uninstantiated variables act as either join or return value variables. Similarly, the tuple non-existence tests are represented in the following form: $\sim R(A_1 = V_1, \ldots, A_k = V_k)$. A test of the non-existence of a work center record in MRP II is shown in the above example as:

  ```
  ~Mwc(wcid = Wcid)
  ```

- *Comparisons* of the form, $X\theta Y$, where θ is a comparison operator ($<, \leq, =, \geq, >$) and X and Y are constants or variables. A comparison evaluates to true if the algebraic relation θ holds between X and Y.
- The *empty* condition. It always evaluates to true.
- *Negative or positive variable instantiation tests* with the form, var(V_i) or nonvar(V_i), where V_i, $1 \leq i \leq n$, are variables introduced in the head of the procedure. The negative instantiation test evaluates to true if the variable V_i is not supplied in the invocation of the current procedure. The positive instantiation test evaluates to true if the variable V_i is supplied in the invocation of the current procedure. In the above example, `var (Wcid)` and `nonvar (Wcid)` are used to test the negative and positive instantiation of the variable Wcid.
- *Existential qualification, exists* $V_1 \ldots V_n C$. An existential qualification evaluates to true if there is at least one substitution of values V_i, $1 \leq i \leq n$ that satisfies the subcondition C, which cannot contain any instantiation tests. There must be at least one occurrence of each V_i that is free in C.

Procedure invocations have one of the following forms:

- an *application procedure invocation* has the form (e_k and f_k are values of the respective attribute):

  $$\langle \textit{user defined name} \rangle\ R(A_1 = e_1, \ldots, A_k = e_k\ [;\ A_1 = f_1, \ldots, A_k = f_k])$$

In the above example, the application procedure involved is:

```
release Mwc (wcid = Wcid, des = Des, dep = Dep,
             cap = Cap)
```

- *insertion and deletion procedure invocations* have the forms:

$$insert\ R(A_1 = e_1, \ldots, A_k = e_k) \text{ and } delete\ R(A_1 = e_1, \ldots, A_k = e_k)$$

respectively. In the above example, the insertion procedure involved is:

```
insert Pwc (wcid = Wcid, des = Des, dep = Dep,
cap = Cap, sts = w)
```

- *update procedure invocations* have the form:

$$update\ R(A_1 = e_1, \ldots, A_k = e_k; A_1 = f_1, \ldots, A_k = f_k)$$

In the above example, the update procedure involved is:

```
update Mwc (wcid = Wcid, sts = h; wcid = Wcid,
cap = Cap, sts = r)
```

- *physical insertion, deletion and update invocations* have the forms:

$$ins\ R(A_1 = e_1, \ldots, A_n = e_n)$$
$$del\ R(A_1 = e_1, \ldots, A_n = e_n)$$

and

$$upd\ R(A_1 = e_1, \ldots, A_n = e_n; A_1 = f_1, \ldots, A_n = f_n)$$

- primitive I/O operations for *read* and *write* and the operation *fail* are also included in the update dependency formalism.

In the above example, the primitive I/O operations involved include:

```
write ('Enter wcid')
read (Wcid)
```

The procedure abstraction/encapsulation hierarchy enforced by the syntax of the update dependency formalism is illustrated in Fig. 5.33. There are three levels in the hierarchy. The bottom level corresponds to the physical operations; the middle level corresponds to the modification procedures; and the top level corresponds to the application procedures. Notice that physical insertion, deletion and update invocations on a base relation **R** are only allowed from insertion, deletion and update procedures on **R**, respectively.

Notice that physical insertion, deletion and update, *ins*, *del* and *upd*, respectively, on a relation **R** can only be invoked from within insertion, deletion and update procedures on the relation **R**, respectively. Furthermore, physical insertion, deletion and update, and not available on views;

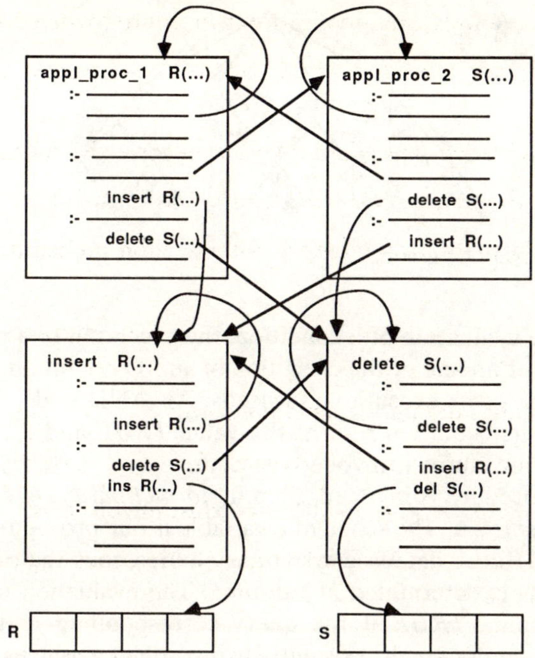

Figure 5.33 Procedure abstraction/encapsulation hierarchy.

procedures for views are specified through the invocation of insertion, deletion and update procedures on the base relations the views are defined from. Finally, procedures may call each other and may call themselves recursively.

In the algorithm and the examples presented later in this section, we utilize the procedures at the application and modification procedure levels only; we assume that the DBMS has provided the implementation of modification procedures, which work as the corresponding physical operations. In other words, we have assumed that procedures *insert*, *delete* and *update* act as operations *ins*, *del* and *upd*, respectively.

(c) UDL semantics

The execution of a procedure can be depicted by an AND/OR graph (Fig. 5.34). The *AND* nodes are those whose executions are tied together by an arc; the *OR nodes* are those whose executions are not tied together by an arc. Each execution of an OR node represents the execution of *one procedure alternative.* The ordered sequence (left-to-right) of executions of an AND node represents the execution of the *elements of one procedure alternative*; the first represents the evaluation of the condition and the following represents the executions of the invoked procedures. A *ROOT node* represents the execution of a user-invoked procedure. A *LEAF node*

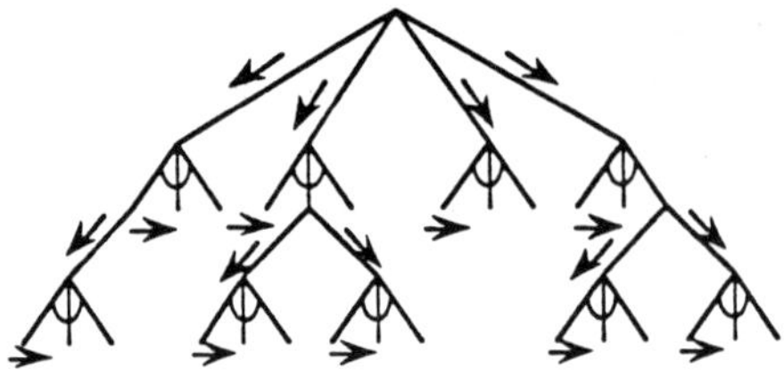

Figure 5.34 AND/OR graph.

represents the evaluation of a condition, the execution of a physical insertion, deletion or update or the execution of an I/O operation. An OR node *succeeds* if one of its executions succeeds. An AND node *succeeds* if the evaluation of its condition returns the value *TRUE* and the execution of each of the procedures it invokes succeeds.

When a procedure is invoked, then its *formal parameters are bound to the actual parameters*. The scope of a variable is one procedure. Conditions are submitted to the database system as queries, thus the order of evaluation of atoms is determined at run-time. The evaluation of a condition returns the value *TRUE* if the query corresponding to the condition returns a non-empty result; existentially quantified variables are bound to values that satisfy the query.

The execution of a physical insertion, deletion or update and the execution of an I/O operation always succeed.

The selection of execution of procedure alternatives is non-deterministic and executions of procedure alternatives may be done in parallel. However, the effects of only one of the alternatives will be seen when the procedure succeeds. Furthermore, while an alternative is executing, it will only see database updates that have occurred on its execution path; it will not see database updates from other alternatives that might be executing in parallel. If a procedure execution fails, i.e. none of its alternatives succeed, then the database is left completely unchanged by the procedure invocation. Conditions are submitted to the database system as queries, as mentioned above.

(d) Feature correspondence between UPN and UDL

This section describes the translation of particular features of UPN to UDL.

(i) Data in UPN as UDL relations

The information flowing through a UPN net can be atomic data, although this atomic information can be aggregated into more complex data structures.

Atomic data and its data set can be translated to UDL as domains. For example, the data set of a work center status in MRP II (which can have only two different values, *h* for hold and *r* for released: $STS = \{r, h\}$) is represented in UDL by a domain of character type.

In UDL, data structures are defined by a relation name and a tuple of data, which correspond to specific attributes specified in UPN:

$$R(A_1 = V_1, \ldots, A_k = V_k)$$

An example of a work center record in MRP II in the form of a UDL relation is shown below. It represents a work center `lt101` (wcid) which is a `lathe` (des), located in the `machining` (dep) department, having `h` (*hold*) status (sts), `na` (*not available*) state (ste), `null` (*unknown*) capacity (cap), `M12` resource code (res) and `null` (*unknown*) affectivity start date (esd). (It is reminded that a general work center record in MRP II is represented as `Mwc (wcid, des, dep, cap, sts, ste, res, esd)`:

```
Mwc (wcid = lt101, des = lathe, dep = Tmachining,
cap = null, sts = Th, ste = na, res = M12, esd = null)
```

(ii) Facts in UPN as UDL conditions

In order to verify whether a rule is enabled or not, it is necessary to verify that the precondition part of the rule matches with the status information in the system. Status information is represented by UPN places and their marking. Access to that information is specified in UPN by means of arcs and arc expressions.

Two different types of status information can be distinguished: information about the database status and information about the reasoning process status.

Database status: requires access to a database record and reading the values of its attributes. This is implemented by using the UDL relational form where the record is identified by the record id number. For example: the MRP II user starts releasing a work center with *wcid* = *lt*101 following the rules specified in Fig. 5.14. The database check of work center *lt*101 with a *hold* status corresponds in UPN to an arc from the place *EMwc* of the MRP II database, with the function *wcid* = *lt*101, *sts* = *h*. This is translated into UDL in the same form: Mwc `(wcid = lt101, sts = h)`. On the other hand, the non-existence of the work center *lt*101 corresponds to the UPN place *NMwc* of the MRP II database, with the function *wcid* = *lt*101; this can be translated into the UDL form: `~Mwc (wcid = lt101)`.

Reasoning process status: generally corresponds to the states of a UDL application procedure. For example, places p_{init}, p_2, p_3, p_4, p_5, p_{ret} in Fig. 5.14.

(iii) Database-related arc conditions in UPN as UDL checking and modification procedures

The next step in the translation process is to identify UPN elements, which correspond to arc conditions directly relating to database places, in order to translate them into UDL elements. They are translated into UDL checking conditions or modification procedures to access or modify the database. These elements are identified as follows:

- *Checking* a record. In UPN form, the database check is represented by a pair of input and output arcs, which have the same arc expression, linked between a transition and a database place. The check is implemented, as mentioned before, for database access. The case of a database place representing the non-existence of the record is implemented using the UDL negative form. For example, transition t_3 in Fig. 5.14 has two arcs to and from place *EMwc* (in the MRP II database) with the same arc expression: *wcid* = *wcid#*, *sts* = *r*. This is translated into UDL form as:

```
Mwc (wcid = Wcid, sts = r)
```

- *Inserting* a record occurs when there is an arc from a database place to a transition which represents non-existence of a record and another arc from the transition to a database place representing the existence of the same record. It is implemented using the UDL modification procedure `insert` (⟨*relation name*⟩(⟨*tuple spec*⟩)). For example, transition t_5 in Fig. 5.14 has one arc from place *NPwc* and one to place *EPwc* (in the CAPP database) with the arc expression *Pwc*(*wcid* = *wcid#*, *des* = *des#*, *dep* = *dep#*, *cap* = *cap#*, *sts* = *w*). This is translated into UDL form as:

```
insert Pwc (wcid = Wcid, des = Des, dep = Dep,
cap = Cap, sts = w)
```

- *Deleting* a record from the database can be recognized when an arc stems from a database place representing the existence of a record to a transition and another arc stems from the transition to a database place representing the non-existence of the same record. It is implemented using the UDL modification procedure: `delete` (⟨*relation name*⟩(⟨*tuple spec*⟩)).
- *Updating* a record in the database can be recognized when an arc stems from a database place representing the existence of a record to a transition and another arc, in the reverse direction, but with a different function. It is implemented using the UDL modification procedure `update` (⟨*relation name*⟩(⟨*old tuple spec*⟩;[⟨*new tuple spec*⟩])). For example, transition t_5 in Fig. 5.14 implies an update to the record *Mwc* (in place of *EMwc*) in the MRP II database that is translated into UDL form as:

```
update(Mwc(wcid = Wcid; wcid = Wcid, cap = Cap, sts = r))
```

(iv) Requesting/printing information in UPN as UDL primitive I/O operations

The next step is to identify UPN elements, which correspond to arc conditions directly relating to information input/output, to translate them into UDL I/O primitives operations. Thus, requesting information from or printing information to the user can be achieved. The primitive operations are identified as follows:

- *Requesting information* from the user. This is detected when a transition is a source transition, where some information that is leaving the transition through the outgoing arc(s) did not enter through any incoming arc(s). This information must be requested from the user. It is implemented using the UDL primitive operation `read` (⟨*domain variable*⟩). For example, transition t_1 of Fig. 5.14 does not receive information from place p_1. Instead one needs to provide a work center identification number in the variable *wcid#*. This information must be provided by the user and is implemented by:

  `read` (Wcid)

  For better legibility, a message like the following can be printed to prompt the user:

  `write` ('Input the value for the variable wcid')

- *Printing* a message to the user. This is detected when sink places appear in the net. Some information arrives at such a place through the incoming arc(s), but does not leave the place through any outgoing arc(s), generally because it has no outgoing arcs. This information must be shown to the user. It is implemented using the UDL primitive `write` ('⟨*place label text*⟩', ⟨*domain variable*⟩). If there is no domain variable, the label identifying the place is shown as `write` ('⟨*place label text*⟩'). The last option may be used to show single-error messages. For example, place p_4 in Fig. 5.14 is translated as an error message for the work center identification provided in variable *wcid#*:

  ```
  write ('Output in P4 for data: ' wcid#)
  ```

  or, if the place has an associated label:

  ```
  write ('work center already exists: 'wcid#)
  ```

(v) Rules and metarules as UDL procedures

The following step corresponds to the translation of the transition set itself. UDL procedures provide a very powerful mechanism to represent if-then rules (transitions). As a first approach, each transition of a UPN net could be easily implemented by a separate UDL procedure. This approach for the translation of transitions is general and simple, but it presents several problems. An important problem is that some additional local variables are required to execute a series of transition firings without user interruptions. For example, if n number of transitions are designed to be fired sequentially, without any user input and they are implemented into individual procedures, at least $n - 1$ new variables representing the completion of the first $n - 1$ transition firings have to be created. Secondly, this approach does not make use of some important programming capabilities available in UDL, such as the use of procedures (application rules) and recursion. This would result in an inefficient implementation.

An actual example is shown below: in order to implement transition t_3 from Fig. 5.14 as a composed operation, we need a new variable, varP₂, to test the value Wcid in place P_2 ⇒

```
Release-Transition-t3  Mwc(wcid = Wcid, sts = Sts)
→ (Wcid = varP2) Ÿ Mwc(wcid = Wcid, sts = r),
write('Work center already has "r" status in MRP II')
```

On the other hand, UDL provides a way to implement a set of related rules in the form of a composed rule. Also, procedures and procedure calls are typical decomposition mechanisms used in UDL programs and recursion is also available, as mentioned above. Therefore, we decided as a general rule to implement a *set* of related rules as *one* UDL procedure. UDL procedures are used to represent subnets at any level of abstraction. It is no longer necessary to use additional local variables (other than the formal parameters of the procedures involved) to implement the transition status of the net execution.

To take advantage of these features, we follow what we call an *information-driven* approach. According to the UDL syntax, only one procedure alternative (transition) of a procedure call can be successfully executed. This presents a limitation for the UPN execution syntax, according to which one transition can be enabled and fired automatically following the firing of its preceding transition. To eliminate this limitation, we use a recursive call to the same procedure, with the necessary parameters carried over. An example is shown below which has three procedure alternatives. The first as well as the second alternative of this procedure will invoke a recursive call, which will execute another copy of the same procedure with initiated parameter values, after all its preconditions are evaluated to be true and all its operations are successfully executed. This enables the continuous execution of more than one alternative within one UDL procedure. The need for recursive calls in UPN models is identified if the outgoing arcs of a transition go to places internal to the subnet that do not belong to database places (i.e. places of local scope – section 5.5.3). This means that other transitions within the same subnet may continue firing uninterrupted. It is obvious when the recursive call sequence has finished. That happens when the procedure call reaches an alternative which does not have recursive call. In this example, the third alternative will end the recursive call sequence.

$$
\begin{aligned}
&OR(A_1 = V_1, \ldots, A_n = V_n)\\
\rightarrow\ & C_1\\
& O_{1,1}\\
& O_{1,2}\\
& OR(A_1 = V_1, \ldots, A_n = V_n)
\end{aligned}
$$

$$\begin{aligned}
\rightarrow\ & C_2 \\
& O_{2,1} \\
& OR(A_1 = V_1, \ldots, A_n = V_n) \\
\rightarrow\ & C_3 \\
& O_{3,1} \\
& O_{3,2}
\end{aligned}$$

Procedure parameters transfer data from one call (transition execution) to the following one. The actual parameters sent in each recursive call correspond to the information that is transmitted to the postconditions of the transition being executed.

(vi) Translation procedure

The translation of UPN to UDL can be seen as another special 'implementation' of Petri nets, specific for this application domain. This implementation of UPN is simpler than the implementation of a generic colored Petri net, due to the added constraints imposed by UPN over the general Petri net formalism. Examples of such added constraints include: the variety of preconditions that are highly constrained, rules that are supposed to be well structured in metarules and specifications that are related to a manufacturing database domain. The overall purpose of the translation procedure is to generate an efficient code in UDL, the language in which the specifications will be executed. To start the translation procedure, the UPN model must be provided. The procedure for translating one subnet into a piece of UDL code is detailed as follows.

Generate a UDL procedure heading, based on the UPN metarule name ($\langle O \rangle$) and its corresponding database relation ($\langle R \rangle$). The set of attribute names to be included in the formal parameter list of the procedure is defined by the set of all attribute names that appear in the arc expressions of the subnet ($A_1, \ldots, A_m$). The procedure head is:

$$\langle O \rangle \langle R \rangle (A_1 = V_1, \ldots, A_n = V_m)$$

where ($V_1, \ldots, V_m$) is the set of formal variables for which the values of attributes, $A_1, \ldots, A_m$, from the relation $\langle R \rangle$ are bound (these variable names can be the same as those in the UPN model).

One UDL procedure is composed by several alternatives, one for each transition in the metarule subnet. The following steps must be taken for each transition.

1. Conditions for alternatives (preconditions of transitions) are defined by incoming arc(s) to a transition:

   (a) Recognize *checking* UDL elements, as explained previously in this section. The conjunction of these checking elements is a precondition for the procedure alternative:

   $$\langle R \rangle (A_m = V_m, \ldots, A_m = V_n)$$

(b) Find positive variable instantiations by looking at the variables in the arc expressions from the incoming arcs, which do not belong to the database checking recognized above ($Var_i, \ldots, Var_j$) and generate a positive variable instantiation test for each one. The conjunction of these tests is another precondition:

$$\texttt{nonvar}(V_i) \wedge \ldots \wedge \texttt{nonvar}(V_j)$$

(c) The rest of the formal variables have negative instantiations. Only variables representing attributes that provide information to the output places and are not coming from the input places ($V_x, \ldots, V_y$) must be checked. A negative variable instantiation test must be generated for each of them. The conjunction of these tests is another precondition:

$$\texttt{var}(V_x) \wedge \ldots \wedge \texttt{var}(V_y)$$

2. Operations for alternatives (postconditions of transitions) are defined by outgoing arcs from a transition. Each one of the following steps can produce new operations:

(a) Recognize *input and output* UDL elements, as explained in section 5.5.3. For each variable that needs to be provided from the user, generate the appropriate input sequence (⟨*Text* V_p⟩ that corresponds to the interpretation of the attribute name bound by V_p in the database record tables):

`write`('`Enter` ⟨*Text* V_p⟩'), `read` (V_p)

For each output variable, generate:

`write` ('⟨*place label text*⟩')}

(b) Recognize *deletion*, *insertion* and *update* UDL modification procedures, as explained in section 5.5.3 and generate the appropriate invocations:

`delete` (⟨*relation name*⟩(⟨*tuple spec*⟩))
`insert` (⟨*relation name*⟩(⟨*tuple spec*⟩))
`update` (⟨*relation name*⟩(⟨*old tuple spec*⟩);[⟨*new tuple spec*⟩]))

(c) Write the calls for all UDL application procedures associated with the transition. The recognition of UDL application procedure calls is done by reading the content (name) associated to the transition, in the case of a *compound* transition.

(d) Generate a recursive call, if any of the transition's output places, which is not a database place, is an input place to any other transition within the subnet. Only the variables ($V_i, \ldots, V_j$) which are used in the outgoing arc expressions, that connect to the output places mentioned above, are used in the parameter list of the procedure call:

$$\langle O \rangle \ \langle R \rangle (A_i = V_i, \ldots, A_j = V_j)$$

(vii) Generation of UDL code

The implementation of the knowledge-based system is based on the translation from the UPN subnets (which are designed, validated and refined according to the system specifications collected) into UDL code. There are two types of UPN subnets to be translated: the first, a single-procedure subnet, which involves only one application procedure; the second, a multi-procedure subnet, which involves more than one application procedure with procedural calls among subnets. Each application procedure has to be translated into one UDL code, following the translation procedure discussed above, including the application procedure calls in the second case. Examples of translations for both single-procedure UPN subnets and multi-procedure UPN subnets are detailed in the following sections.

(viii) Example of translating a single-procedure UPN subnet into one UDL procedure

In order to clarify the translation procedure, we return to the example shown in Fig. 5.14, which was used to illustrate the creation of UPN models described in section 5.5.3. This net is *simple* because it does not require further refinement to create additional subnets. The goal now is to translate the UPN representation to the respective UDL code.

The name of the UPN is 'release Mwc' and the corresponding database records – work center record in MRP II and work center record in CAPP – are described below (a more detailed description of *Mwc* is given in Table 5.4):

Work center record in MRP II:
`Mwc(wcid, des, dep, cap, sts, ste, res, esd)`
Work center record in CAPP:
`Pwc(wcid, des, dep, cap, sts)`

(ix) Translation procedure

1. Procedure heading generation:

   ⟨*O*⟩ ← `release` (metarule name)
   ⟨*R*⟩ ← `Mwc` (corresponding database record)

   Attribute names that appear in the arc expressions are: *wcid*, *des*, *dep*, *cap*, *sts* and their corresponding variables (*wcid#*, *des#*, *dep#*, *cap#*) are modified into the following UDL variable syntax: `Wcid`, `Des`, `Dep`, `Cap`. The procedure heading becomes ⇒

```
release Mwc (wcid = Wcid, des = Des, dep = Dep,
cap = Cap)
```

2. Conditions for the alternatives:
   Transition t_1: there is no connection with database places (rows *NMwc* and *EMwc* in I^- and I^+ are 0). This means there is no checking of the database. The column of transition t_1 in I^- shows that there is only one incoming arc connected with place p_1 with no variables attached to the arc expression. This means that no positive variable instantiations are needed. The rest of the variables (`Wcid`, `Des`, `Dep` and `Sts`) have negative instantiations; however, the column of transition t_1 in I^+ shows that there is only one outgoing arc connected with place p_2 with an arc expression *wcid#*. This means that in the incoming arcs to this transition a work center identification number (variable `Wcid`) was not provided, but will be provided to the outgoing arc. In order to reduce the code, only this test is really needed: `var (Wcid)`. The complete condition is ⇒

```
var (Wcid)
```

   Transition t_2: this has incoming and outgoing arcs to *NMwc* (MRP II database) with the same arc expression *Mwc*(*wcid* = *wcid#*). This is a checking for the non-existence of `Mwc` with that specific work center identification number ⇒ `~Mwc (wcid = Wcid)`. It has another incoming arc with *wcid#* from p_2 providing the work center id. This information must be checked for positive instantiation ⇒ `nonvar (Wcid)`. There is no more outgoing information for the arc because the arc expression to place p_1 has no variables. This means that no negative instantiation test is necessary. The complete condition part is the conjunction of these two conditions ⇒

```
nonvar (Wcid) ∧ ~Mwc (wcid = Wcid)
```

   Transition t_3: similarly, the complete condition is ⇒

```
nonvar (Wcid) ∧ Mwc (wcid = Wcid, sts = r)
```

   Transition t_4: similarly, the complete condition is ⇒

```
nonvar (Wcid) ∧ var (Cap) ∧
Mwc (wcid = Wcid, des =  Des, dep = Dep,
sts = h, ste = na)
```

   Transition t_5: similarly, the complete condition is ⇒

```
nonvar (Wcid) ∧ nonvar (Des) ∧ nonvar (Dep) ∧
nonvar (Cap) ∧ Mwc (wcid = Wcid, sts = h)
```

3. Operations for the alternatives:
   Transition t_1: column t_1 from I^- and I^+ shows that variable *wcid#* needs to be requested (there are no incoming variables and variable *wcid#* is outgoing) ⇒

```
write ('Enter  wcid')
read (Wcid)
```

No other UDL elements (output, deletion, creation or update) can be recognized. However, transition t_1 has an output place, p_2, which is an input place to transitions, t_2, t_3 and t_4. This means that the reasoning process is not completed yet and a recursive call is required. The parameters of this call are the ones required by the outgoing arcs (in this case only `Wcid`) ⇒

```
release Mwc (wcid = Wcid)
```

Transition t_2: an output primitive can be easily recognized here: place p_3 is an output place (or a sink place), thus the information in the arc expression, *wcid#* and the text associated with the interpretation of p_3 must be displayed ⇒

```
write ('Work center ID does not exist in MRP II,
enter again', Wcid)
```

As before, a recursive call is required, in this case with no call parameters (arc expression outgoing to place p_1 has no variables) ⇒

```
release Mwc (  )
```

Transition t_3: only an output statement is needed to display the information in the arc expression, *wcid# and* the text associated to the interpretation of p_4 ⇒

```
write ('Work center already has "r" status in
MRP II', Wcid)
```

No new call is needed because the output place p_4 is not connected to any other transition.

Transition t_4: the input for variable `Cap` is required and then a recursive call is made with the information for the `wcid`, `des`, `dep`, `cap` and `sts` parameters ⇒

```
write('Enter capacity')
read(Cap)
release Mwc(wcid = Wcid, des = Des, dep = Dep,
cap = Cap)
```

Transition t_5: an update modification procedure can be identified because there is an arc coming from the database place *EMwc* with a different function (*Mwc*(*wcid* = *wcid#*)) to the one that is going back to *EMwc* (*Mwc*(*wcid* = *wcid#*, *des* = *des#*, *dep* = *dep#*, *cap* = *cap#*, *sts* = *r*)) ⇒

```
update Mwc (wcid = Wcid, sts = h; wcid = Wcid,
cap = Cap, sts = r)
```

```
release Mwc(wcid=Wcid,des=Des,dep=Dep,cap=Cap)
→       var(Wcid),
        write('Enter wcid'),
        read(Wcid),
        release Mwc(wcid=Wcid).

→       nonvar(Wcid) ∧ ~Mwc(wcid=Wcid),
        write('Work center ID does not exist in MRP II, enter again',
Wcid), \\
        release Mwc().

→       nonvar(Wcid) ∧ Mwc(wcid=Wcid,sts=r),
        write('Work center already has ''r'' status in MRP II', Wcid),

→       nonvar(Wcid) ∧ var(Cap) ∧
        Mwc(wcid=Wcid,des=Des,dep=Dep,sts=h,ste=na),
        write('Enter capacity'),
        read(Cap),
        release Mwc(wcid=Wcid,des=Des,dep=Dep,cap=Cap).

→       nonvar(Wcid) ∧ nonvar(Des) ∧ nonvar(Dep) ∧ nonvar(Cap),
        update Mwc(wcid=Wcid,sts=h;wcid=Wcid,cap=Cap,sts=r),
        insert Pwc(wcid=Wcid,des=Des,dep=Dep,cap=Cap,sts=w).
```

Figure 5.35 UDL code for the scenario 'Release of a work center in MRP II'.

It also has an associated procedure call ⇒

```
insert Pwc(wcid = Wcid, des = Des, dep = Dep,
cap = Cap, sts = w)
```

The final UDL code resulting from this translation is shown in Fig. 5.35.

(x) Example of translating a multi-procedure UPN subnet into UDL procedures

A UPN subnet, which has been designed using a top-down refinement technique into a set of subnets, each representing one UDL application procedure, has to be translated into more than one UDL code segments. An example of this kind is the removal of a work center record from MRP II presented here. When the removal operation is invoked in MRP II, the following system checks are initiated. A check is made to see that the work center being removed exists in MRP II. The status of the work center is not relevant to the operation. In addition, all routings maintained by the MRP II routing module are checked. If any routings utilizing this work center exist and are on 'hold' or 'release' status in CAPP, the operation fails and a message to this effect is displayed. The reason is that work centers which are utilized by active routings cannot be removed. If the above checks are satisfied, the work center is removed from the databases of MRP II, CAPP and SFC. The above specification is first modeled in UPN at the abstract level as shown in Fig. 5.36 and then further refined down to a more detailed level. The complete net involves three subnets, which are translated to three UDL procedures: one major procedure (procedure

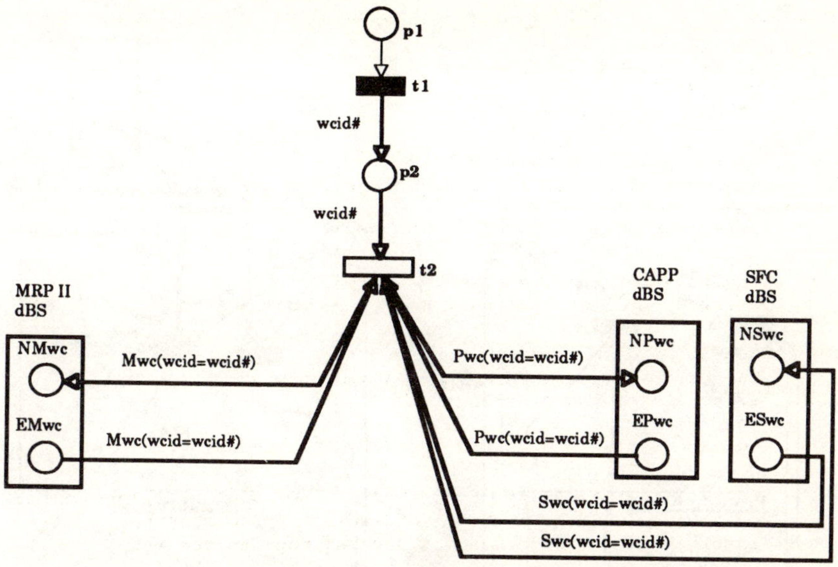

Figure 5.36 Subnet of the scenario 'Deletion of a work center in MRP II'.

no. 1) removes the work center via MRP II and two other procedures check the MRP II and the CAPP databases. The top-down refinement technique used was discussed in section 5.5.3.

The goal now is to translate the UPN representations to the respective UDL codes. Following the same translation procedure for all the subnets involved, three UDL application procedures are generated as shown below.

(i) UPN subnet no. 1 During the translation of operations t_2, the following two application procedures called by it have to be satisfied before any other modification procedures can be implemented ⇒

```
check.1  Mwc(wcid = Wcid)
check.2  Prout(wcid = Wcid, psts = Psts)
```

Three deletion modification procedures can be identified: an arc coming from the database place *EMwc* with the expression *Mwc*(*wcid* = *wcid*#) and another one going to the database place *NMwc* with the same expression (same is the case for CAPP and SFC) ⇒

```
delete Mwc(wcid = Wcid)
delete Pwc(wcid = Wcid)
delete Swc(wcid = Wcid)
```

(ii) UPN subnet no. 2 During the translation of operation $t_{2,4}$, we observe that the output is a place $p_{2,3}$ which represents the interface with the

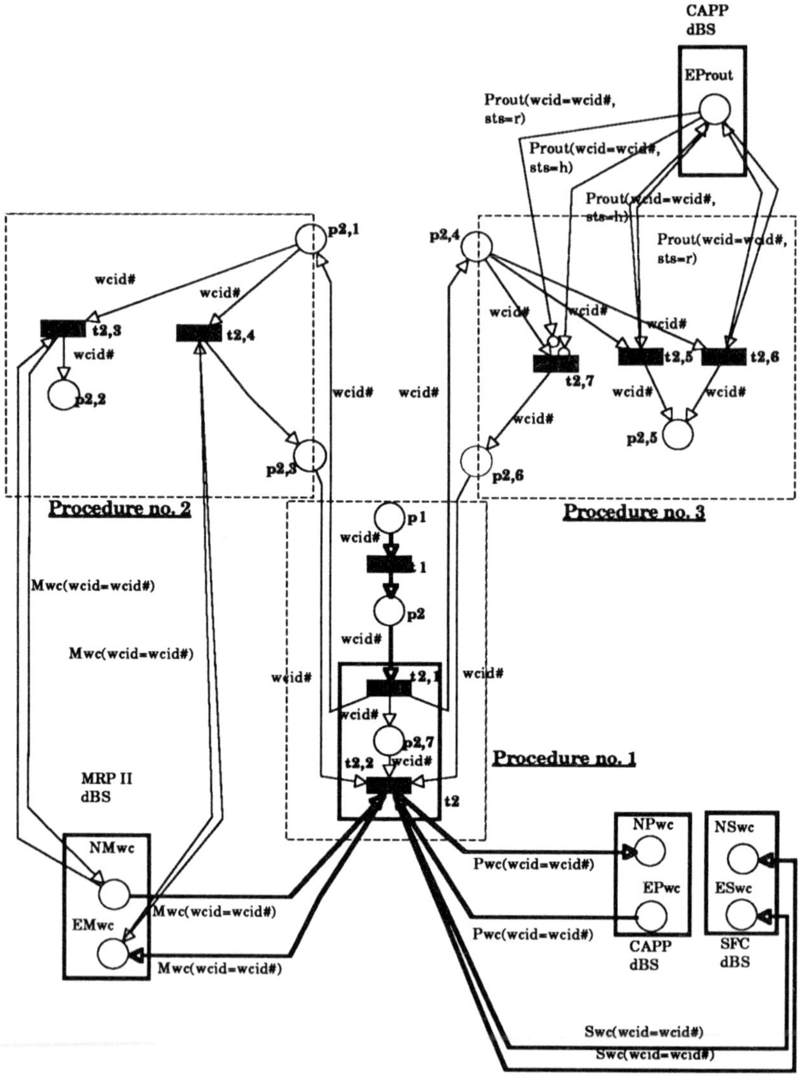

Figure 5.37 Subnet of the scenario 'Deletion of a work center in MRP II'.

higher-level subnet. This place will receive a token as long as all the preconditions are satisfied. Therefore, no operation is required here.

(iii) UPN subnet no. 3 During the translation of operation $t_{2,7}$, there are two negative checkings, which are represented by the inhibitor arcs, for the non-existence of any routing `EProut(wcid = Wcid, psts = h)` using that specific work center identification number and bearing an *h* or *r* status Π

`~ EProut(wcid = Wcid, psts = h)`

```
remove Mwc(wcid=Wcid)
→     var(Wcid),
      write('Enter wcid'),
      read(Wcid),
      remove Mwc(wcid=Wcid).

→     nonvar(Wcid),
      checkl Mwc (wcid=Wcid),
      check2 Prout (wcid=Wcid,psts=Psts),
      delete Mwc(wcid=Wcid),
      delete Pwc(wcid=Wcid),
      delete Swc(wcid=Wcid).


checkl Mwc(wcid=Wcid)
→     nonvar(Wcid) ∧ ~Mwc(wcid=Wcid),
      write('Work center ID does not exist in MRP II', Wcid).

→     nonvar(Wcid) ∧ Mwc(wcid=Wcid).


check2 Prout(wcid=Wcid)
→     nonvar(Wcid) ∧nonvar(Wcid) ∧ EPwc (wcid=Wcid,psts=h),
      write('Work center is in use by active process plans', Wcid).

→     nonvar(Wcid) ∧ nonvar(Wcid) ∧ EPwc (wcid=Wcid,psts=r),
      write('Work center is in use by active process plans', Wcid).

→     nonvar(Wcid) ∧ ~EProut (wcid=Wcid,psts=h) ∧

~EProut(wcid=Wcid,psts=r).
```

Figure 5.38 UDL code for the scenario 'Deletion of a work center in MRP II'.

and

$$\sim \text{EProut(wcid = Wcid, psts = r)}$$

Similarly, the complete condition is the conjunction of all related conditions ⇒

$$\text{nonvar(Wcid)} \wedge \sim \text{EProut(wcid = Wcid, psts = h)} \wedge$$
$$\sim\text{EProut(wcid = Wcid, psts = r)}$$

The full UDL code for this multi-procedural subnet is presented in Fig. 5.38.

5.6 CONCLUSIONS

The purpose of this chapter was to illustrate the use of Petri nets for the design, analysis and formal specifications of various aspects of manufacturing information systems. Obviously, Petri nets, and especially high-level

Petri nets, are efficient models to represent and analyze company policies and the behavior of information systems in addition to classical data models used to represent the static properties of information systems.

When manufacturing information system analysis is concerned with logical design of local databases to be implemented by classical or object-oriented data models, the *M** methodology with the PDN model is recommended. When the analysis focuses more on knowledge-base design using existing local databases to implement manufacturing policies, the second methodology with the UPN model is preferred.

To make the techniques presented in this chapter more practical for industrial case studies, there is the need for formal techniques and computer tools, which still have to be developed in the form of commercial tools.

BIBLIOGRAPHY

Albano, A., De Antollenis, V. and Di Leva, A. (eds) *Computer-Aided Database Design: The DATAID Project*. North-Holland, Amsterdam, 1985.

Appleton, D. S. The CIM database. *SME Computer-Integrated Manufacturing Series*, 1(4). SME, Dearborn, MI, 1984.

Atkinson, M., Bancilhon, F., DeWitt, D., Dittrich, K., Maier, D. and Zdonik, S. The object-oriented database system manifesto. In *Deductive and Object-Oriented Databases* (W. Kim, J. M. Nicolas and S. Nishio, eds), North-Holland, Amsterdam, pp. 223–240, 1990.

Beck, C. L. and Krogh, B. H. (1986) Models for simulation and discrete control of manufacturing systems. *Proc. IEEE Int. Conf. on Robotics Automation*, pp. 305–310, 1986.

Biemans, F. and Blonk, P. On the formal specification and verification of CIM architectures using LOTOS. *Computers in Industry*, 7, 491–504, 1986.

Ceri, S. (ed.) *Methodology and Tools for Database Design*, North-Holland, Amsterdam, 1983.

Chen, P. P. S. The entity-relationship model – toward a unified view of data. *ACM Transactions on Database Systems*, 1(1), 9–36, 1976.

Coad, P. and Yourdon, E. *Object-Oriented Analysis*. Prentice-Hall, Englewood Cliffs, NJ, 1990.

Codd, E. F. A relational model of data for large shared data banks. *Communications of the ACM*, 13(6), 377–386, 1970.

Date, C. J. *An Introduction to Database Systems*. 4th edition, Addison-Wesley, Reading, MA, 1986.

Di Leva, A. and Giolito, P. *I Sistemi Informativi: Analisi e Progetto*, Dipartimento di Informatica, Università di Torino, 1989.

Di Leva, A., Giolito, P. and Vernadat, F. Executable models for the representation of production systems. *Proc. IMACS-IFAC Symposium MCTS-91*, Lille, France, pp. 561–566, 7–10 May 1991.

Di Leva, A., Vernadat, F. and Bizier, D. Information system analysis and conceptual database design of production environments with *M**. *Computers in Industry*, 9, 183–217, 1987.

Famili, A. and Vernadat, F. Integrity constraints of manufacturing data bases. *Engineering with Computers*, 2(1), 41–52, 1986.

Gane, C. and Sarson, T. *Structured System Analysis: Tools and Techniques*. Prentice-Hall, Englewood Cliffs, NJ, 1979.

Genrich, H. J. Predicate/transition nets. In *Petri Nets: Central Models and their Properties* (W. Brauer, W. Reisig and G. Rozenberg, eds.) Lecture Notes in Computer Science, 254, Springer-Verlag, Berlin, pp. 208–247, 1987.

Genrich, H. J. and Lautenbach, K. System modelling with high level Petri nets. *Theoretical Computer Science*, 13, 109–136, 1981.

Gerelle, E. G. R. and Stack, J. *Integrated Manufacturing Strategy, Planning, and Implementation*. McGraw-Hill, New York, 1988.

Hardwick, M. and Spooner, D. The ROSE data manager: using object technology to support interactive engineering applications. *IEEE Transactions on Knowledge and Data Engineering*, pp. 285–289, 1989.

Hsu, C. and Rattner, L. Information modeling for computerized manufacturing. *IEEE Transactions on Systems, Man, and Cybernetics*, 20(4), 758–776, 1990.

Huber, P., Jensen, K. and Shapiro, R. M. Hierarchies in colored Petri nets. In *Advances in Petri Nets 1990* (G. Rozenberg, ed.), Lecture Notes in Computer Science. Springer-Verlag, Berlin, 1990.

Hull, R. and King, R. Semantic database modeling: survey, applications and research issues, *ACM Computing Surveys*, 19(3), 201–260, 1987.

Jeng, M. D. and DiCesare, F. A review of synthesis techniques for Petri nets. *Proceedings of IEEE Computer Integrated Manufacturing Systems Conference*, 1990.

Jensen, K. Coloured Petri nets and the invariant-method. *Theoretical Computer Science*, 14, 317–336, 1981.

Jensen, K. Computer tools for construction, modification and analysis of Petri nets. *Advances in Petri Nets, Part II*, pp. 4–19, 1986.

Jensen, K. Coloured Petri nets. In *Petri Nets: Central Models and their Properties*, (W. Brauer, W. Reisig and G. Rozenberg, eds), Lecture Notes in Computer Science, 254, Springer-Verlag, Berlin, pp. 248–299, 1987.

Jensen, K. Coloured Petri nets: a high level language for system design and analysis. In *Advances in Petri Nets 1990* (G. Rozenberg, ed.), Lecture Notes in Computer Science. Springer-Verlag, Berlin, 1990.

Lee, K. H. and Favrel, J. Hierarchical reduction method for analysis and decomposition of Petri nets. *IEEE Transactions on Systems, Man, and Cybernetics*, SMC-15(2). 1985.

Lin, C. P. Design, verification and implementation of rule based information system for integrated manufacturing. Ph.D. thesis, Department of Mechanical Engineering, University of Maryland, College Park, Maryland, 1991.

Lopez, B., Meseguer, P. and Plaza, E. Knowledge based systems validation: a state of the art. *AI Communications*, 3(2), 58–72, 1990.

Mark, L. and Roussopoulos, N. Operational specification of update dependencies. *Systems Research Center Technical Report No. SRC TR-87-37*, University of Maryland, 1987.

Martin, J. and McClure, C. *Structured Techniques for Computing*. Prentice-Hall, Englewood Cliffs, NJ, 1985.

Martinez, J. and Silva, M. A simple and fast algorithm to obtain all invariants of a generalized Petri net. *Second European Workshop on Application and Theory of Petri Nets*, pp. 301–310, 1982.

Meyer, B. *Object-Oriented Software Construction*. Prentice-Hall, New York, 1988.

Muro, P. R., Villarroel, J. L., Martinez, J. and Silva, M. A knowledge representation tool for manufacturing control systems design and prototyping. *INCOM '89, 6th IFAC/IFIC/IFORS/IMACS Symposium on Information Control Problems in Manufacturing Technology*, Madrid, Spain, 1989.

Narahari, Y. and Viswanadham, N. A Petri net approach to the modeling and analysis of flexible manufacturing systems. *Annals of Operations Research*, 3, 381–391, 1985.

Ngu, A. H. H. Conceptual transaction modelling. *IEEE Transactions on Knowledge and Data Engineering*, 1(4), 508–518, 1989.

Nguyen, T. A., Perkins, W. A., Laffey, T. J. and Pecora, D. Knowledge base validation. *AI Magazine*, pp. 67–75, summer 1987.

Peterson, J. L. *Petri Net Theory and the Modeling of Systems*, Prentice-Hall, Englewood Cliffs, NJ, 1981.

Pressman, R. S. *Software Engineering: A Practitioner's Approach*. 2nd edition, McGraw-Hill, New York, 1987.

Sartori, L. G. *Manufacturing Information Systems*, Addison-Wesley, Reading, MA, 1988.

Sibertin-Blanc, C. Object-oriented structuring using high-level Petri nets. In *Advances in Petri Nets '91* (G. Rozenberg, ed.), Springer-Verlag, Berlin, 1991.

Smith, J. M. and Smith, D. C. P. Data abstraction: aggregation and generalization. *ACM Transactions on Database Systems*, 2(2), 1977.

Spooner, D., Hardwick, M. *et al.* The evolution of ROSE: an engineering object-oriented database system, *Proc. of IEEE Conf. on CIM*, RPI, pp. 16–23, 1990.

Stotts, P. D. and Furuta, R. Petri net based hypertext: document structure with browsing semantics. *ACM Transactions on Information Systems*, 7(1), 3–29, 1989.

Su, S. Y. W. Modeling integrated manufacturing data with SAM*. *Computer*, 19(1), 34–49, 1986.

Suzuki, I. and Murata, T. A method for stepwise refinement and abstraction of Petri nets. *Journal of Computer System Science*, 27, 51–76, 1983.

Tardieu, H., Rochfeld, A. and Colletti, R. *La méthode MERISE: principes et outils*, Les Editions d'Organisation, Paris, 1983.

Teorey, T. J., Yang, D. and Fry, J. P. A logical design methodology for relational databases using the extended entity-relationship model. *ACM Computing Surveys*, 18(2), 197–222, 1986.

Tsichritzis, D. C. and Lochovsky, F. H. Hierarchical data base management: a survey. *ACM Computing Surveys*, 8(1), 105–123, 1976.

Ullman, J. D. *Principles of Database Systems*. 2nd edition, Computer Science Press, Rockville, MD, 1982.

Valette, R. Analysis of Petri nets by stepwise refinements. *Journal of Computer and System Sciences*, 18, 35–46, 1979.

Vernadat, F. A conceptual schema for a CIM database, *Proc. AUTOFACT VI Conference*, Anaheim, CA, 1–4 October 1984.

Vernadat, F., Di Leva, A. and Giolito, P. Organization and information system design of manufacturing environments: the new *M** approach. *Computer-Integrated Manufacturing Systems*, 2(2), 69–81, 1989.

Wirfs-Brock, R. J. and Johnson, R. E. Surveying current research in object-oriented design. *Communications of the ACM*, 33(9), 105–124, 1990.

Index

Page numbers in **bold** refer to figures and those in *italics* refer to tables.